Chipless and Conventional Radio Frequency Identification:

Systems for Ubiquitous Tagging

Nemai Chandra Karmakar
Monash University, Australia

Managing Director:	Lindsay Johnston
Senior Editorial Director:	Heather A. Probst
Book Production Manager:	Sean Woznicki
Development Manager:	Joel Gamon
Development Editor:	Hannah Abelbeck
Acquisitions Editor:	Erika Gallagher
Typesetter:	Jennifer Romanchak
Cover Design:	Nick Newcomer, Lisandro Gonzalez

Published in the United States of America by
    Information Science Reference (an imprint of IGI Global)
    701 E. Chocolate Avenue
    Hershey PA 17033
    Tel: 717-533-8845
    Fax: 717-533-8661
    E-mail: cust@igi-global.com
    Web site: http://www.igi-global.com

Library of Congress Cataloging-in-Publication Data

Chipless and conventional radio frequency identification: systems for ubiquitous tagging / Nemai Chandra Karmakar, editor.
    p. cm.
  Includes bibliographical references and index.
  Summary: "This book explores the use of conventional RFID technology as well as chipless RFID technology, which provides a cheaper method of implementation, opening many doors for a variety of applications and industries"-- Provided by publisher.
  ISBN 978-1-4666-1616-5 (hardcover) -- ISBN 978-1-4666-1617-2 (ebook) -- ISBN 978-1-4666-1618-9 (print & perpetual access) 1. Radio frequency identification systems. 2. Radio frequency--Identification. 3. Wireless communication systems. 4. Radio--Antennas. I. Karmakar, Nemai Chandra, 1963-
  TK6570.I34C45 2012
  621.3841'92--dc23
                        2011050428

British Cataloguing in Publication Data
A Cataloguing in Publication record for this book is available from the British Library.

All work contributed to this book is new, previously-unpublished material. The views expressed in this book are those of the authors, but not necessarily of the publisher.

To the Memory of My PhD Supervisor, Professor Marek Bialkowski

Table of Contents

Section 3
Chipless RFID Tags

Section 4
RFID System and Detection of RFID Tags

Detailed Table of Contents

Section 1
Introduction to RFID

Chapter 1

    Nemai Chandra Karmakar, Monash University, Australia

The book provides a comprehensive coverage on most recent developments in chipless and conventional RFID. It covers a wide range of topics from component level design, analysis, and development, to system integration, middleware, anti-collision, and security protocols. The chipless RFID will bring revolutionary impacts on low-cost item tagging in this millennium. The RFID based sensors and RF sensors play a vital role in real time condition monitoring of objects. The designs of various chipless RFID tags and printing techniques to achieve a goal toward less than a cent tag are presented. The reading methods of RFID tags of various types, system perspective design, and analyses, detection techniques, sensor nodes for RFID system, security risk and vulnerability of the RFID technology and their remedies, anti-collision protocols, middleware and enterprise software implementation, and innovative applications of RFIDs in various fields are all presented in the book. The book will make a valuable reference in the RFID field, which has been growing exponentially.

Section 2
RFID Antennas and Amplifiers

Chapter 2

    Zhonghao Hu, Auto-ID Lab, The University of Adelaide, Australia
    Peter H. Cole, Auto-ID Lab, The University of Adelaide, Australia
    Christophe Fumeaux, The University of Adelaide, Australia
    Yuexian Wang, The University of Adelaide, Australia

A simple analytic formula, found in the literature for calculating the resonant frequency of a meander line dipole antenna (MDA) in free space from its physical parameters is described. The formula is modified to calculate the resonant frequency of an MDA on a dielectric substrate and for use as an RFID tag antenna by taking two factors into account: (i) the effects of dielectric material underneath the MDA, (ii) the special needs of an impedance matching condition in RFID tag antenna design. The parameter of relative effective permittivity for an MDA on a dielectric board, and the method for deriving this parameter, are introduced. Experiments to verify the modified formula are reported. Test results such as input impedance and reading range of an RFID tag antenna design based on an MDA on a dielectric board are provided. Following that, the radiation pattern and efficiency of an MDA either in free space or on a board are investigated.

Chapter 3

Tapas Chakravarty, Tata Consultancy Services Ltd (TCS), India
P. Balamuralidhar, Tata Consultancy Services Ltd (TCS), India

This chapter addresses the various design issues, requirements, and specifications in designing reader antennas for UHF RFID applications. In a typical UHF system, the RFID reader antennas are geography specific; that is, there are different antennas for different geography, namely North America, Europe, et cetera. The discussion on the design challenges and performance expectations lead to a new form of compact and broad band antenna, which will be applicable for the global UHF RFID band of 860MHz to 960 MHz. In addition, this chapter also provides potential future trends in RFID reader antenna design. The issues and challenges discussed in this chapter are envisioned to provide a roadmap to the potential challenges to be faced by a chipless RFID system.

Chapter 4

Christina Junjun Wang, Beihang University, China

The emphasis of this chapter is to introduce the design of the integrated circuit package antenna (ICPA), which is a compact and cost-effective antenna design method for RFID reader. The concept, the architecture, and the characterizations of the ICPA will be discussed in details. As differential circuitry dominates in RF transceiver integrated circuit design due to its good performance, microstrip antennas can be seen for use in radio systems with differential signal operation. In this chapter, the improved theory of single-ended microstrip antennas based on the cavity model is expanded to analyze the input impedance and radiation characteristics of the differentially-driven microstrip antennas and ICPA. The occurrence of the resonance for the differentially-driven microstrip antennas, which can be tuned by adjusting the ratio of the separation of the dual feeds to the free space wavelength, will be analysed. Furthermore, the frequency band selection capability of the differential ICPA will be presented.

Chapter 5

A. K. M. Baki, Monash University, Australia
Nemai Chandra Karmakar, Monash University, Australia
Uditha Wijethilaka Bandara, Monash University, Australia
Emran Md Amin, Monash University, Australia

It is possible to achieve higher BE and lower SLL of array antenna by implementing different amplitude or phase distribution technique in the array antenna. The phase errors of the system should also be kept to a minimum in order to maintain lower SLL and higher BE. The phase errors can come from any of the stages: signal detection, MW/RF generation, amplifier/attenuator, phase synchronization, phase shifter, et cetera. The phase error can be reduced by using non-uniform element spacing. In this chapter some methods of SLL reduction and increase of BE by adopting some edge tapering concepts and minimization of phase errors by implementing non-uniform spacing of array elements are discussed. The spectrum below 10 GHz frequency will likely be congested, and the spreading of millimetre wave technology in different emerging wireless applications as well as associated increase in energy consumption will be witnessed in the near future. In this chapter some new and better beam forming techniques for optimization between side lobe levels and beam efficiency are discussed. Different frequency bands of RFID systems are also focused on in this chapter.

Chapter 6

Shivali G. Bansal, Deakin University, Australia
Jemal Abawajy, Deakin University, Australia

In this chapter the authors discuss the physical insight of the role of wireless communication in RFID systems. In this respect, this chapter gives a brief introduction on the wireless communication model followed by various communication schemes. The chapter also discusses various channel impairments and the statistical modeling of fading channels based on the environment in which the RFID tag and reader may be present. The chapter deals with the fact that the signal attenuations can be dealt with up to some level by using multiple antennas at the reader transmitter and receiver to improve the performance. Thus, this chapter discusses the use of transmit diversity at the reader transmitter to transmit multiple copies of the signal. Following the above, the use of receiver combining techniques are discussed, which shows how the multiple copies of the signal arriving at the reader receiver from the tag are combined to reduce the effects of fading. The chapter then discusses various modulation techniques required to modulate the signal before transmitting over the channel. It then presents a few channel estimation algorithms, according to which, by estimating the channel state information of the channel paths through which transmission takes place, performance of the wireless system can be further increased. Finally, the Antenna selection techniques are presented, which further helps in improving the system performance.

Chapter 7

Liming Gu, Nanjing University of Science and Technology, People's Republic of China
Yang Yang, Monash University, Clayton Campus, Australia
Shichang Chen, City University of Hong Kong, Hong Kong
Nemai Chandra Karmakar, Monash University, Clayton Campus, Australia

In RFID reader systems, the power amplifier plays a critical rule for efficiency enhancement. A high efficiency power amplifier may not only increase the life expectancy of portable RFID devices but also reduce the reliance on heat sinks. Heat sinks usually occupy plenty of space and lead to packing difficulties. A well designed power amplifier with high efficiency and output power may also increase the reading range of RFID and system reliability, especially for the applications requiring long reading range (e.g. vehicle tagging in complicated traffics) or in a lossy environment (e.g. in sensing in rainy weather). This chapter systematically introduces the typical power amplifiers classified as Class A, AB, B, E, and F. The principles of Class F are emphasized due to its outstanding performance in efficiency enhancement. A practical design example is also presented, and also some recent typical techniques for improving the performances of Class F power amplifier are summarized.

The rapid development in wireless identification devices and subsequent applications is at the origin of intensive investigations in order to fulfill various constraints that can exist when implementing applications in practice. Chipless technologies have many advantages. They are fundamentally wireless and powerless devices, and can be all passive components, which potentially means infinite lifetime. However, chipless technology is still in its infancy age, even if it is the most effective for cost reduction. One of the most important features of chipless is coding capacity and ways to imprint it into the device. This chapter will review and discuss various coding techniques. It will address a comparison of the most relevant coding techniques. For sake of clarity some global parameters that can be used as figure of merit will be introduced and applied to compare different practical chipless tags.

Radio Frequency Identification (RFID) is an emerging technology playing a vital role in modern automatic identification system. Chipless RFID is a new dimension in the field of radio-frequency application systems with immense potential to manufacture low-cost, multi-bit RFID tags for potential barcode replacement on polymer, paper, and other flexible substrates. In this chapter, the authors present a detailed overview of the printing methods, substrates, and materials used for printing chipless RFID tags. Based on the available literature, an attempt is made to review the printing and performance related issues of printed RF devices that are currently published. The basic aspects of printing of chipless tags with conductive inks are discussed in brief.

The problems of radio frequency identification are discussed. It was shown that the use of passive transponders is preferable, but weak energy of system in this case reduces the operation distance and decreases the noise-immunity of the system. The problems of traditional radio-frequency identification systems are discussed. In this chapter the use of homodyne method of useful signal selecting was proposed. The augmentation signal of transponder was obtained by means of frequency shift with the help of controlled phase shifter. This solution allows to increase the energy and the noise-immunity of the system (the operation distance is increased). Furthermore, the interrogator can treat several transponders simultaneously in this case. Additionally the use of one-port transistor amplifier for increasing of operation range was proposed. The energy consumption of such amplifier and its cost are very low, but the gain of amplifier can reach 20 dB and more.

Chapter 11

Prasanna Kalansuriya, Monash University, Australia
Nemai Chandra Karmakar, Monash University, Australia
Emanuele Viterbo, Monash University, Australia

This chapter presents a different perspective on the chipless RFID system where the chipless RFID detection problem is viewed in terms of a digital communication point of view. A novel mathematical model is presented, and a novel approach to detection is formulated based on the model. The chipless RFID tag frequency signatures are visualized as points in a signal space. Although data bits are stored in the tags using unconventional techniques, the proposed model enables the detection of these data bits through conventional robust detection methods. Through simulations it is shown that the proposed detection method has better performance compared to contemporary detection approaches.

Chapter 12

Gour C. Karmakar, Monash University, Australia
Laurence S. Dooley, The Open University, UK
Nemai C. Karmakar, Monash University, Australia
Joarder Kamruzzaman, Monash University, Australia

Object analysis using visual sensors is one of the most important and challenging issues in computer vision research due principally to difficulties in object representation, segmentation, and recognition within a general framework. This has motivated researchers to investigate exploiting the potential identification capability of RFID (radio frequency identification) technology for object analysis. RFID however, has a number of fundamental limitations including a short sensing range, missing tag detection, not working for all objects, and some items being just too small to be tagged. This has meant applying RFID alone has not been entirely effective in computer vision applications. To address these restrictions, object analysis approaches based on a combination of visual sensors and RFID have recently been successfully introduced. This chapter presents a contemporary review on these object analysis techniques for localisation, tracking, and object and activity recognition, together with some future research directions in this burgeoning field.

Chapter 13

A. K. M. Azad, Monash University, Australia
Joarder Kamruzzaman, Monash University, Australia
Nemai C. Karmakar, Monash University, Australia

Radio Frequency Identification (RFID) systems and Wireless Sensor Networks (WSNs) are believed to be the two most important technologies in realizing the ubiquitous computing vision of Future Internet. RFID technology provides much cheaper solution for object identification and tracking based on radio wave. On the other hand, data on various parameters about the physical environment can be acquired using WSNs. Integration of the advantages of both RFID systems and WSNs would benefit many application domains. In RFID system, either an active RFID tag itself or an RFID reader (reading passive or semi-passive tags) consisting of an RF transceiver poses communication capability similar to that for nodes in WSNs. Therefore, instead of using single hop RFID protocol, RFID networks can take advantage of WSN-like multihop communication, and in this regard a number of WSN protocols can be useful for such RFID systems. In this chapter we present possible scenario of the integration of RFID system and WSNs and study a number of wireless sensor network protocols suitable to use in RFID system.

Foreword

Radio Frequency Identification (RFID) is one of the most significant technologies in recent decades. RFID has added many value added services in businesses, processes, and services. In the era of information and communication explosion, acquisition, processing, retrieval, and utilisation of data in a timely manner are important for business processes, security, surveillance, tracking, and tracing. RFID has been playing significant roles in these endeavours of capturing, processing, and utilisation of data without much human intervention. Compared to optical barcodes, which have been dominating tagging and tracking industries for the last four decades, RFID offers much flexibility, capacity, and efficiency. In recent demands, operations and processes need more data handling capabilities than the optical barcodes can offer. Due to its flexibility, higher data handling capacity, and efficiency, RFID is favoured more in businesses compared to optical barcodes. However, mass deployment of RFID for low cost item tagging and tracking has been hindered by the cost of RFID tags. The chipless RFID tag technology will influence the new millennium with higher operability, flexibility, and data processing power, compared to optical barcodes. It may create huge impact in our day-to-day life if it can be made cheaply and comes with sensing capabilities.

Significant amount of research work has been conducted on the development of the chipless RFID. This book has addressed the chipless RFIDs and RF sensing devices, related physical layer development, and software protocols. The book aims to provide a complete solution of chipless and conventional tags and reader systems and their emergence as a significant technology in tagging and sensing. The topics include the physical layer developments of components such as antennas for both tags and readers, high efficiency power amplifiers for RFID readers, chipless RFID tags —design and printing techniques, detection techniques of chipless and chipped tags, security issues of RFID, network protocols, and finally, emerging applications of the RFID and RF sensor technology. The book covers seven distinctive sections on active and passive RF components, chipless RFID tag design, reader systems, detection and discriminations of tagged objects, security issues, middleware, anti-collision protocols, and physical implementations of various emerging applications. The book presents most recent developments in the field specially the chipless RFID.

The book will be a significant resource for engineers, managers, and researchers working in the RFID industry and postgraduate students who are doing research and implementation in both active and passive RF and microwave design.

Leena Ukkonen
Tampere University of Technology, Finland

Preface

Radio Frequency Identification (RFID) is a wireless data capturing technique for automatic identification, tracking, security surveillance, logistic, and supply chain management. In the modern era RFID is one of the top ten technologies that have tremendously impacted society. RFID offers flexibility in operation and higher data capacity compared to that for optical barcodes. Therefore, RFID has gained momentum to be used in all possible applications. The most visible application of RFID is Electronic Article Surveillance (EAS) in superstores. Expensive items are tagged so that the unpaid items give warning signals at the entry and exit points of stores. EAS is a 1-bit tag, can only respond to *yes or no* situations. More expensive and high capacity tags carry much more useful information that an optical barcode can offer. Therefore, RFIDs offer not only flexibility and capacity but also item level tagging, tracking, and surveillance. However, the bottleneck of mass deployment of RFIDs for low cost item tagging is the cost of the tag. The cost of the conventional RFIDs has been decreasing day by day. However, there is a limit due to the silicon chip attached to the tag. These chips are application specific integrated circuits (ASICs) and the price of the chip can be tens of cents. To alleviate this cost problem, researchers are envisaging alternatives such as chipless tags. Thin film transistors (TFTs) on organic substrates and fully printable tags are the two commercially viable solutions that can compete with optical barcodes in mass implementations. If the cost of the tags can be reduced to less than a cent, the tags will find many potential applications. This book has addressed the most recent development of chipless and conventional tags—their systems and applications.

In 1995 when the author of this preface was a full-time PhD student at the University of Queensland (UQ), Brisbane, Australia, a few researchers were working on quasi-optical power combiners (QOPCs) in microwave and antenna research laboratories. In the 90s, QOPC was a hot topic. The development of an RFID or a radio frequency transponder was a topic of research. The RFID resembled the principle of harmonic mixing in space similar to that for QOPC. TO this author's memory, the researchers were referring to the seminal paper "Progress in Active Integrated Antennas and Their Applications" by Yongxi Qian and Tatsuo Itoh (1998). For the author, that was the first introduction to the technical details of a physical layer development of a transponder—a bow-tie antenna connected with a sub-harmonically pumped quasi-optical mixer (Stephan & Itoh, 1984). The UQ group designed effectively a passive RFID tag with a few lumped components. During the '90s, there was very little known about the RFID tags to common people. Toll roads, intelligent transportation systems, car immobilizers, and personalized door bells were in the inception phase and just had been introduced in other parts of the world. As for an example, European insurance companies enforced car immobilizers to protect cars from theft in the early 90s. After the author's PhD thesis in 1998, he moved to Nanyang Technological University in Singapore. There, he and his colleagues applied for a nationally strategic large research grant to the National Science and Technology Board (NSTB) of Singapore. The project was the development of a micro-electromechanical-system identification tag (MEMSIT) in the mm-wave frequency band. The objective of the project is to develop a MEMSIT for smart cards. Professor Choi Look Law was the lead chief investigator and Dr. Karmakar was the chief investigator of RF Design. That was his formal intro-duction to research with the RFID. In the project they designed many interesting circuits and gears for the characterization of the designed circuits. The developed circuits were 24 GHz MEMS slot antennas, microstrip antenna arrays for RFID readers, filters, and their measuring jigs. Later on after the completion of the project, when Karmakar moved to Monash University, Melbourne, Australia in 2004, his CSIRO colleague Dr. Gerhard Swiegers invited him to apply for the Australian Research Council Discovery Grant on Chipless RFID. Dr. Sweigers had been developing a one-bit tag using printable pn-junction diode on a polymer substrate. He wanted to design a fully printable tag on a thin polymer substrate. Dr. Karmakar did comprehensive literature review on the topic and found that not much work was done on the chipless RFID tags at that time. A few developments on chipless RFIDs were: chemical nano-structures, radio frequency Surface Acoustic Wave (RF-SAW), and RFID tattoos. Except RF-SAW, the proposed and available chipless tags offered very short reading distance in the range of a few millimetres. The author found a gap in research on fully printable chipless tag using passive microwave circuits and antennas. The inception of chipless idea at Monash University germinated with a multi-resonator circuit and two antennas—one for transmission and one for reception. The proposed multi-resonator circuit is a planar wheel filter developed in 2001 (Karmakar & Padhi). The antennas are broadband fractal antennas to cover the multi-bit frequency signatures. Later on, a former PhD student of the author's defined this method as the *Re-transmission* based tag. The concepts of chipless tags have gained momentum from 2006 with the successful ARC Discovery grant *DP0665523: Chipless RFID for Barcode Replacement*. For the last five years three successful PhD completions with high distinction and more than one hundred referred journal and conference proceedings papers, book chapters, one edited book on *Smart Antennas For Radio Frequency Identification Systems* (2010) with Wiley InterScience, two Australian Provision Patents, two international patent applications and many regional ones on chipless RFIDs generated from the research work. The author's research group is highly cited on chipless RFIDs. This book is his first initiative to combine most recent works from prominent researchers in the field.

A few works on chipless tags are well cited: (i) Jalaly and Robertson (2005), (ii) McVay, Hoorfar, & Engheta (2006), (iii) Zhang, Tenhunen, Zheng (2006), (iv) Mukherjee (2007), and (v) Preradovic *et al.* (2007). Both frequency and time domain chipless tags were addressed in these papers. In the recent years many research teams from France, Japan, Germany, Korea, Spain, Thailand, and USA have been investigating chipless RFID tags and reader systems.

According to the respected research institute IDTechEx, chipless RFID tags will occupy more that 60% market share of RFID markets within a few years time if the tags can be made less than a cent (IDTechEx, 2009). The market of RFID technology surpassed $5bn in 2009 and is projected to be more than $25bn in 2018. The accelerated pace of RFID tags, middleware and reader development will address many technological challenges as well as provide many new solutions in printing techniques, algorithm and reader architectures. Anti-collision will also play an important role in mass deployments of chipless RFID technology in multi-faceted applications. Chipless RFID will change the culture of the way we do transactions in our businesses and livings. Like chipped RFIDs, chipless RFIDs have the capability to provide flexibility in operations with its salient features of non-line-of-sight (NLOS) and all-weather reading capability without much human intervention. It has the potential to replace trillions of optical barcodes printed each year. Therefore, many research activities on chipless RFID tags have been con-ducted not only in academia but also in industries. In this regard, printed electronics technologies shall play the vital role. Again according to the market analyses by IDTechEx, a few hundred industries are engaged in printed electronics for identifications, tagging and telecommunications markets. As quoted by IDTechEx, "This organic and printed electronics is growing to become a $300 billion market and, in 2007 alone, many factories came on stream to make "post silicon" transistors, displays and solar cells using thin films and, increasingly, printing. Most of the action is taking place in East Asia, Europe and North America.......*There is also the prospect of replacing 5-10 trillion barcodes yearly with printed RFID that is more versatile, reliable and has a lower cost of ownership*" (Harrop, Reuter, & Das, 2009).

Every new technology goes through a cycle of developmental phases. From the inception of the technology and its conceptual development, many doubts, confusions and technical challenges hinder the progress of the development. Once these barriers are overcome, investors and funding bodies come forward to support the research and development activities. New applications and investment on returns motivate the end users to embrace the technology for growth and expansion of their businesses. Once the development reaches to the maturity, huge commercial implementation happens to all possible sectors. The deep penetration of the technology in mass market changes the business and transactions cultures. A few examples of such matured technologies are optical barcodes, emails, internets and mobilephones. The evolution of RFID has no exception. Starting from the Identification, Friend or Foe (IFF) in the World War II, the seminal paper by Stockman (1948) on RFID, and strong patronization of RFID technology by Walmart, US Department of Defence and similar consortia from the rest parts of the world, the RFID technology has grown to a mammoth technology. However, as the RFID finds applications in every wake of our lives and businesses, the challenges to cater the need for these applications are also enormous. The main challenge of mass deployment of RFID is the cost of the tag and reliable reading processes. Significant momentum has been gained on to develop various technologies to address the issues of the cost and reliable reading process of the RFID tag. The book presents the recent technical developments in chipless and conventional tags, their components and reading methods.

Significance of Chipless RFID Tags: Removal of application specific integrated circuits (ASIC) microchips from the conventional RFID tags can only provide viable commercial solution to mass deployment of RFID for low cost item tagging (Karmakar, 2010). The author's industry partner FE Technologies Pty Ltd., based in Geelong, Victoria, Australia has been marketing their Smart Library® (FeTech Group, 2010) RFID system in Australia and overseas. In February 2009 FE Technologies demonstrated their automated library database management system in front of a group of librarians from Monash University. Smart Library® comprises an automatic check out kiosk, a smart trolley and a magic wand as a handheld reader for inventory checking and misplaced items. Monash University's library possesses more than 3 million books to cater about 10,000 staff and 50,000 students in Australia and overseas campuses in Malaysia and South Africa. With a book tag costing 50¢ each; Monash University immediately needs to invest about $2m to implement their RFID system. While the existing optical barcodes for books cost less than 10¢ per unit and the existing library database management system based on the optical barcode works very well within the existing infrastructure and operational culture, there is a question always remains about the return on investment of more than $2m to implement the RFID system for the library database system for Monash University's libraries. This is a big question mark and an uphill battle to persuade the management to finance the implementation of RFID for the library. This is only one example. The huge potential of the RFID in such many other applications is hindered by the high price of the chipped tags. The viable solution is the low cost printable chipless RFID tag which will cost less than 1¢ and can complete with the optical barcode. The chipless RFID tags developed by the author's research group at Monash University are simple in concept. It is a fully printable passive microwave electronic circuit, which can be printed with conductive inks using an inkjet printer or other printing tools. Some conductive inks are invisible. Therefore, RFID tags can be made invisible. This technology will open up a fully new spectrum of applications starting from Australian polymer banknotes, library books, apparels, shoes, and tagging of low cost and perishable items such as apples and bananas. Now imagine the market volume if low cost printed tags can be delivered and reliably read. To make the tag chipless and simple in operation, the bulk of the operation will be bestowed on the reader electronics. Certainly the reader should be built more powerful that the conventional chipped tag readers to process the returned echoes of the tags and encode the unique identification and location of the tag. The smart signal processing and detection algorithm and smart antennas in the reader will play a major role to improve the reading of the tags.

RFID is an emerging technology that has been going through various developmental phases in terms of technological developments and businesses (applications), the potential as well as the challenges are huge. As for the example of the implementation of the RFID in Monash University's Library above, the bottleneck is the cost of the tag and its mass deployment. The answer to the problem lies in the development of new materials and printing technologies which can appropriately address the problem and bring forth a sustainable solution in terms of economy and technological advancements. A full chapter is dedicated to the development of printing techniques, polymer substrate, and flexible design of chipless tags on the polymer using various printing techniques.

Where the tags become dumb, the reader should be smart. The smartness will come from the smart signal capturing capabilities from the dumb tags and the post-processing of the returned echoes which are the signals from the uniquely identifiable tags. Significant advancement has been made in the new design of RFID systems and detection techniques of RFID tags, discrimination of tagged items and protocols developed for wireless sensors network applied to RFID systems. The book includes a full section on these topics.

As an enabling technology, RFID encompasses multiple disciplines. Similar to radar technology, RFID is a multi-disciplinary technology which encompasses a variety of disciplines: (i) RF and microwave engineering, (ii) RF and digital integrated circuits, (iii) antenna design, and (iv) signal processing software and computer engineering. The latter encodes and decodes analog signals into meaningful codes for identification. According to Lai et al (2005), "The fact that RFID reading operation requires the combined interdisciplinary knowledge of RF circuits, antennas, propagation, scattering, system, middleware, server software, and business process engineering is so overwhelming that it is hard to find one single system integrator knowledgeable about them all. …. In view of the aforesaid situation, this present invention (RFID system) seeks to create and introduce novel technologies, namely redundant networked multimedia technology, auto-ranging technology, auto-planning technology, smart active antenna technology, plus novel RFID tag technology, to consolidate the knowledge of all these different disciplines into a comprehensive product family." The book has incorporated these multi-disciplinary contents in three different sections: (i) RFID antennas and Amplifiers, (ii) Chipless RFID Tags, and (iii) RFID System and Detection of RFID Tags.

Due to the flexibility and numerous advantages of RFID systems compared to barcodes and other identification systems available so far, RFIDs are now becoming a major player in retail sectors and government organisations. Patronization of the RFID technology by organisations such as Wal-Mart, K-Mart, the USA Department of Defense, Coles in Australia and similar consortia in Europe and Asia has accelerated the progress of RFID technology significantly in the new millennium. As a result, significant momentum in the research and development of RFID technology has developed within a short period of time. The RFID market has surpassed the billion dollar mark recently (Das & Harrop, 2006), and this growth is exponential, with diverse emerging applications in sectors including medicine and health care, agriculture, livestock, logistics, postal deliveries, security and surveillance and retail chains. The book includes application in tracing systems on the integrity of pharmaceutical products, near field authentication, monitoring system for sleep apnoea diagnosis in wireless body sensor network (WBSN) using active RFID and MIMO technology, chipless RFID based temperature and partial discharge (PD) detection sensors and finally, wireless sensors network and their applications in RFIDs.

Today, RFID is being researched and investigated by both industry and academic scientists and engineers around the world. Recently, a consortium of the Canadian RFID industry has put a proposal to the Universities Commission on the education of fresh graduates with knowledge about RFID (GTA, 2007). The Massachusetts Institute of Technology (MIT) has founded the AUTO-ID centre to standardize RFID, thus enabling faster introduction of RFID into the mainstream of retail chain identification and asset management (McFarlane & Sheffi, 2003; Karkkainen, & Ala-Risku, 2003). The synergies of implementing and promoting RFID technology in all sectors of business and day to day life have overcome the boundaries of country, organisation, and discipline.

As a wireless system, RFID has undergone close scrutiny for reliability and security (EPCglobal, Inc., 2006). With the advent of new anti-collision and security protocols, efficient antennas and RF and microwave systems, these problems are being delineated and solved. Smart antennas have been playing a significant role in capacity and signal quality enhancement for wireless mobile communications, mobile ad-hoc networks and mobile satellite communications systems. Smart antennas are used in RFID readers where multiple antennas and associated signal processing units are easy to implement (Lai et al, 2005). Even multiple antennas are proposed in RFID tags to improve reading rate and accuracy (Ingram, 2003).

Besides the contributions from outside, the author's research group at Monash University have contributed significantly in the physical layer development of RFID reader architectures for chipped and chipless RFID tag systems, RFID smart antennas, wireless sensor network protocols for RFID, and anti-collision algorithm. The research group has been supported by the Australian Research Council's Discovery Project Grants *DP665523: Chipless RFID for Barcode Replacement* and *DP110105606: Electronically Controlled Phased Array Antenna for Universal UHF RFID Applications*; the Australian Research Council's Linkage Project Grants *LP0989652: Printable, Multi-Bit RFID for Banknotes*; *LP0776796: Radio Frequency Wireless Monitoring in Sleep Apnoea (particularly for paediatric patients)*; LP0669812: *Investigation into improved wireless communication for rural and regional Australia*; *LP0991435: Back-scatter based RFID system capable of reading multiple chipless tags for regional and suburban libraries*; and *LP0989355: Smart Information Management of Partial Discharge in Switchyards using Smart Antennas*; and finally, Victorian Department of Innovation, Industry & Regional Development (DIIRD) Grant: *Remote Sensing Alpine Vehicles Using Radio Frequency Identification (RFID) Technology* within the Department of Electrical and Computer Systems Engineering, Monash University from 2006 to date. The dedication of former postgraduate students Drs. Sushim Mukul Roy, Stevan Preradovic and Isaac Balbin and current research staff and PhD students under the author's supervision has brought the chipless RFID tag and reader system as the viable commercial products for Australian polymer banknotes, library database management systems, access cards, remote sensing of faulty power apparatuses in switchyards and the wireless monitoring of sleep apnoea patients. The RFID and smart sensor related research projects supported by Australian Research Council's Discovery and Linkage Projects and Victoria Government are worth approximately three million dollars. More than twenty researchers have been working in various aspects of these projects. The book contains six chapters on our research findings in the above research topics.

The dramatic growth of the RFID industry has created a huge market opportunity. Patronization from Wal-Mart alone has triggered their more than two thousand suppliers to implement RFID system for their products and services. The motto is to track the goods, items, and services from their manufacturing point until the boxes are crushed once the goods are sold. Thus industries can track every event in their logistics and supply chain management and make sound plan for efficient operations and business transactions. The RFID system providers are searching all possible technologies that can be implemented in the existing RFID system (Gen2 becomes a worldwide standard) that can be made cheap, can be implemented to provide high accuracy in multiple tags reading with minimum errors and extremely low false alarm rate, location finding of tags for inventory control and asset tracking. Employing smart information management system in the reader presents an elegant way to improve the performance of the RFID system. The book has covered many technical aspects of these requirements.

Deploying smart antennas in the reader architecture and network, smart antennas may bring outstanding improvement in throughput, high speed reading and position detection of tagged items. These facilities can be obtained with an efficient beamforming scheme and diversity techniques. Positioning of tagged items has many applications in industry thanks to the direction finding ability of the smart antennas. The RFID technology is moving in higher and higher frequencies to incorporate more data-bits. Frequency bands in upper microwave and mm-wave such as 5.8, 24, and 60 GHz are less occupied by conventional telecommunications technologies. Therefore, new developments in RFID (Ingram, 2003), and especially in chipless RFID are happening in these frequency bands. In the author's lab he and his colleagues are developing ultra-wide band (UWB) microwave and mm-wave chipless RFID tags, reader architectures and smart antennas. The book includes two chapters dedicated to a low side lobe microwave smart antenna design for RFID readers at 5.8 GHz and UWB chipless tags on thin polymer substrates from the author's research group. Besides these, UWB chipless RFID developments in Grenoble–INP of France and Georgia Institute of Technology of USA are also presented in the book.

Smart antennas can also be used in handheld RFID readers making the reading more efficient and long range. The beamforming and interference suppression abilities of the smart antenna make the reader capable to increase throughput. In a networked environment of the RFID readers where each reader represent a node, the smart antenna in a node with packet routing protocols, the direction finding and suppression of interference abilities from the neighboring nodes provides the optimum routing relaying between nodes. A chapter dedicated to smart sensor network protocols for RFID system is included in the book.

A MIMO wireless communication channel can be built by installing antenna arrays that provide uncorrelated signal outputs at both readers and tags. The MIMO system provides many number of channels with the number of antenna elements in both transmit and receive chains. The MIMO system enhances the channel capacity hence the throughput of the RFID reader. Even Multiple antennas are proposed in the RFID tags by pushing the operating frequency at 5.8 GHz frequency band to incorporate multiple antennas in a credit card size tag (Griffin & Durgin 2010). The benefit is the high speed tag reading and significant throughput improvement. MIMO also enhances the anti-collision capability and capturing effect of the tag when the reader reads multiple tags in close proximity. Antenna selection and channel estimation play an important part for throughput improvement of RFID system. A couple of chapters are dedicated to address the implementation issues of MIMO antennas in RFID system. The book includes a couple of chapters on anti-collision protocols for RFIDs.

The book aims to provide the reader with comprehensive information with the recent development of chipless and conventional RFID systems both in the physical layer development and the software algorithm and protocols. To serve the goal of the book, it features fourteen chapters authored by the leading experts in both academia and industries. They offer in depth descriptions of terminologies and concepts relevant to the RFID components and systems—antennas related to the RFID, physical layer development including the printing techniques of chipless RFID tags, the system development and various detection techniques for both chipless and conventional tags, the security issues of chipped and chipless tags, development of chipless RFID tags, and reader system to address authentications.

The author has about fifteen books on RFIDs in his personal library collections. He continuously collects and reads books on RFIDs. These books are readily available from online book shops such as Amazon.com. Every scientific book publishers have a series of book on RFIDs and their applications in governance, pharmaceuticals, logistics, supply chain managements, retail, and original part manufacturing. These books mainly report specific applications, introduce fundamental issues, and gather information on RFIDs, specific technical details that are commonly available from other resources. This book aims to come out of the convention approach of reporting the technology. The book presents the most recent technological development from renowned researchers and scientists from academia and industries. Therefore, a comprehensive coverage of definitions of important terms of RFID systems and how the RFID technology is evolving into a new phase of development can be found in the book. The book covers the state-of-the-art development on RFID in recent years. Seven scientists from five large to medium size industries including Microsoft Research Center, USA, Securency Intl. Pty. Ltd, Australia, Unique Microwave Design, Australia, Tata Consultancy Services, India, and fifty academic researchers from Australia, Chain, France, Italy, Mexico, Taiwan, Ukraine, UK, and USA, have contributed chapters in the book. Therefore, the book not only delivers the emerging development in a total package of chipless and conventional RFIDs, but also provides diversities in topics. The rich contents of the book will benefit the RFID technologist, planners, policy makers, educators, researchers, and students. Many universities and tertiary educational institutions teach RFID in certificate, diploma, undergraduate, and graduate levels. This book can be served as a textbook or a companion book and a very useful reference for students and researchers in all levels.

The book can be best used as a complete reference guide if an expert wants to design a complete RFID system using either a chipless or a conventional radio frequency identification system. The beneficiaries of the book are the specialists of specific disciplines such as antennas and RF designs for both tags and readers, chipless RFID tag designs, system aspects on detection, discrimination, sensor network protocols, security issues, and design of security protocols and systems. The readers of the book can maximize their knowledge on a systematic middleware and enterprise software planning, anti-collision protocol designs for multiple tag and reader scenarios such as warehouses, manufacturing plants, supply chain managements, and pharmaceuticals. If some experts and executives want to implement RFID in a particular system in their organizations, they are encouraged to read the last few chapters on design and implementation of RFIDs and RFID based sensors in various emerging applications. Each section is rich with new information and research results to cater for the needs of specialists in system as well as specific components of the RFID.

In the book, utmost care has been paid to keep the sequential flow of information related to the various aspects as mentioned above on the RFID system and its emerging development. The hope is that the book will serve as a good reference of RFID and will pave the ways for further motivation and research in the field.

Nemai Chandra Karmakar
Monash University, Australia
February 6, 2012

REFERENCES

Das, R., & Harrop, P. (2006). *RFID forecasts, players & opportunities 2006 – 2016.* London, UK: IDTechEx. Retrieved September 11, 2007, from http://www.idtechex.com/products/en/view.asp?productcategoryid=93

EPCglobal. (2005, January). *EPCTM radio-frequency identity protocols class-1 generation-2 UHF RFID protocol for communications at 860 MHz - 960 MHz, Version 1.0.9.*

FeTech Group. (n.d.). Retrieved September 1, 2011, from http://www.fetechgroup.com.au/

Griffin, J., & Durgin, G. D. (2010). Fading statistics for multi-antenna RF tags . In Karmakar, N. C. (Ed.), *Handbook of smart antennas for RFID systems.* Wiley Book Series in Microwave and Optical Engineering. doi:10.1002/9780470872178.ch18

GTA. (2007). *RFID industry group RFID applications training and RFID deployment lab.* Request Background Paper January, 2007.

Harrop, P., Reuter, S., & Das, R. (2009). *Organic and printed electronics in Europe.* IDTechEx.

IDTechEx. (2009). *RFID forecasts, players and opportunities 2009-2019, executive summary and collusions.*

Ingram, M. A. (2003, January 21). *Smart reflection antenna system and method* (US patent no. US 6,509,836, B1)

Jalaly, I., & Robertson, I. D. (2005). Capacitively tuned microstrip resonators for RFID barcodes. *Proceedings of the 35th European Microwave Conference 2005,* (pp. 1161-1164).

Karkkainen, M., & Ala-Risku, T. (2003). *Automatic identification – Applications and technologies.* Logistics Research Network 8th Annual Conference, London UK, September 2003.

Karmakar, N. C. (2010). *Handbook of smart antennas for RFID systems.* Hoboken, NJ: Wiley Microwave and Optical Engineering Series. doi:10.1002/9780470872178

Karmakar, N. C., & Padhi, S. K. (2001, March 1). A novel study of an electrically small printed charka (wheel) antenna. *IEE Electronics Letters, 37*(5), 269–271. doi:10.1049/el:20010187

Lai, K. Y. A. Wang, O. Y. T., Wan, T. K. P., Wong, H. F. E., Tsang, N. M., Ma, P. M. J., Ko, P. M. J., & Cheung, C. C. D. (2005, 22 July). *Radio frequency identification (RFID) system.* (European Patent Application EP 1 724 707 A2).

McFarlane, D., & Sheffi, Y. (2003). The impact of automatic identification on supply chain operations. *International Journal of Logistics Management, 14*(1), 407–424. doi:10.1108/09574090310806503

McVay, J., Hoorfar, A., & Engheta, N. (2006). Theory and experiments on Peano and Hilbert curve RFID tags. *Proceedings of the Society for Photo-Instrumentation Engineers, 6248,* 624–808.

Mukherjee, S. (2007). *System for identifying radio-frequency identification devices* (patent). (Patent No. US20070046433, 03/01/2007).

Preradovic, S., Balbin, I., Karmakar, N. C., & Swiegers, G. (2008). A novel chipless RFID system based on planar multiresonators for barcode replacement. *The 2008 IEEE International Conference on RFID, 2008*, (pp. 289-96).

Qian, Y., & Itoh, T. (1998). Progress in active integrated antennas and their applications. *IEEE Transactions on Microwave Theory and Techniques, 46*(11), 1891–1900. doi:10.1109/22.734506

Stephan, K. D., & Itoh, T. (1984). Inexpensive short-range microwave telemetry transponder. *Electronics Letters, 20*(21), 877–878. doi:10.1049/el:19840595

Stockman, H. (1948). Communication by means of reflected power. *Proceedings of the Institute of Radio Engineering*, (pp. 1196–1204).

Zhang, L., Tenhunen, R. S., & Zheng, L. R. (2006). An innovative fully printable RFID technology based on high speed time-domain reflections. *Conference on High Density Microsystem Design 2006*, (pp. 166 – 170).

Acknowledgment

I would like to thank IGI Global for the invitation to edit a book on RFID. Special thanks go to the editorial assistants, Ms. Emily E. Golesh and Hannah Abelbeck of IGI Global. Hannah's continuous support and cooperation made the total editing process smooth. Ms. Hannah Abelbeck was always beside me with her continuous support and patience throughout editing and writing process of the manuscript. Generous support from the authors of the chapters of the book and their timely responses for submission of chapters are highly acknowledged. Special thanks to those authors who submitted their chapters on time, but had to wait for a long time until the completion of the manuscript.

Special thanks to my beloved wife, Mrs. Shipra Karmakar, for collating the huge depository of email addresses of prospective authors and drafting email invitations for the authors. Her dedication and effort to initial the book are highly acknowledged. My current and former students and research staff gave me moral support to take the mammoth task to edit the manuscript. I thank them for their generous support and chapter contributions. I would like to thank the members of advisory editorial board for their inspiration, guidance, and contributions to the book. My special thanks go to Prof. Smail Tedjini of Grenoble-inp/LCIS, Valance, France. His continuous suggestions and guidance helped me shaping the book title and chapter organized. The special contribution from his research group on the chipless RFID tags and detection systems truly impacted the book. Authors and many expert reviewers reviewed the chapters of the book. I acknowledge their supports. I must acknowledge and thank Prof. Subhas Chandra Mukhopadhyay of Massey University, New Zealand for his special contribution on RF sensors. He has accepted the invitation of writing a unique chapter on the RF sensors in his very busy schedule. This invited chapter has unique significance in the book.

Special thanks to Dr. AKM Azad for his contribution. He reviewed and edited a few chapters besides his own contributions in the book. Yang Yang has prepared the index term of the book. Special thanks go to Yang for his dedication and hard work to make the index of the book. Thanks to Wan Muhammad Imran, Wan Mohd Zamri, and Ka Seng Chan for their assistance in compiling the book.

During the preparation of the manuscript, my family, wife Mrs. Shipra Karmakar, and daughters, Antara and Ananya Karmakar, experienced my absence. With their continuous moral support and motivational attributes, I completed the huge task of editing and preparing the manuscript. They brought my moral strength back when I felt exhausted in completing the big volume of this book. I highly value their love and care during the preparation of the manuscript.

Finally, the research funding supports from Australian Research Council's Discovery Project Grants, Linkage Project Grants, and Monash University's internal research grants are highly acknowledged.

Nemai Chandra Karmakar
Monash University, Australia October 2011

Section 1
Introduction to RFID

Chapter 1
Introduction to Chipless and Conventional Radio Frequency Identification System

Nemai Chandra Karmakar
Monash University, Australia

ABSTRACT

The book provides a comprehensive coverage on most recent developments in chipless and conventional RFID. It covers a wide range of topics from component level design, analysis, and development, to system integration, middleware, anti-collision, and security protocols. The chipless RFID will bring revolutionary impacts on low-cost item tagging in this millennium. The RFID based sensors and RF sensors play a vital role in real time condition monitoring of objects. The designs of various chipless RFID tags and printing techniques to achieve a goal toward less than a cent tag are presented. The reading methods of RFID tags of various types, system perspective design, and analyses, detection techniques, sensor nodes for RFID system, security risk and vulnerability of the RFID technology and their remedies, anti-collision protocols, middleware and enterprise software implementation, and innovative applications of RFIDs in various fields are all presented in the book. The book will make a valuable reference in the RFID field, which has been growing exponentially.

INTRODUCTION TO CHIPLESS AND CONVENTIONAL RFIDS

Radio frequency identification (RFID) is an emerging wireless technology for automatic identifications, access controls, tracking, security and surveillance, database management, inventory control and logistics. The RFID has two main components: a tag and a reader. The reader sends an interrogating radio signal to the tag. In return the tag responds with a unique identification code to the reader. The reader processes the returned signal from the tag into a meaningful identification code. Some tags coupled with sensors can also provide data on surrounding environment such as temperature, pressure, moisture contents, acceleration and location. The tags are classified into active, semi-active and passive tags based

DOI: 10.4018/978-1-4666-1616-5.ch001

on their on-board power supplies. An active tag contains an on-board battery to energize the processing chip and to amplify signals. A semi-active tag also contains a battery, but the battery is used only to energize the chip, hence yields better longevity compared to an active tag. A passive tag does not have a battery. It scavenges power for its processing chip from the interrogating signal emitted by a reader, hence lasts forever. However, the processing power and reading distance are limited by the transmitted power of the reader.

The main constraint of mass deployment of RFID tags for low-cost item tagging is the cost of the tag. The main cost comes from the application specific integrated circuit (ASIC) or the micro-chip of the tag. If the chip can be removed without losing functionality of the tag, then the tag can have the potential to replace the optical barcode. The optical barcode has several limitations in operation, including: (a) each barcode is individually read; (b) needs human intervention; (c) has less data handling capability; (d) soiled barcodes cannot be read; and (e) barcodes need line of sight operation. Despite these limitations, the low cost benefit of the optical barcode makes it very attractive as it is printed almost without any extra cost. Therefore, there is a pressing need to remove the ASIC from the RFID tag to make it competitive in deployment to co-exist or replace trillions of optical barcodes printed each year. The solution is to make the RFID tag chipless. Similar to the optical barcodes, the tag should be fully printable on low cost substrates such as papers or plastics. A reliable prediction by the respected RFID research organization IDTechEx (2009) advocates that 60% of the total tag market will be occupied by the chipless tag if the tag can be made less than a cent. However, removal of ASIC from the tag is not a trivial task as it performs many RF signal and information processing tasks. It needs tremendous investigation and investment in designing low-cost but robust passive microwave circuits and antennas using conductive ink on low-cost substrates. However, obtaining high

fidelity response from low cost lossy materials is very difficult. In the interrogation and decoding aspects of the RFID system is the development of the RFID reader, which is capable to read the chipless RFID tag.

Currently, only a few chipless RFID tags, which are in the inception stage, are reported in the literature. They are: a capacitive gap coupled dipole array (Jalaly & Robertson, 2005); a reactively loaded transmission line (Zhan, 2006); a ladder network (Mukherjee, 2007); and finally, a piano and a Hilbert curve fractal resonators (McVay et al., 2006). These tags are in prototype stage and no further development in commercial grade is reported so far. To the best of the author knowledge, there is no chipless RFID reader reported in open literature besides the work reported from the author's group (Preradovic and Karmakar, 2010). To fill up the gap in the literature for the potential chipless RFID field, the author's chipless RFID research team has been working on the paradigm chipless RFID tag since 2004 (Karmakar, 2010, Preradovic, 2011). The designed tag has mainly targeted to tag chipless RFID tags as replacement of optical barcodes. Significant strides have been achieved to tag not only the polymer banknotes but also many low cost items such as books, postage stamps, secured documents, bus tickets and hanging cloth tags. The technology relies on encoding spectral signatures and decoding the amplitude and phase of the spectral signature. The other methods are phase encoding of backscattered spectral signals and time domain delay lines. So far as many as more than ten varieties of chipless RFID tags and three generations of readers are designed by this team. The proof of concept technology is being transferred to the banknote polymer and paper for low cost item tagging. These tags have potential to co-exist or replace trillions of optical barcodes printed each year. To this end it is imperative to invest on low loss conducting ink, high resolution printing process and characterization of laminates on which the tag will be printed. The design needs to push in higher frequency bands to accommodate and increase the number of bits in the chipless tag to compete with the optical barcode. The reader design needs to accommodate large distance and high-speed reading, multiple tag reading in close proximity, error correction coding and anti-collision protocols. Also wide acceptance of RFID technology by consumer and business requires strong privacy and security protection. The book aims to address all these issues mentioned above to make the chipless RFID system a viable commercial product for mass deployment.

SPECIAL FEATURES OF THE BOOK

The book provides readers with comprehensive coverage on chipless and conventional RFID tags and reader systems. As mentioned above the chipless RFID will impact low-cost item tagging in this millennium. The designs of various chipless RFID tags and the dedicated reader to read the developed chipless tags are presented. The book will report most recent developments and challenges ahead in the chipless and conventional RFID field, which has been growing exponentially, and will further accelerate research and development of this field leading to successful mass commercialization of RFID technologies. Recommended Topics are organized in four distinct sections. They are:

1. Component design for RFID tags and readers. Various antennas are designed for both tags and readers, and an efficient class-F power amplifier for potential use in handheld readers is presented in the section.
2. Chipless RFID tag design, encoding data bits and decoding techniques for the tags and finally, the printing issues for a viable printed tag are presented in section 3.
3. RFID systems and detection techniques. Both active passive and chipless RFID tags are discussed in section 4.

The book features:

1. A single comprehensive and most recent source of the chipless and conventional RFID technologies applied to low cost item tagging for large scale deployment to replace/co-exist trillions of optical barcodes printed each year;
2. Provides broad, integrated coverage of subject matters not usually covered in a single reference of RFID open literature;
3. Presents the latest achievements in the designs and applications of chipless and conventional RFID tags and RF sensors;
4. Provides basic concepts, terms, protocols, systems, architectures and case studies for chipless RFID tags and readers;
5. Identifies the fundamental problems, key challenges, future directions in designing chipless RFID tag and reader systems;
6. Covers a wide spectrum of topics, including signal processing algorithms, hardware architectures, test-bed evaluations and practical applications of chipless RFID tag and reader systems.

The above topics are covered in four distinct sections (the first of which is this introductory chapter) with thirteen chapters. Executive summaries of the chapters are produced below.

SECTION 2: RFID ANTENNAS AND AMPLIFIERS

This section includes six very high quality chapters on the RFID antennas and amplifiers. Antennas and RF designs such as amplifiers, mixers, down-conversion circuits and RF detectors play significant roles in the infrastructure development of the RFID system. Antennas are used in both tags and readers. Antennas are like eyes to electronics that make the wireless communications viable (Balanis, 1988). Therefore, efficient antenna design is paramount in the RFID systems. A plethora of work on antennas on RFID tags is a tremendously independent domain on its own right. Every commercial tag has its own antenna design that makes the form factor of the tag. Squiggle antennas of various shapes are mainly the printed dipole antennas and variants of them.

In Chapter 2, *Design and Analysis of a Meander Line Dipole Antenna* is presented. The effect of the dielectric substrate on frequency response and the need for impedance matching are the two main objectives of the analysis. These two important parameters of the antenna play important roles in antenna efficiency. Also importance of form factors to the total tag layout is also given in the chapter.

In Chapter 3, *UHF RFID Reader Antenna,* the authors present a universal UHF tag reader antenna in a compact package. The antenna is a planar monopole patch as radiating element with a z-shaped capacitive tuning post and a compact ground plane. The antenna covers 860- 960 MHz frequency band with a gain of 3 dBi and a 3-dB beamwidth of approximately 75°. The antenna is designed for handheld readers to be used in healthcare applications.

Chapter 4 presents a differential feed integrated circuit package antenna for RFID readers. A differential circuitry dominates the RF transceiver integrated circuits design. Therefore, a differential feed antenna is compatible with the conventional circuitry. The differentially fed antennas also have some added advantages. The advantages are higher integration density, reduced cross talk, removed grounding problems, and ease of testing of system performance. An integrated circuit package antenna (ICPA), which is fed differentially for better performance compared to the conventional patch antennas, is the main topic of the chapter. This ICPA is a compact and cost-effective antenna for compact and handheld RFID reader. Detailed mathematical model of the antenna input impedance, differential input voltage, radiation patterns

are also presented along with the design guideline in an integrated package RFID device module. The antenna has better cross-polar and co-polar components and higher input impedance compared to the conventional single feed antenna.

Chapter 5, *Beam Forming Algorithm with Different Power Distribution for RFID Reader,* presents phased array antenna architectures for various RFID bands. The generic smart antenna has adaptive beamforming capability with prescribed low sidelobe levels. A set of voltage variable attenuators and digital phase shifters are the heart of beamforming devices. The basic theories of the antenna array—analysis and synthesis of array antennas and adaptive beamforming algorithm - are presented. Demonstration prototypes of 8-element linear array of rectangular patch antennas at 900 MHz and 5.8 GHz ISM bands are also demonstrated as the proof of concept. The antenna architecture has the capability to cater the services from several MHz up to 60 GHz ISM band for future RFID applications (Pursula et al 2011).

Chapter 6 *Multi-Input-Multi-Output Antennas for Radio Frequency Identification Systems* presents a comprehensive study on the channel for RFID communications in a communication theoretic perspective. Introduced with the most recent developments and research works of multiple input and multiple output (MIMO) antenna systems applied to the RFID systems, the authors discuss the physical insight to the role of wireless communication in RFID systems. In this respect, this chapter gives a brief introduction on the wireless communication model followed by various communication schemes. Various channel impairments and the statistical modelling of fading channels based on the environment in which the RFID tag and reader may be present is also discussed. By using multiple antennas at the reader transmitter and receiver the performance can be improved compared to the conventional single input and single output (SISO) system. Use of transmission diversity at the reader's transmitter to transmit multiple copies of the signal is a proposed technique. Multiple copies of the signal arriving at the reader's receiver from the tag are combined to reduce the effects of fading. Performance of the wireless system can be further increased by implementing a few channel estimation algorithms. Finally, the antenna selection techniques are presented which further helps in improving the system performance.

Chapter 7 *Design of High Efficiency Power Amplifiers for RFID* presents a highly efficient class-F power amplifier (PA) for RFID reader applications. In RFID reader systems, a high efficiency PA not only increases the efficiency and battery life to enhance the life expectancy of portable RFID devices but also reduces the reliance on heat sinks. A well designed power amplifier with high efficiency and high output power also increases the read range of a RFID reader. The PA also increases system reliability, especially for the applications requiring long reading range. The author's research group has been developing reader systems for both microwave and mm-wave frequency bands. Conventional voltage controlled oscillators cannot provide enough transmission power levels required for long range reading. In this regard, PAs play crucial roles in the overall system performance and reliability. This chapter systematically introduces the typical power amplifiers classified as Class A, AB, B, E, and F. The principles of Class F are emphasized due to its outstanding performance in enhancing efficiency. A practical design example of a class F PA with a measured peak power added efficiency (PAE) of 80% at 3.5 GHz with an output power of 38.7 dBm and power gain of 15.5 dB is also presented. Finally, some recent typical techniques for improving the performances of Class F power amplifier are also summarized.

Overall, the section presents various fundamentals components for various RFID systems, antenna designs for RFID tags and reader systems, advanced level smart antenna designs, channel modelling and enhanced system performance using MIMO antennas and antenna selection, and

finally, a highly efficient class-F power amplifier. The chapters of the section are very useful for designers, postgraduate students and practitioners to comprehend advanced and state-of-the-art component designs for RFID tags and readers.

SECTION 3: CHIPLESS RFID TAGS

In this section, two very crucial chapters on the chipless RFID are included. Chapter 8 *Mastering the Electromagnetic Signature of Chipless RFID Systems* presents comprehensive accounts for chipless RFID tag development. Starting with the history of the RFID and especially chipless RFID, the authors present a beautiful classification of chipless RFID tags and encoding systems. The authors then move to the technical details of frequency signature based, group delay and phase encoded chipless RFID tags. They also present a very crucial issue of how to increase the bit capacity of the chipless tag. The techniques are very similar to the conventional pulse position modulation (PPM) and on-off keying (OOK). Finally, the authors emphasise the bit density per surface area of the chipless tag and recommend pushing the design in THz region. Chapter 9 *Printing Techniques and Performance of Chipless Tag Design on Flexible Low-Cost Thin-Film Substrates* presents various printing techniques of conducting inks on thin film polymer substrates. Various printing techniques include flexography, off-set lithography, screen printing, inkjet and gravure. The resolution, accuracy, longevity, conductivity and continuity of the conducting tracks vary with the printing techniques and curing temperature. Finally, a fully printed multi-bit chipless RFID tag on thin film polymer substrate using screen printing is presented. The challenging features on printed chipless tags are discussed in details.

In conclusion, this section presents a comprehensive overview, design and design challenges of various chipless RFID tags—time domain reflectometry, frequency signature, reactively loaded transmission lines and radiating patches and group delay structures. Transferring the chipless RFID technology from the conventional printed circuit boards (PCBs) to low cost thin film polymers and paper imposes serious challenges on the performance of the tag. To the best of the author's knowledge, this is the first time so comprehensive investigation on printing technology is presented in the section.

SECTION 4: RFID SYSTEM AND DETECTION OF RFID TAGS

After the comprehensive coverage on component level designs and chipless RFID tags, some fundamental system level issues of the conventional and chipless RFID tag systems and their detection methods are presented in the section. Chapter 10 *The Multi-tag Microwave RFID System with Extended Operation Range* presents a unique approach to the low powered passive RFID for improved performance. Passive transponders are the most popular tags for its low cost and moderately high data handling capability and reading range. However, weak energy of the system reduces the reading distance and increases the system's vulnerability to noise. The chapter proposes a homodyne method of useful signal selection. The augmentation signal of transponder was obtained by means of frequency shift with the help of controlled phase shifters. This solution allows increasing the energy and the noise-immunity of the system. Consequently, the reading distance is also increased. Furthermore, the interrogator can treat several transponders simultaneously in this case. Additionally, the use of one-port transistor amplifier to increase the operating range is proposed. The energy consumption of such an amplifier and its cost are very low but the gain of the amplifier can reach 20 dB and more.

Chapter 11 *A Novel Approach in Detection of Chipless RFID* presents a novel detection technique based on signal space representation and maximum likelihood detection based on minimum Euclidian distance. This is an approach to the chipless RFID tag detection based on a digital communication point of view. A novel mathematical model is presented and a new approach to detection is formulated based on the model. The chipless RFID tag frequency signatures are visualized as points in a signal space. Although data bits are stored in the tags using unconventional techniques (frequency signatures), the proposed model enables the detection of these data bits through conventional robust detection methods. Through simulations it is shown that the proposed detection method has better performance compared to contemporary detection approaches.

Chapter 12 *Object detection and discrimination with and without RFID* presents a combinational approach to object detection base on computer vision augmented by RFID technology. Object analysis using visual sensors is one of the most important and challenging issues. An object representation, segmentation and recognition in a general framework are always challenging. Researchers use the potential identification capability of RFID technology for object analysis along with computer vision. However, due to the fundamental limitations of RFID such as short reading range, missing detection for small objects, collisions in proximity tagged objects, RFID alone is not an effective tool in computer vision applications. To bridge this gap, recently researchers are introducing object analysis approaches based on a combination of visual sensors and RFID. This chapter provides a contemporary review on such approaches for object analysis for localization and tracking of objects and activity recognition along with future research directions in this field.

Chapter 13 *Wireless Sensor Network Protocol Applicable to RFID System* bridges the gap between the two technologies RFID and WSN. Firstly, the comprehensive reviews of the two technologies have been presented followed by the technical details and similarity in these two technologies. Various protocols that are invented for WSN, are being proposed for the RFID technology. A synergy between the two technologies shall provide ubiquitous computing of future Internet.

CONCLUSION

RFID is an emerging technology for automatic identification, tracking and tracing of goods, animals and personnel. In recent decades, the exponential growth of RFID market signifies its potentials in numerous applications. The advantageous features and operational flexibility of RFID have attracted many innovative applications areas. Therefore, there is a need for tremendous development and open literature on RFID to report new results. The book is an initiative to publish most recent results of research and development on chipless and conventional RFID systems. The book aims to serve the needs for a broad spectrum of readers. The book has started with component level design and development followed by chipless tag design and printing techniques of the chipless tag like optical barcodes. The book also cover system aspects of the two types of tags—chipless and chipped tags, their detection techniques and system architectures. Integrations of RFID with computer vision and WSN have opened up new applications and revolutionised ubiquitous computing in new dimensions. Overall, the book has become a one-stop-shop for a broad spectrum of readers who have interests in RFID and sensor technologies.

REFERENCES

Balanis, C. A. (1988). *Antennas: Theory, design and analysis* (2nd ed.). New York, NY: Wiley.

IDTechEx. (2009). *RFID forecasts, players and opportunities 2009-2019*. Executive Summary and Collusions.

Jalaly, I., & Robertson, I. D. (2005). Capacitively tuned microstrip resonators for RFID barcodes. *Proceedings of the 35th EUMC* (pp 1161-1164)

Karmakar, N. C. (2010). *Handbook of smart antennas for RFID systems*. New Jersey: Wiley Microwave and Optical Engineering Series. doi:10.1002/9780470872178

McVay, J., Hoorfar, A., & Engheta, N. (2006). Theory and experiments on Peano and Hilbert curve RFID tags. *Proceedings of the Society for Photo-Instrumentation Engineers, 6248*(1).

Mukherjee, S. (2007). Chipless radio frequency identification by remote measurement of complex impedance. *Proceedings of Wireless Technologies, European Conference* (pp. 249-252).

Preradovic, S. (Ed.). (2011). *Advanced radio frequency identification design and applications in RFID Tags*. Rijeka, Croatia: INTECH.

Preradovic, S., & Karmakar, N. C. (2010). Multiresonator based chipless RFID tag and dedicated RFID reader. *Digest of 2010 International Microwave Symposium*, Anaheim, California. (CD-ROM)

Pursula, P., Karttaavi, T., Kantanen, M., Lamminen, A., Holmberg, J., & Lahdes, M. (2011). 60-GHz millimeter-wave identification reader on 90-nm CMOS and LTCC. *IEEE Transactions on Microwave Theory and Techniques, 59*(4), 1166–1173. doi:10.1109/TMTT.2011.2114200

Zhang, L., Rodriguez, S., Tenhunen, H., & Zheng, L. R. (2006). An innovative fully printable RFID technology based on high speed time-domain reflections. *HDP'06 Conference*, (pp. 166-170).

KEY TERMS AND DEFINITIONS

Anti-Collision Protocols: Protocols or software algorithm that helps avoid collision in proximity tags

Back-Scatter Chipless RFID Tag: A chipless RFID tag which exploits returned echoes as a data encoding method

Chipless RFID: Radio Frequency tags without a microchip

Detection: Techniques to detect objects or signals

Error Correction Coding: A coding technique which can correct errors in symbols or bits or signals

Microwave Transceiver: An microwave frequency electronic device that combines both transmitter and receiver in a package

Middleware: A software program that interface technical hardware and software services

Multi-Bit RFID Tags: RFID tags those are capable of generating multiple bits

Multi-Resonator Circuits: A passive microwave circuit that generates multiple resonances in specific frequency bands

Polymer Tags: RFID tags which are printed on polymer substrates

Printing Techniques: Various methods of printing microwave electronic circuits

Reader Antennas: Antennas used for RFID readers

RFID Antennas: Antennas for both RFID readers and RFID tags

RFID Applications: Various emerging applications for RFID technology

RFID Reader: An electronic device to read an RFID tag

Security: Security features of RFID tags

Spectral Signature: Distinctive frequency spectrum with on-off keying features

Section 2
RFID Antennas and Amplifiers

Chapter 2
Analysis and Design of Meander Line Dipole Antennas

Zhonghao Hu
Auto-ID Lab, The University of Adelaide, Australia

Peter H. Cole
Auto-ID Lab, The University of Adelaide, Australia

Christophe Fumeaux
The University of Adelaide, Australia

Yuexian Wang
The University of Adelaide, Australia

ABSTRACT

A simple analytic formula, found in the literature for calculating the resonant frequency of a meander line dipole antenna (MDA) in free space from its physical parameters is described. The formula is modified to calculate the resonant frequency of an MDA on a dielectric substrate and for use as an RFID tag antenna by taking two factors into account: (i) the effects of dielectric material underneath the MDA, (ii) the special needs of an impedance matching condition in RFID tag antenna design. The parameter of relative effective permittivity for an MDA on a dielectric board, and the method for deriving this parameter, are introduced. Experiments to verify the modified formula are reported. Test results such as input impedance and reading range of an RFID tag antenna design based on an MDA on a dielectric board are provided. Following that, the radiation pattern and efficiency of an MDA either in free space or on a board are investigated.

1. INTRODUCTION

Nowadays the meander line dipole antenna (MDA) is used widely in UHF RFID tag antenna design (Choi & Shin et al., 2006), (Marrocco, 2008) because of its size reduction property and relative high radiation efficiency. The MDA is actually a dipole loaded with meander lines. One example of an MDA loaded with six meander lines is shown in Figure 1. Significant research work has been done on MDAs. Nakano, Tagami, Yoshizawa & Yamauchi (1984) inspired by the appearance of the meander monopole antenna (Rashed & Tai, 1982), firstly proposed the meander line dipole antenna (MDA). Nakano, Tagami, Yoshizawa & Yamauchi (1984) not only proposed the MDA, but

DOI: 10.4018/978-1-4666-1616-5.ch002

also analysed its radiation pattern, input imped-ance and size reduction ratio relative to the half wavelength dipole, when they are both resonant at the same frequency. The radiation efficiency of the MDA has been studied by Marrocco (2003). According to his research, an approximate current distribution on an MDA loaded with six meander lines is also shown in Figure 1. Since the currents on the adjacent vertical segment of each meander cell are opposite, these currents actually do not contribute to the radiation but bring losses. The radiation resistance is mainly determined by the horizontal segments of the MDA. Moreover, be-cause the large currents occur near the centre of the MDA as shown in Figure 1, one should not place meander lines, especially long meander lines (with large vertical dimension), near the centre. Genetic algorithms (GA) are also used to obtain a gain-optimised MDA within a fixed maximum available area (Marrocco, 2003). However, most analyses of MDAs are based on numerical meth-ods, and to arrive at the optimal MDA design, such numerical computations have to be iterated, with the result that the calculations are extensive. As a result, a simple analytic formula was proposed by Endo, Sunahara, Satoh & Katagi (2000) to calculate the resonant frequency from an MDA's geometrical parameters.

All the literature introduced above including the formula proposed by Endo, Sunahara, Satoh & Katagi (2000) assumes that MDA is working in free space. However, for the purpose of an-tenna protection, mechanical stability, size reduc-tion (Michishita, Yamada & Nakakura, 2004) and high radiation efficiency (Yamada & Michishita,

2005), large numbers of MDAs are commonly fabricated on dielectric substrates. Although nu-merical electromagnetic methods such as MOM (Method Of Moments) and FEM (Finite Element Method) can provide us reliable and accurate characteristics of MDAs on dielectric substrates, a simple method is still needed to avoid the bur-dens of the numerical methods in analysing the MDA.

This chapter aims, by summarising the existing literature, to give a complete analysis of resonant frequency, radiation pattern and radiation effi-ciency of an MDA and by contributing original thoughts to modify the formula proposed by Endo, Sunahara, Satoh & Katagi (2000) to calculate the resonant frequency of an MDA not only in free space but also above a dielectric board for RFID tag antenna design.

The outline of this chapter is as follows. Section 2 introduces and validates the formula proposed by Endo, Sunahara, Satoh & Katagi (2000). Section 3 gives the limitations of the formula proposed by Endo, Sunahara, Satoh & Katagi (2000) for calculating the resonant frequency of an MDA on a dielectric board and used for an RFID tag antenna. In order to overcome these limitations, several modifications are made to the formula. One significant modification is to add a new factor named as relative effective permittivity ε_{reff} in the formula so that the effects of the dielectric substrate can be taken into account. The method for deriving the factor ε_{reff} is given and validated by comparison with results from electromagnetic simulation software Ansoft HFSS. An RFID tag antenna based on the MDA pattern is investigated

Figure 1. A sample of meander line dipole antenna with approximate representation of the instantaneous current distribution. The arrows on the antenna represent the current flow direction and the number of the arrows illustrates the magnitude of the current distribution.

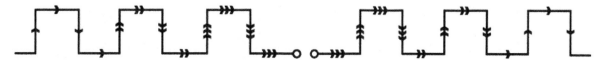

and tested in Section 4 to verify the modified formula. Following that, the radiation pattern and radiation efficiency of MDA are discussed in Section 5. Finally, in Section 6, conclusions are drawn.

2. INTRODUCTION AND VALIDATION OF THE FORMULA FOR CALCULATING RESONANT FREQUENCY OF AN MDA IN FREE SPACE

In this section, the closed-form expression for calculating the resonant frequency of an MDA in free space is derived, following the available literature, and subsequently validated with several examples.

2.1. Formula Derivation

This subsection introduces the derivation of the formula proposed by Endo, Sunahara, Satoh & Katagi (2000) for calculating the resonant frequency of an MDA in free space, from its geometrical parameters.

Figure 2 shows a dipole antenna with a single meander line on each dipole arm. The two parallel vertical lines are treated as twin lines with a short circuited termination. In addition, the horizontal part of the MDA with total length s is considered as a straight conducting wire with a diameter b.

The derivation of MDA's resonant frequency is derived as follows. The characteristic impedance of twin lines can be expressed by Balanis (2005):

$$Z_0 = \frac{\eta}{\pi} \log \frac{2w}{b}, \tag{1}$$

where, η is the wave impedance in free space, w is the centre-to-centre distance between the twin lines, b is diameter of the conducting line. Z_{in} is the input impedance of the twin lines, which is given by the following transmission line equation (Balanis, 2005):

$$Z_{in} = Z_0 \frac{Z_L + jZ_0 \tan \beta h}{Z_0 + jZ_L \tan \beta h}. \tag{2}$$

where β is equal to $2\pi/\lambda$, h is the height of the twin lines and Z_L is the load impedance. Now suppose that all twin lines are terminated in a short circuit. Thus, the load impedance of the twin lines is zero ($Z_L=0$), and (2) becomes

$$Z_{in} = jZ_0 \tan \beta h . \tag{3}$$

Following Endo, Sunahara, Satoh & Katagi (2000), $\tan\beta h$ can be approximately by a Taylor expansion with the third order on condition that $\beta h \ll 1$

Figure 2. Meander line dipole antenna loaded with two meanders

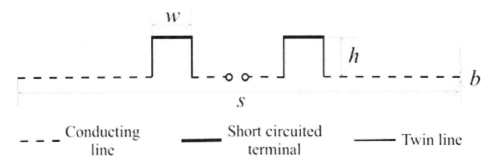

$$\tan \beta h \approx \beta h + \frac{1}{3}(\beta h)^3. \qquad (4)$$

Then a new expression of input impedance is obtained

$$Z_{in} = j\omega L = jZ_0(\beta h + \frac{1}{3}(\beta h)^3]. \qquad (5)$$

If we insert (1) into (5), the reactance formed by twin lines can be shown to be:

$$L = \frac{\mu_0 h}{\pi}[1 + \frac{1}{3}(\beta h)^2]\log\frac{2w}{b}. \qquad (6)$$

On the assumption that the number of meanders is m, therefore the total reactance obtained by the twin lines should be $L_p = m \cdot L$. The straight conducting line, which length is s, also results in a self-inductance. It is given by the following equation (Endo, Sunahara, Satoh & Katagi, 2000):

$$L_s = \frac{\mu_0 s}{2\pi}(\log\frac{4s}{b} - 1), \qquad (7)$$

where μ_0 is vacuum permeability. Then (6) and (7) can be solved to obtain the total inductive reactance of the MDA,

$$L_T = L_s + m \times L. \qquad (8)$$

Thus

$$L_T = \frac{\mu_0 s}{2\pi}(\log\frac{4s}{b} - 1) + m \cdot \frac{\mu_0 h}{\pi}[1 + \frac{1}{3}(\beta h)^2]\log\frac{2w}{b}. \qquad (9)$$

The self-inductance of a half wave-length dipole antenna can also be derived by (7)

$$L_H = \frac{\mu_0 \lambda}{4\pi}(\log\frac{2\lambda}{b} - 1) \qquad (10)$$

Following Endo, Sunahara, Satoh & Katagi (2000), we consider that the inductance of the MDA and a corresponding half wave-length dipole antenna must be the same when they resonate at the same frequency, thus $L_H = L_T$:

$$\frac{\mu_0 \lambda}{4\pi}(\log\frac{2\lambda}{b} - 1) = \frac{\mu_0 s}{2\pi}(\log\frac{4s}{b} - 1) + m \cdot \frac{\mu_0 h}{\pi}[1 + \frac{1}{3}(\beta h)^2]\log\frac{2w}{b}. \qquad (11)$$

As predicted by (11), the resonant frequency c/λ is decided by the physical dimension of the meanders. In detail, the resonant frequency declines as some features increase such as the meander lines height h, number of folds m, the ratio of meander lines width to the conducting line's diameter w/b and the conducting line length s. Moreover, the loaded position of meander lines does not appear to affect significantly the resonant performance.

2.2. Validation of the Closed-Form Expression for the MDA in Free-Space

For examining the validity of the method introduced in Subsection 2.1, the simulation software Ansoft HFSS is employed. MDAs loaded with different numbers of meanders are modeled by HFSS as shown in Figure 3, where the number of meanders $m = 2, 8, 14$, the length of the MDA $s = 129$mm, and the gap between dipole arms is 3mm. In order to establish the influence brought by each parameter, the following methodology is used. First, the length of MDA s and diameter b remain 129mm and 1mm respectively in all the following cases. Secondly, two of the three geometrical parameters of an MDA, (i) m, (ii) h, (iii) w, are fixed. Then, the third unfixed one is varied over a range. Therefore, the resonant frequencies of MDA for different shapes can be obtained by (11) and compared to the reference obtained by the simulation software Ansoft HFSS respectively as shown in Figure 4 (a-c) in which all the grey

Figure 3. Three models of MDA with various numbers of meander lines

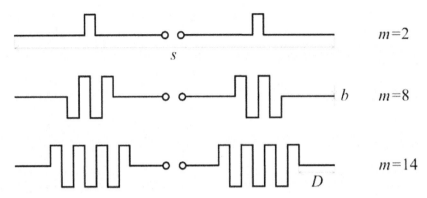

curves are derived by (11) and all the black curves are derived by simulation.

The number of meander lines m is firstly varied from 2 to 14 on condition that each meander line has a width $w = 6$mm and a height $h = 10$mm. The resonant frequencies calculated by (11) and HFSS are derived, as shown in Figure 4(a). The resonant frequency of the MDA as a function of the meander line height h is shown in Figure 4(b) when the meander line width w is 6mm and the number of the meander lines m is equal to 2, 8 and 14 respectively. Similarly, the resonant fre-

Figure 4. The resonant frequency of MDA as a function of its physical parameters

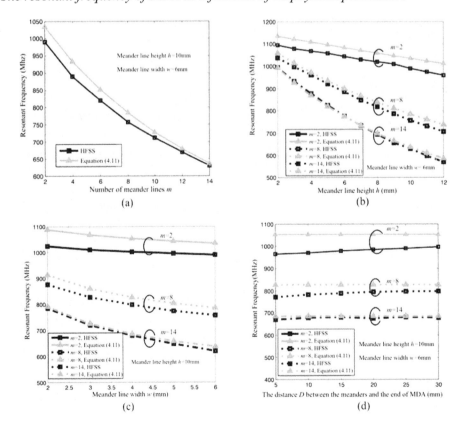

quency of MDA as a function of the meander line width w is shown in Figure 4(c) when the meander line height h is 10mm and the number of the meander lines m is equal to 2, 8 and 14 respectively. All meander lines discussed until now were loaded in the middle of each arm of the dipole.

The influence on resonant frequency resulting from the loaded position of the meander lines on the dipole arms can be studied by moving the meander lines close to the end of the dipole step by step, which corresponds to decreasing the distance D in Figure 3 step by step. In this validation example, the values of the other features remain unchanged (w = 6mm, h = 10mm, m = 2, 8 or 14). The resonant properties analysed by (11) and HFSS are illustrated in Figure 4(d).

All in all, the resonant characteristics of MDA predicted by the closed-form expression (11) have a satisfactory qualitative agreement with simulation results. The meander line's width w, height h and number m are the features which influence to the resonant characteristics whereas and the meander lines' loaded position D does not. However, this parameter affects the radiation efficiency which will be introduced in Section 5. Interestingly, the simulation results and the calculated results are getting closer with the increase of the number of meander lines. Clearly, the two curves almost superpose, when the number m reaches fourteen. As a result, this simple expression can be applied to calculate the MDA's resonant frequency with better accuracy when the number of the meander lines is large. Compared to HFSS, this method does not require building models for analysis. Moreover the calculation efficiency of a closed-form expression is of course much higher than the traditional full-wave numerical methods such as Moment of Method (MOM) and Finite Element Method (FEM), so that the designers can quickly find an optimal shape for the MDA corresponding to their requirements and make resonant at a desired frequency.

3. MODIFICATIONS ON EQUATION (11) FOR RFID TAG ANTENNA DESIGN

3.1. Limitations of Equation (11) in RFID Tag Antenna Design

Section 2 has demonstrated that (11) can derive a reasonably accurate resonant frequency from the geometrical parameters of an MDA or in other words, (11) can estimate an approximate MDA's shape for a given resonant frequency. However, (11) is not suitable to estimate the MDA's shape on a dielectric substrate for RFID tag antenna design, because of the following reasons: 1) When the MDA is placed or manufactured on a dielectric substrate, the effects of the substrate should be considered. 2) An MDA on a board is composed of metal strip instead of the wire assumed in (11), so an electrical equivalent diameter of a dipole made from a strip should be calculated from b = $0.5a$ (Balanis, 2005), where b is the equivalent wire diameter and a is the width of the strip. 3) The size of each meander could be different, whereas the meanders in (11) are uniform. 4) Compared with antennas employed in other areas, the RFID tag antenna design is special in its impedance matching condition. The tag antenna input impedance is not required to be real but complex, since the RFID tag antenna should be connected to a chip which typically present a small real component and a ten times larger imaginary impedance (e.g. approximately $12 - j130\Omega$ around 1GHz). Hence, the needed inductive load reactance X_a should be taken into account in designing a tag antenna. After considering all the four issues, (11) is modified to become

$$\frac{\mu_0 \lambda_{eff}}{4\pi} (\log \frac{4\lambda_{eff}}{a} - 1) + L_a =$$
$$\frac{\mu_0 s}{2\pi} (\log \frac{8s}{a} - 1) + \sum_{n=1}^{N} \frac{\mu_0 h_n}{\pi} [1 + \frac{1}{3}(\beta_{eff} h_n)^2] \log \frac{4w_n}{a}$$

$$(12)$$

where $\lambda_{eff} = \dfrac{\lambda_0}{\varepsilon_{reff}}$, $\beta_{eff} = \dfrac{2\pi}{\lambda_{eff}}$, ε_{reff} is the relative effective permittivity representing the effects of the dielectric substrate to the resonant frequency of an MDA. $L_a = X_a/\omega$ is the inductance brought by the extra needed inductive reactance X_a, and ω is angular frequency. N is the total number of the meander lines loaded on the dipole, h_n, w_n are the nth meander line's height and width respectively. The meander line height and width are measured between mid-lines of the strips. If the geometrical parameters of the MDA on a dielectric substrate and the impedance of the chip which is going to be mounted on an MDA are known, the only unknown factor in (12) is ε_{reff} which is discussed in the next subsection.

3.2. Method for Calculating the Relative Effective Permittivity of an MDA on a Dielectric Substrate

Before introducing the method for calculating the ε_{reff} of an MDA on a dielectric substrate, the method for calculating the ε_{reff} of the coplanar strip (CPS) transmission line structure is given, since the former is developed according to the latter.

Relative Effective Permittivity of CPS on a Substrate

The two coplanar strips (CPS) lying on a dielectric substrate are shown in Figure 5. The grey rectangle denotes an infinite substrate with dielectric constant ε_r and thickness t. The two dark rectangles represent the metal strips, with width *(d-c)/2* and a gap with width c between them.

The transverse electric field configurations of the CPS for quasi-static approximation are shown in Figure 6 in which the electric field travels from one strip into the air and ends at the other strip.

The determination of the distributed capacitance between the strips or the effective dielectric constant requires a solution of Laplace's equation. Hanna (1980) proposed an equation for estimating the ε_{reff} of the CPS shown in Figure 5 by a conformal mapping technique. A conformal mapping is a function that transforms curves in one complex plane to other curves in another complex plane, preserving angles between intersecting curves as it does so. It has the property that if a potential satisfies Laplace's equation in the original coordinates it will continue to satisfy it in the transformed coordinates. Conformal mapping is useful for solving problems in physics involving incon-

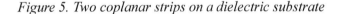

Figure 5. Two coplanar strips on a dielectric substrate

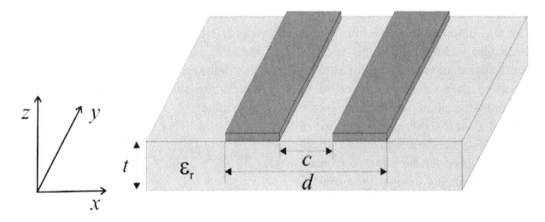

Figure 6. Illustration of the transverse electric field distribution in the cross section of a CPS

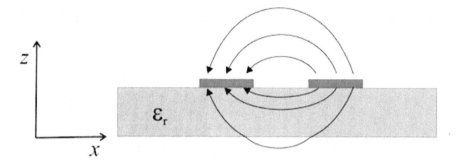

venient geometries. By making use of a conformal mapping technique, the original coordinate system shown in Figure 5 may be transformed to a new one in which the electric field or potential is parallel to one of the axes, and the solution to Laplace's equation is easily found. Hanna (1980) does not disclose much of this technique; only the final equations which are expressed as follows.

$$\varepsilon_{reff} = 1 + \frac{\varepsilon_r - 1}{2} \frac{K(k^{'})K(k_1)}{K(k)K(k_1^{'})} \tag{13}$$

where

$$k = \frac{c}{d} \tag{14}$$

$$k^{'} = \sqrt{1 - k^2} \tag{15}$$

$$k_1 = \frac{\sinh(\frac{\pi c}{4t})}{\sinh(\frac{\pi d}{4t})} \tag{16}$$

$$k_1^{'} = \sqrt{1 - k_1^2} \tag{17}$$

and $K(k)$ is the complete elliptic integral of the first kind.

Relative Effective Permittivity of MDA on a Substrate

As mentioned previously, each meander line on an MDA can be regarded as a CPS transmission line shortened an its end. Therefore, we first infer that the method for calculating the relative effective permittivity of a CPS may be useful for calculating the counterpart of an MDA. Hence, it is assumed that the ε_{reff} of an MDA on a dielectric substrate as a whole is approximately equal to the ε_{reff} of each meander line on the same dielectric substrate which is actually a CPS model. Then, the former can be calculated according to the latter. For example, an MDA loaded with four identical meander lines on a dielectric substrate is shown in Figure 7. The grey rectangle denotes the FR4 substrate which has dielectric constant ε_r equal to 4.4 and thickness t equal to 1.6mm. The dark strips represent the copper tape building the meander lines. The strip width is 1mm and the distance between the adjacent strips is 1.7mm. Hence, by inserting the geometrical parameters of the meander line into (13), i.e. $c = 1.7$mm, $d = 3.7$mm, $t = 1.6$mm, $\varepsilon_r = 4.4$, the ε_{reff} is derived as 2.4.

In order to test whether this calculation is valid, simulation was conducted on the MDA shown in Figure 7 by HFSS. The method for deriving the ε_{reff} by HFSS is described as follows. First, the model in Figure 7 with the dielectric substrate, which thickness is 1.6mm and dielectric

Figure 7. An MDA loaded with four identical meander lines

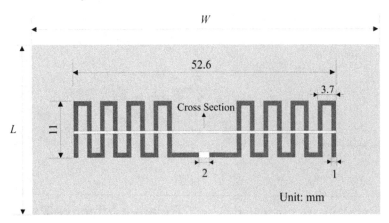

constant is 4.4, is simulated and its resonant frequency with the board, defined as the lowest frequency at which the reactance is equal to zero, is obtained and is denoted by f_b. Then, the substrate is removed from the model, so that the MDA is placed in free space, and the resonant frequency f_f of the MDA in free space is obtained. The relative effective permittivity brought by the board is then derived by (18).

$$\varepsilon_{reff} = (\frac{f_f}{f_b})^2 \qquad (18)$$

For the model shown in Figure 7, W and L are set to be large enough in the simulation so that the substrate can be regarded as infinite. This infinite extent is chosen as we intend to compare the simulated results with the calculated results derived by (13) which assumes the dielectric substrate is infinite. f_b and f_f obtained from HFSS are 1047MHz and 1279MHz respectively, so that ε_{reff} of the MDA is obtained by (18) to be 1.49. This value is much smaller than 2.4 previously derived by (13).

To understand this discrepancy, it is necessary to realize that the ε_{reff} is not simply dependent on the shape resemblance but ultimately dependent on the electric field distribution. If we look into the transverse field distributions of an MDA and

a CPS on a dielectric substrate, we still find a resemblance of the transverse field distribution which makes (13) useful in calculating the ε_{reff} of an MDA after making some modifications. The transverse electric field in the cross section of a CPS has been shown in Figure 6. It is clear that the electric field travels from one strip to the other. In terms of strength the field will be stronger if the observing point is closer to the strip and the field attenuates if the observing point is moved away from the strip.

The electric field magnitude distribution of the MDA shown in Figure 7 can also be obtained by the simulation software Ansoft HFSS. When the dielectric substrate thickness is 1.6mm and the relative permittivity is 4.4, the resonant frequency of the MDA is about 1047MHz as mentioned previously. At this frequency, the electric field distribution at the cross section marked in Figure 7 when the antenna is excited by a 1A r.m.s. current at the feed point is given in Figure 8.

According to Figure 8, the strongest electric field densities concentrate on the two strips at the ends of the MDA and diminishes, when the fields go further away from these two strips. This field distribution is similar to that of CPS model. Based on this observation, a new assumption can be made that the electric field radiated from an MDA can be represented by the electric field radiated from a CPS model by treating the two strips at

Figure 8. Cross section view of electric field magnitude distribution of the MDA shown in Figure 7

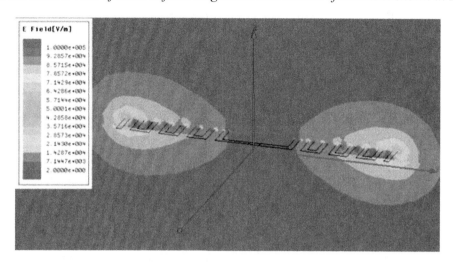

the ends of the MDA as coplanar strips (CPS) and ignoring the other components of the MDA. In addition, the electric field distribution ultimately decides the ε_{reff}. In other words, if the electric field distributions are similar, the ε_{reff} should be similar. The ε_{reff} of the MDA can thus be calculated by Equation (13) after adjusting the values of c and d in that equation according to the geometry of the MDA. In the case of the MDA in Figure 7, d should be the whole width of the MDA which is 52.6mm and c is the gap between the two strips at the ends of the MDA which is 50.6mm. Substituting $d = 52.6$mm, $c = 50.6$mm, $t = 1.6$mm and $\varepsilon_r = 4.4$ into (13), the ε_{reff} is derived as 1.61 which is very close to the simulation result of 1.49.

In order to further verify the method for calculating the ε_{reff} of an MDA by (13) as introduced above, more simulations have been done on the MDA of Figure 7 by varying the thickness t and the relative permittivity ε_r of the dielectric substrate. The ε_{reff} based on the simulation results and derived by (13) are given in Figure 9. In detail, the ε_{reff} as a function of the ε_r is illustrated in Figure 9(a), when the substrate thickness remains 1.6mm. Similarly, the ε_{reff} as a function of the dielectric substrate thickness t is illustrated in Figure 9(b), when the dielectric constant ε_r is held to be constant at 4.4. In addition, the black

curves in Figure 9 represent the results derived by (13) and the grey curves represent the results derived by the simulation.

There is a relative agreement in Figure 9(a) between the simulation results and the calculated results obtained by (13), at least for low values of the relative permittivity. This demonstrates that the method of simplifying the MDA to the CPS according to their resemblance in field distribution to calculate the MDA's ε_{reff} is feasible. But, it is also noted in Figure 9(a) that the higher the relative permittivity ε_r is, the less the agreement between the simulation results and the calculated results.

We then investigated the effects of various relative permittivity ε_r on the electric field distribution of the MDA. In order to observe the effects conveniently, the relative permittivity ε_r of the substrate underneath the MDA shown in Figure 7 is varied over a large range from 1 to 50. Meanwhile the thickness of the substrate remains 1.6mm. Simulations were conducted on the MDA respectively when the ε_r are 1, 10, 20, 30, 40 and 50. The electric field magnitude of the MDA at its resonant frequency defined earlier and depending on the ε_r was obtained by the simulation. The electric field distribution at the cross section marked in Figure 7 when the antenna is excited

Figure 9. The relative effective permittivity of the MDA in Figure 7 as a function of the dielectric constant ε_r for sub-figure (a) and as a function of the substrate thickness t for sub-figure (b)

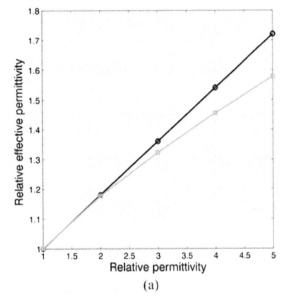

(a)

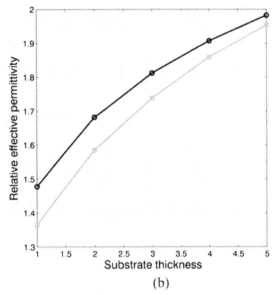

(b)

by a 1A r.m.s. current at the feed point is shown in Figure 10. The subtitle under each figure gives the values of the relative permittivity ε_r and the resonant frequency f_b.

According to Figure 10, it is found that along with the increase of the ε_r, the electric field magnitude near the two strips at the two ends of the MDA decreases more rapidly than the electric field magnitude in other places does. Hence, when the ε_r becomes large, besides the two strips at the ends, the effects of other components of the MDA should be taken into account for calculating the ε_{reff}. As the method for calculating ε_{reff} of an MDA proposed here only considers the two strips at the ends, this explains the divergence between the curves in Figure 9(a). However, this does not impedes the usage of (13) in designing MDA in RFID applications, since the dielectric materials used in fabricating or packaging RFID tags are usually low dielectric constant materials ($1 < \varepsilon_r < 4$), and the divergence between the curves of Figure 9 is then acceptable.

Another notable aspect of this method is the dielectric substrate has been assumed infinite but

in most of the RFID applications, the dielectric substrate size usually is equal to the footprint of the MDA or a little bit larger. However, simulation results not shown have confirmed that the variations of the substrate size do not affect the results of ε_{reff} significantly. This can be understood by considering that most transverse fields associated to propagation in the meanders are confined in-between the meander lines.

3.3. Further Numerical Validation of the Method

In the last subsection, the method for calculating the ε_{reff} of an MDA on a dielectric substrate was proposed and this method was initially examined by simulations on the MDA shown in Figure 7. In order to further confirm this method, we examine this method by simulations on another type of MDA which is shown in Figure 11. It is noted that the shape of the MDA in Figure 11 is quite different from that in Figure 7 in terms of the number of the meander lines and the shape of each meander line.

Figure 10. The variation of the MDA's electric field magnitude distribution at the resonant frequency along with the variation of the ε_r

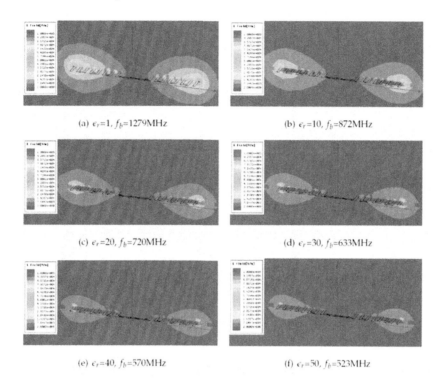

(a) $\varepsilon_r=1$, $f_b=1279$MHz

(b) $\varepsilon_r=10$, $f_b=872$MHz

(c) $\varepsilon_r=20$, $f_b=720$MHz

(d) $\varepsilon_r=30$, $f_b=633$MHz

(e) $\varepsilon_r=40$, $f_b=570$MHz

(f) $\varepsilon_r=50$, $f_b=523$MHz

According to the geometry of the MDA in Figure 11 and the method for calculating the ε_{reff} of an MDA by (13), introduced in Subsection 3.2, the values of c and d which should be substituted in (13) are 39.8mm and 41.8mm as marked in Figure 11. The values of the ε_{reff} of the MDA on a dielectric substrate obtained by (13) and simulations are shown in Figure 12. In detail, Figure 12(a) describes the variation of the ε_{reff} along with the variation of the ε_r at three different substrate thicknesses. Similarly, Figure 12(b) describes the variation of the ε_{reff} along with the variation of the substrate thickness at three different ε_r. The black curves in Figure 12 represent the results derived by (13) and the grey curves represent the results derived by the simulation.

The agreement between the simulation results and the calculated results in Figure 12 demonstrates again that the method of simplifying the MDA to the CPS according to their resemblance

in electric field distribution to calculate the MDA's ε_{reff} is feasible.

Again, it is observed that with the increase of the ε_r, the divergence between the two curves increases in Figure 12(a). The observation of this divergence has been explained previously in Subsection 3.2.

The experimental validation of this method and Equation (12) is discussed in the next section.

4. EXPERIMENTAL VALIDATION OF THE MODIFIED EQUATION

In order to test the validity of (12), the MDA shown in Figure 11 was fabricated and its input impedance was measured. The dielectric substrate of the MDA is FR4 with overall size $W \times L = 43.8$mm$\times 28.8$mm, thickness $t = 1.6$mm and dielectric constant $\varepsilon_r = 4.4$. The dimensions

Figure 11. An MDA loaded with three different meander lines

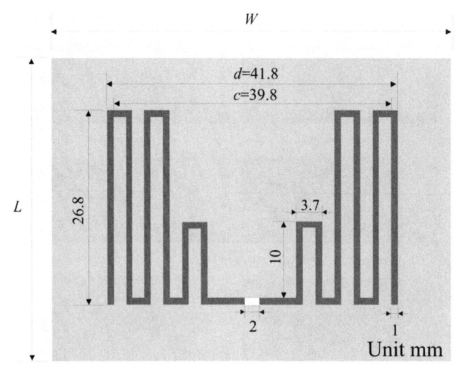

Figure 12. The relative effective permittivity of the MDA in Figure 11 as a function of the dielectric constant ε_r for sub-figure(a) and as a function of the substrate thickness t for sub-figure(b)

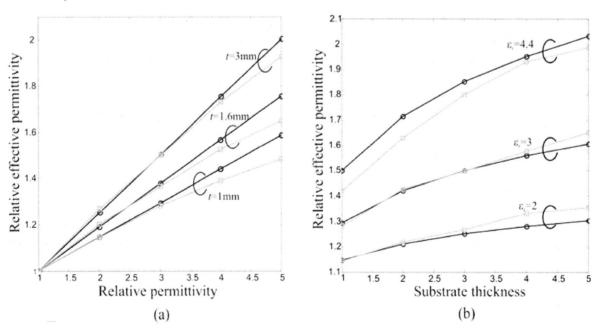

Figure 13. The half MDA on a ground plane being tested

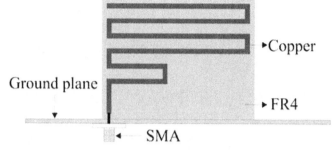

(a) Picture of the half MDA (b) Details and side view of the half MDA

of the tested MDA have been optimised by the simulation software HFSS in order to make its input impedance a conjugate match to the chip impedance (13.6 - j142Ω) at 923MHz. The frequency of 923MHz was chosen because it is the centre frequency of a 6MHz band according to Australian RFID standards.

The input impedance is tested by the method which is shown in Figure 13. The unbalanced version of the MDA is soldered on to an SMA connector which is mounted on a ground plane. An image of the half MDA is created by the ground plane, which can complete the whole MDA. A half unbalanced antenna on a ground plane will have half of the input impedance of a complete balanced antenna. The input impedance of the half MDA on a ground plane at 923MHz measured by the network analyser 8714C was 5.7+j60Ω, as is shown in the Smith Chart in Figure 14. Hence, the input impedance of the complete MDA is 11.4+j120Ω, which is very close to the target design impedance. The complete MDA mounted with a Higgs-2 chip manufactured by Alien Technology (Higgs™-2, 2008) is shown in Figure 15. The reading range under the 4W EIRP of the tag is 5.2m.

Inputting the physical dimensions and ε_r equal to 4.4 of the MDA on the substrate being examined into (13) to derive the ε_{reff} gives the result 1.64. Then, substituting ε_{reff} = 1.64 into (12) with the physical dimension of the MDA to calculate the resonant frequency of this MDA with a known

complex chip impedance, gives the result 910MHz. Comparing the experimentally verified resonant frequency of 923MHz with the calculated resonant frequency of 910MHz reveals that they are reasonably close to each other which validates the feasibility of using (12) for design.

5. RADIATION PATTERN AND EFFICIENCY

As is verified by Nakano, Tagami, Yoshizawa & Yamauchi (1984), the radiation pattern of an MDA loaded with uniform meanders in free space is similar to that of the half wavelength dipole. We also found by using the Ansoft HFSS simulation software that the radiation pattern of an MDA on a dielectric substrate, either loaded with uniform meanders or different size meanders, is similar to that of the half wavelength dipole. Two factors which affect MDA's radiation efficiency are discussed in the two following subsections respectively. The two factors are the physical dimension of the MDA and the properties of the dielectric substrate underneath.

5.1. Physical Dimension of MDA

The radiation efficiency of an antenna is defined as (19)

Figure 14. Smith chart derived by the network analyser 8714C showing input impedance of the half MDA on a ground plane. The mark is at the frequency of 923MHz.

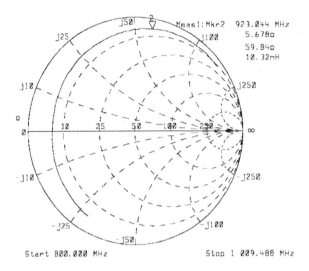

$$\eta_r = \frac{P_r}{P_r + P_L} \, . \qquad (19)$$

where P_r is the radiation power and P_L is the loss power.

Following Endo, Sunahara, Satoh & Katagi (2000), we assume that the distribution of the current along the conductor forming an MDA is in sinusoidal form, and the peak value of the current

occurring in the middle of the MDA is I_0. This current diminishes to zero at the two ends of the MDA. The current can be expressed as

$$I(z) = I_0 \sin k'(\frac{l}{2} - |z|) \, . \qquad (20)$$

where z is the coordinate along the conductor and the MDA's feed point sits on the origin of

Figure 15. A tag based on the MDA in Figure 11

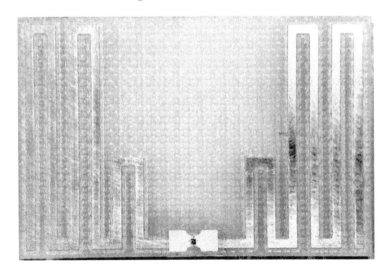

this coordinate. Different from the symbols used in the paper published by Endo, Sunahara, Satoh & Katagi (2000), our l is the whole length of the conductor either in the form of strip or wire. Our k_0 which has the same value as Endo, Sunahara, Satoh & Katagi (2000) is equal to π/l.

The loss resistance per unit length of the conductor can be expressed using the skin depth as

$$R_L = \frac{1}{P}\sqrt{\frac{\omega\mu_0}{2\sigma}}. \tag{21}$$

where ω is the angular frequency, μ_0 is the permeability of free space, σ is the conductivity, and P is the perimeter of the cross section of the conductor ($P = \pi b$ for a circular wire of diameter b; $P = 2a$ for a very thin strip of width a).

Hence, the loss power of the MDA can be expressed in (22).

$$P_L = \int_{-\frac{l}{2}}^{\frac{l}{2}} I^2(z) R_L dz \tag{22}$$

Substituting (20) and (21) into (22), the loss power can be derived as

$$P_L = \frac{1}{2} I_0^2 \frac{l}{P}\sqrt{\frac{\omega\mu_0}{2\sigma}}. \tag{23}$$

Equation (23) states that the loss power of the MDA is proportional to the whole length of the strip or wire comprising the MDA and the square root of the angular frequency, and it is inversely proportional to the strip width and square root of conductivity.

The radiation power can be obtained by (24).

$$P_r = \frac{1}{2} I_0^2 R_r \tag{24}$$

where R_r is the radiation resistance defined at the feed point. The factor of 1/2 results from our use of peak value phasors.

Endo, Sunahara, Satoh & Katagi (2000) assumed and experimentally verified that the ratio of the horizontal length of an MDA to the length of a half wavelength dipole is equal to the ratio of the radiation resistance of the MDA to the radiation resistance of the half wavelength dipole, if the two antennas are resonant at the same frequency. The equality is expressed in (25).

$$\frac{R_r}{R_d} = \frac{s}{\lambda/2} \tag{25}$$

where s is the horizontal length of the MDA as is shown in Figure 3. R_d is the radiation resistance of the half wavelength dipole which is readily known as 73Ω. Hence, the radiation resistance of the MDA can be expressed as

$$R_r = \frac{2sR_d}{\lambda}. \tag{26}$$

Therefore, after inserting (26) into (24), the radiation power becomes

$$P_r = \frac{1}{2} I_0^2 \frac{2sR_d}{\lambda}. \tag{27}$$

Substituting (23) and (27) into (19), the other form of radiation efficiency of the MDA can be obtained:

$$\eta_r = \frac{2sR_d}{2sR_d + \dfrac{l}{P}\sqrt{\dfrac{\omega\mu_0}{2\sigma}}} \tag{28}$$

where l (the total length of the wire or strip comprising the MDA) is equal to $s + 2mh$, if the MDA is loaded with m uniform meander lines which height is h. Generally speaking, loading more and longer meander lines on a dipole will decrease the resonant frequency or decrease the ratio of the horizontal length of the antenna to the

Figure 16. Radiation efficiency comparison between two types of MDA

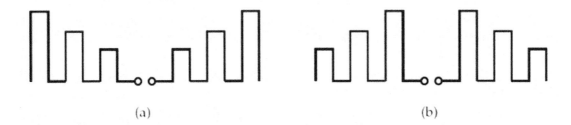

(a)　　　　　　　　　　　　　　(b)

wavelength according to the study in the previous sections. However, more or longer meanders result in the increase of the whole length of the wire or strip comprising the MDA (*l* goes up) and that increase according to (28) lowers the radiation efficiency. Therefore, a tradeoff has to be made between the horizontal length of the MDA and radiation efficiency at a known resonant frequency.

In addition, according to the assumption of the sinusoidal current distribution on an MDA, the current is large in the middle and declines as the current flows further away from the centre of the MDA. Hence, the meander line loaded in the middle leads to more losses than the same meander loaded at the end of the MDA. For example, in Figure 16, there are two types of MDA. According to the study in Section 2 and Section 3, the two MDAs will have similar resonant frequencies whether they are in free space or on a dielectric substrate. However, for achieving better radiation efficiency, type (a) is preferable. For the MDA loaded with uniform meanders, leaving more horizontal segments near the feed port before loading the meander lines is preferable.

5.2. Dielectric Substrate

The effects on an MDA's radiation efficiency by a dielectric substrate have been studied by Yamada & Michishita (2005). The radiation efficiency of two types of MDAs is compared. Both of them have the same footprint and are loaded with uniform meander lines. One of them is placed in free space, $\varepsilon_r = 1$. The other MDA is sandwiched between two dielectric substrates of thickness $t = 0.1$mm and $\varepsilon_r = 10$. By adjusting the number of meander lines m loaded on the MDA, these two antennas are made resonant at the same frequency. It is found that the MDA sandwiched between high dielectric constant material needs fewer meander lines than the other does and the smaller number of loaded meander lines results in higher radiation efficiency as analysed above.

Yamada & Michishita (2005) stated that deploying higher dielectric constant material will result in fewer loaded meander lines. This can be explained by (12). By deploying higher ε_r material, the factor on the left side of (12) becomes smaller, so that on the right side of (12), fewer meander lines are needed to make the equation valid.

6. CONCLUSION

In this chapter, a simple analytic equation (11) from the literature (Endo, Sunahara, Satoh & Katagi, 2000) has been used as a basis for calculating the resonant frequency of an MDA in free space. The validity has been verified by simulations. However, this equation has some limitations in analysing the MDA on a dielectric substrate for RFID tag antenna design. In order to overcome those limitations, a modified closed-form Equation (12) as been derived starting from equation (11). One significant modification is to consider a relative effective permittivity ε_{reff} in Equation

(12) so that the equation can include the effects brought by the dielectric substrate. A method of deriving an approximate effective permittivity ε_{reff} is given and verified by simulations. In addition, the special needs of an impedance matching condition in designing RFID tag antennas is also taken into account in Equation (12). The modified Equation (12) is experimentally examined. Following that the radiation pattern and efficiency of an MDA are discussed.

It should be noted that the method for obtaining the factor ε_{reff} has its own limitation. It has been demonstrated that the method can only give reasonably accurate result when the relative permittivity ε_r of the substrate is in the range from 1 to 4. However, this does not impede the usage of this simple method in analysing the MDA made for RFID tags, since the relative permittivity ε_r of the materials usually used for packaging and manufacturing the RFID tags are within that range. Extensive of the model to account for the deviation on substrate with large permittivity are feasible, but were not considered because of this reason.

In summary the content in this chapter provides a thorough analysis of meander line dipole antennas which are commonly used for designing RFID tag antennas because of their size reduction and relatively high radiation efficiency properties. By making use of the modified Equation (12), the shape of the MDA designed as an RFID tag antenna can be approximately estimated. The process of the estimation is much more efficient than that by simulation software based on numerical methods, but to some extent at the expense of accuracy. As a result, we recommend that an antenna designer should use Equation (12) in the primary design of the MDA for RFID tag and then finalise the design by simulation software. The combination of the analytic equation and the simulation can both shorten the design cycle and maintain the accuracy.

REFERENCES

Balanis, C. A. (2005). *Antenna theory* (3rd ed.). Hoboken, NJ: John Wiley.

Choi, W., Son, H., Shin, C., Bae, J. H., & Choi, G. (2006, July). *RFID tag antenna with a meandered dipole and inductively coupled feed*. Paper presented at IEEE Antennas and Propagation Society International Symposium.

Endo, T., Sunahara, Y., Satoh, S., & Katagi, T. (2000). Resonant frequency and radiation efficiency of meander line antennas. *Electronics and Communications in Japan (Part II Electronics)*, *83*, 52–58. doi:10.1002/(SICI)1520-6432(200001)83:1<52::AID-ECJB7>3.0.CO;2-7

Hanna, V. F. (1980). Finite boundary corrections to coplanar stripline analysis. *Electronics Letters*, *16*(15), 604–606. doi:10.1049/el:19800419

Higgs™-2. (2008). *Product overview*. Retrieved from http://www.alientechnology.com/docs/products/DS H2.pdf

Marrocco, G. (2003). Gain-optimized self-resonant meander line antennas for RFID applications. *IEEE Antennas and Wireless Propagation Letters*, *2*, 302–305. doi:10.1109/LAWP.2003.822198

Marrocco, G. (2008). The art of UHF RFID antenna design: Impedance-matching and size-reduction techniques. *IEEE Antennas and Propagation Magazine*, *50*(1), 66–79. doi:10.1109/MAP.2008.4494504

Michishita, N., Yamada, Y., & Nakakura, N. (2004). *Miniaturization of a small meander line antenna by loading a high ε_r material*. Paper presented at the 5th International Symposium on Multi-Dimensional Mobile Communications and The 2004 Joint Conference of the 10th Asia-Pacific Conference.

Nakano, H., Tagami, H., Yoshizawa, A., & Yamauchi, J. (1984). Shortening ratios of modified dipole antennas. *IEEE Transactions on Antennas and Propagation*, *32*(4), 385–386. doi:10.1109/TAP.1984.1143321

Rashed, J., & Tai, C. T. (1982, May). *A new class of wire antennas*. Paper presented at Antennas and Propagation Society International Symposium.

Yamada, Y., & Michishita, N. (2005, March). *Antenna efficiency improvement of a miniaturized meander line antenna by loading a high ε_r material*. Paper presented at IEEE International Workshop on Antenna Technology: Small Antennas and Novel Metamaterials.

KEY TERMS AND DEFINITIONS

CPS: CPS is the abbreviation for coplanar strips. This kind of structure is composed of two parallel strips usually built on a dielectric substrate.

Effective Permittivity: The term is used for describing the effects of a dielectric substrate on the transmission line or antenna above it in terms of permittivity.

Impedance Match: Impedance match is very important in the antenna design, especially in RFID tag antenna design, since the power transmission between the tag antenna and the chip is significantly dependent on the impedance matching condition.

MDA: MDA is the abbreviation for meander line dipole antenna. A meander line dipole antenna is one form of dipole antenna.

Radiation Pattern and Efficiency: Radiation pattern and efficiency are two terms for describing firstly an antenna's radiation directivity and secondly the ratio of the power radiated out to the power lost in the antenna.

RFID: RFID is the abbreviation of radio frequency identification. RFID is a type of automatic identification technology making using of radio waves.

UHF: Ultra high frequency; ultra high frequency is the spectrum usually defined as ranging from 300MHz to 3000MHz, but in RFID systems, it is normally taken to mean from 860MHz to 960MHz.

Chapter 3
UHF RFID Reader Antenna

Tapas Chakravarty
Tata Consultancy Services Ltd (TCS), India

P. Balamuralidhar
Tata Consultancy Services Ltd (TCS), India

ABSTRACT

This chapter addresses the various design issues, requirements, and specifications in designing reader antennas for UHF RFID applications. In a typical UHF system, the RFID reader antennas are geography specific; that is, there are different antennas for different geography, namely North America, Europe, et cetera. The discussion on the design challenges and performance expectations lead to a new form of compact and broad band antenna, which will be applicable for the global UHF RFID band of 860MHz to 960 MHz. In addition, this chapter also provides potential future trends in RFID reader antenna design. The issues and challenges discussed in this chapter are envisioned to provide a roadmap to the potential challenges to be faced by a chipless RFID system.

1. INTRODUCTION

In recent years, RFID technology is seeing tremendous growth as it is been rapidly developed and used in many applications: Healthcare, supply chain management systems, pharmaceuticals, transportation, agriculture etc. RFID Readers can be divided into Fixed/Stationary readers and Mobile/Handheld Readers (Elisabeth, Kemeny, Egri, Monostori. 2006). Currently RFID systems are operating in three frequency bands: Low Frequency (LF), High Frequency (HF) and Ultra High Frequency (UHF). Microwave frequency range (ISM band) is also used in some cases, where the frequency of choice is generally 2.4-2.485GHz. Majority of applications of microwave band relates to "active tag" system, that is, the tag is powered by battery. Globally each country has its own frequency allocation for LF, HF and UHF bands. Each frequency has advantages and disadvantages relative to its capabilities.

An appropriate choice of frequency for a RFID system deployment (LF/HF/UHF) depends on multiple factors namely reading distance, number of assets that can be tracked together, size, cost effectiveness of the system and more importantly,

DOI: 10.4018/978-1-4666-1616-5.ch003

the reliability of reading (since wireless propagation is strongly dependent on environment). It is worth noting that an optimum antenna aperture is half wavelength size. Therefore, if the desired frequency of operation is lowered (longer wavelength), the physical antenna size is required to go up for better efficiency. One can now estimate the size of the efficient antenna if the carrier frequency is in hundreds of KHz! The resultant compromise in size & usability at lower frequency, does indicate that a useful LF system is based on 'near field' reading; generally through magnetic coupling. The mechanism of magnetic coupling (used for both LF & HF systems) assumes transformer like action between the reader antenna (primary coil) and the tag (secondary coil) and are well behaved very close to the reader antenna.

Low-Frequency (LF) RFID systems are typically 125 KHz, while some systems operating at 134 KHz do exist as well (Klaus, 2003) (Barthel 2009). For LF RFID systems read range is smaller(< 0.5m or 1.5 ft) and also suffers from slower read speed than the higher frequencies. LF systems tend to be less sensitive to interference than higher frequency options. But at LF, antenna volume will be large compared to other frequencies. The LF spectrum is not considered a truly global application and such deployments worlwide have got limited reach.

High Frequency (HF) RFID systems operate at 13.56MHz frequency. Working at higher frequencies extend the applicability of RFID systems due to some inherent advantages like the ability to have higher data rates (Klaus, 2003) (Barthel 2009). Similarly higher resonance frequency of the antenna circuit gives lower inductance and capacitance values. The lower antenna inductance means a fewer number of turns (5–10) (compared to 100–200 turns for the LF systems). It also presents the possibility of being produced not only from copper wire but also printed or etched on foil. These antennas are flexible and can be easily laminated in other embedded systems (electronics). These basic advantages lead to a lowered cost for the system.

Ultra high frequency (UHF) RFID systems operate from 860MHz to 960MHz band and globally each country has its own frequency allocation (determined by the respective wireless planning regulators), e.g., 866–868 MHz in Europe, 902–928-MHz band in North and South of America, 865–867 MHz in India,, 952–955 MHz in Japan (Klaus, 2003) (Barthel 2009). Deployment of UHF RFID systems have increased many-folds world wide due to their global standardization and some silent features like longer reading range, high speed reading, more data storage capability and lower cost of passive tags (Karmakar, Zakavi, & Kumbukage, 2010). The antennas used for UHF systems are typically patch, dipole, slot and helix antennas; the exact type is dependent on exact requirement. The coupling mechanism for UHF is no longer Magnetic or Capacitive. UHF RFID systems are based on backscatter principle, where, it is assumed that the passive tag resides in the far-field of the reader antenna. It is also appropriate to discuss about the gain of the UHF reader antenna. At UHF, one can use traditional antenna evaluation methods like gain & effective aperture for evaluating the reading range.

It is to be noted that typical commercial UHF readers have geography specific antennas; this is due to the fact that the reader antennas are expected to have high gain (approximately 6dBi). It is difficult to design an antenna with 5-6dBi gain for the entire frequency bandwidth (100MHz bandwidth at the centre frequency of 900 MHz). The key point in the reader antenna is that each antenna has a specific pattern of coverage. One can opt for an antenna type based on the implementation need, cost, coverage area, reading distance and size. As explained in (Foster & Burberry, 1999).antennas which are omni directional should be avoided and wherever possible, directional antennas should be used because they have the advantage of fewer disturbances to the radiation pattern and the return loss. An important consideration for a reader antenna design occurs due to the restrictions imposed by the respective countries in granting permission to operate wireless systems. Such restrictions are reflected in the respective wireless planning documents like FCC part 15 for North America. As an example, in North America, the maximum permissible radiated power level is 4 Watts EIRP (Effective Isotropic Radiated Power). Since the input to the reader antenna can go as high as 1 Watt; one must not try to connect an antenna with gain in excess of 4 (6dBi).

In this chapter, we intend to discuss some of the important design aspects of UHF RFID, leading to the development of a new and smaller sized UHF RFID reader antenna. To enable a better understanding of the trends in technology development in UHF reader antennas, we begin with discussions on important design issues first. While we trace some recent developments in reader antenna design, we also present a new design methodology for a compact & broadband UHF reader antenna of a moderate gain.

The remainder of the chapter is organized as follows. In section 2, we describe the significance of different radio coupling mechanisms between tag and reader. Performance limitations arising due to small antennas are described in section 3. These are followed by a brief discussion on some important recent works related to UHF RFID reader. We introduce the new design for UHF reader antenna in section 5 and discuss the possible future trends in section 6.

2. SIGNIFICANCE OF COUPLING METHOD FOR RFID SYSTEM DESIGN

An excellent discussion on the differences in coupling type and their implication on the performance is given by Dobkin (2008). For a successful design and implementation of a RFID system (including reader/tag antenna), it is crucial to understand the implication of the coupling mechanism between the reader, which is interrogating and the tag, which is responding with its own unique identification (ID) number. The reader interrogates the tag(s) by emitting electromagnetic energy. This coupling mechanism can be categorized as 'near-field' or 'far-field' (Want, 2006). We have stated in the earlier section that LF and HF RFID systems use near-field (magnetic or capacitive) coupling mechanism, while UHF tags will be operating in predominantly far-field. For antennas whose size is comparable to wavelength (used in UHF RFID), the approximate boundary between the far field and the near field region is commonly given as (Balanis, 2008):

$$r = \frac{2D^2}{\lambda} \qquad (1)$$

where D is the maximum antenna dimension and λ is the wavelength.

On the other hand, for electrically small antennas, the radiating near field region is small and the boundary between the far field and the near field regions is commonly given as $r = \lambda / 2\pi$ (Balanis, 2008). The form of coupling that is used depends upon the intended application. Normally inductive and capacitive couplings are used for short range links and back scattering coupling for long range links

The backscatter coupling mechanism for UHF RFID systems work beyond the near field region; the radio signal propagating away from the RFID reader. As the signal reaches the RFID tag; the tag wakes-up by converting the incoming electromagnetic energy to DC voltage. Subsequently the tag, alternately presents short and matched load at the tag antenna terminals. Such sequence is based on the unique binary number stored in the transponder chip (Note: A passive tag consists of a transponder chip connected to a tag antenna). When the antenna terminals are "short", the incoming signal is scattered back towards the reader and when the terminals are matched (impedance matched), the incoming signal is absorbed by the chip. This difference in amplitude levels for the backscattered signal is detected as alternate '1'(s) and '0'(s); thereby giving the unique code of the tag.

Therefore, the essential difference between the three types of RFID systems is the two-way coupling mechanism between the reader and the tag. While LF/HF systems are limited to 'near-field' only, UHF RFID systems can perform well for both near-field and far-field. For near-field operations (in UHF), an inductive loop is included in the tag in addition to the extended antenna structure.

There is also an inherent similarity in the operation for the three types of RFID techniques. For all the cases, the transponder IC, in the passive tag, require a minimum amplitude level of the incoming interrogating signal just to wake-up and become functional. Such level is defined as the sensitivity of the tag IC. Sensitivity of the tag IC typically determines the tag reading range i.e the farthest distance (from the reader antenna) from where the tag can be detected reliably.

If we now assume that this very chip (or transponder IC) is removed from the tag; leading to the design of *chipless RFID*; the reading range of the tag is determined by two factors namely the sensitivity of the reader and the reflection coefficient of the frequency selective tag antenna (s). The interrogating signal from the reader needs to reach the tag and when reflected back (along with some modulation denoting the unique ID) needs to reach the reader above a minimum threshold set for such detection. The entire detection process, theoretically at least, is similar in nature to Radar based target detection and identifying method

A proper understanding of the coupling mechanism (and its effects on system performance) is an important consideration when it comes to designing the RFID antennas; both reader and the tag. The particular coupling type selected, imposes limits on system performance in terms of detection range, sensitivity to antenna orientations, size of the antennas. In particular, such considerations have a direct bearing on the attempt to design compact reader antennas.

Price has to be paid for using compact antennas (small electrical size) in RFID systems. In the following section, we introduce a few important functional limitations when we deploy a compact antenna. Such applicability is not just limited to RFID but these are valid for other wireless systems as well.

3. PERFORMANCE LIMITATIONS FOR SMALL ANTENNAS

When a high-frequency signal flows through a conductor or a pair of conductors whose length is nearly half-wavelength, a significant part of the source energy escapes into the space as radiation. However, in many cases, it is required to embed an antenna into a physically small device, not leaving much room to incorporate a half-wavelength (or bigger sized) antenna or antenna arrays. For such cases, we are forced (by the end objective) to reduce the physical size of the antenna yet maintain an optimum performance. A reduced size of the antenna has fundamental limitations of performance in terms impedance matching, frequency bandwidth, effective gain and polarization purity.

In the following sub-sections, we briefly highlight some important issues related to limitations imposed by small size on the performance of reader antenna

3.1. Effect of Antenna Volume on Electrical Performance

In 1946, Wheeler (1947) introduced the concept of lumped-element loaded electrically small antennas, and defined the fundamental relationship between Q (quality factor) and physical (occupied) volume of antenna. In 1948, Chu (1948) extended Wheeler's analysis and expressed the fields for an omni-directional antenna in terms of spherical wave functions and found limits for the minimum antenna quality factor (Q), the maximum gain (Gm) and the ratio G/Q. In 1959, Harrington (1960) related the effects of antenna size, minimum Q, and gain for the near and far field diffraction zones for linearly and circularly polarized waves, and also treated the case where the antenna efficiency η is less than 100%. In 1964, Collin and Rothschild (1964) presented a method to find minimum Q without using the equivalent network for both spherical and cylindrical modes. In 1969, Fante (1969) extended Collin work to multimode antennas. In recent years there has been much progress in this area.

A small antenna is defined in terms of its electrical size. Typically, an antenna is called 'small' when the following relationship hold true (Balanis, 2008)

$$ka \leq 0.5 \qquad (2)$$

where, k is the propagation constant in free space and 'a' is the radius of the sphere around the radiating aperture (equal to the maximum dimension of the aperture with respect to the ground). For large ground plane reference, the image of the antenna must be included in the definition. For such antennas, the fundamental limit of performance is defined by Chu-Harrington relationship as:

$$Q = \eta_r \left(\frac{1}{(ka)^3} + \frac{1}{ka} \right) \qquad (3)$$

where, Q is the minimum quality factor related to the smallest volume in which the antenna is enclosed. Thus, eq. 3 imposes a constraint of an upper bound on the achievable bandwidth. Moreover, for a small size of the antenna, the input terminal (antenna) sees a highly reactive component. This reactance leads to high Q and narrow bandwidth. Input reactance also has the effect of storing energy; thus radiation efficiency is drastically reduced

3.2. Effect of Truncated Ground Plane on Antenna Performance

In estimating the performance of an antenna, the size of the ground plane needs to be considered. In general, stable (and therefore insensitive to environment) radiation patterns are produced for much larger ground planes; but this makes the antenna bigger (Chen, 2007). For modern day devices it isn't uncommon for the ground plane to be of nearly the same size as active patch. When a ground plane is of the order of the size of the radiating patch, it can couple and produce currents along the edges of the ground plane resulting in spurious radiation. The resultant antenna pattern is, therefore, a combination of the two patterns, the desired one and the spurious pattern.. On the other hand, if the ground plane is significantly small then tuning becomes increasingly difficult and the overall performance degrades. One very useful antenna type (antenna with small ground plane) is Inverted-F Antenna (IFA). Chen, 2007, provides very important insight into the design and functioning of IFA since this particular antenna type is an excellent choice for small, low-profile wireless designs. One way of increasing antenna (like IFA) bandwidth and efficiency is to use the technique of increasing substrate thickness, (works up to a certain extent). Thereafter, IFA performance will degrade due to the higher cross-polar level and surface-wave excitation. As explained in many texts on microstrip antenna like Chen, (2007); the surface waves get excited and travel along the dielectric substrate (between the ground plane and the dielectric-to air interface due to total internal reflection). Such surface waves get scattered, and diffracted at the edge of a finite comparable sized substrate causing a reduction in gain and an increase in end-fire radiation along with high cross polar levels.

A serious error in judgment may occur when a small antenna with a nearly equal sized ground plane is designed and evaluated in isolation. When this antenna (with the finite ground plane), is connected to a radio transmitter board (RF PCB), the presence of additional ground plane (available on the RF PCB) may completely alter the radiation pattern. Thus, the design methodology followed for such antennas is two pronged: first, design a basic antenna in isolation under the given space constraint then simulate and revalidate the design in presence on the given RF PCB using any standard full-wave numerical methods like Method of Moments (MoM), Finite-Difference Time Domain (FDTD) etc.

4. RECENT TRENDS IN READER ANTENNA DESIGN

In recent years there has been much progress in antenna design for different RFID applications. This section describes the some of the recent developments on reader antenna design found in standard literatures.

4.1. Notable Fixed Reader Antenna Designs

In general, antennas used for fixed readers are planar type and large in size. Two recently published works are worth noting.

An innovative antenna type: broad band circularly polarized antenna, is described in (Chen, Xianming & Chung, 2009) for use with fixed readers for universal UHF RFID applications. The above described antenna consists of four layers of conductor, including two suspended radiating patches; a suspended microstrip feed line, and a finite-size ground plane. Air substrate (to incorporate the advantages of lower dielectric constant value) is used by the authors to achieve higher gain, broader bandwidth, and lower cost. The microstrip feed line of a width of 24 mm is suspended above the ground plane of size 250 mm×250 mm at a height of 5 mm. One end of the feed line is connected to an RF input, while the

other one is open circuited, which is claimed to simplify the antenna structure. By using simple feeding structure and combining several band broadening techniques, the optimized antenna has achieved the desired performance over the UHF band of 840–960 MHz with the gain of more than 3 dBi, AR (Axial Ratio) of less than 3 dB, return loss of better than 15 dB, and a 3dB AR beam width of larger than 75 deg. Therefore, this universal design can be applied to all the UHF RFID applications worldwide. The authors mention that the reading-range measurement has validated the purpose of the proposed antenna to be incorporated into the multi-band RFID readers or/and readers operating at different RFID bands to achieve desired reading ranges. Furthermore, the parametric studies presented by the authors, have addressed the effects of the truncations of the patches, height of the parasitic patch, size of the feeding probes, extension of the open-circuited feed line, and size of the ground plane on the performance of the antenna. The information derived from the above study will be highly helpful for antenna engineers to design and optimize the RFID antennas.

Another antenna design for universal UHF RFID reader is described in (Abo-Elnaga, Abdallah & El-Hennawy, 2010a) Proposed antenna structure is composed of a single loop on the top of FR4 substrate, which is meandered in the form of a rose to obtain compact size. The rose antenna is fed directly through microstrip line connecting the rose with the RF input. Down the microstrip feeding line a finite ground plane placed on the bottom of the substrate. The antenna input impedance is controlled by adjusting the offset length of the ground plane. The described antenna covers the universal UHF frequency band of 840 – 860 MHz with gain of 1.21dBi. Therefore, the proposed fabricated antennas are simple, cheap, and compact and suitable for the universal UHF RFID applications.

4.2. Notable Handheld Reader Antenna Designs

Handheld readers give the advantage of mobility in asset tracking deployment scenarios particularly in warehouse situaions. Antennas for handheld readers need to be compact; small physical size and light weight.

A novel compact high gain antenna design for handheld UHF RFID readers is reported in (Nikitin & Rao, 2010a). The described antenna is designed based on a helical structure operating in backfire mode, with additional reflector. It has a compact form factor, lightweight and simple in construction. The described antenna is alternative to generic three dimensional helical antennas built suspended wires over ground plane and covered with radomes. Due to the compact form factor and performance characteristics comparable to those of large patch antennas used in fixed reader installations, the described antenna can be an attractive solution for handheld readers to maximize tag read range while providing circular polarization and insensitivity to the orientation of tags being read. The antenna is shown to possess maximum linear gain >6dBi and an axial ratio of <2 dB over 60MHz band, which could be centered to cover any desired portion of the global UHF band (860 – 960 MHz).

Another important antenna type is described in (Nikitin & Rao, 2010b). It is a compact linearly polarized and high gain antenna for a handheld UHF RFID reader. The printed Yagi antenna described by Nikitin et al, (2010b) has three elements (driver, director, and reflector) which fit into compact footprint (100 mm x 100 mm). Key parameters in the design are lengths and shapes of antenna elements and their mutual spacing. The authors have claimed to achieve ruggedness, by stamping out the antenna traces out of metal and molding it into 6mm thick ABS plastic (dielectric permittivity of 3.5). Antenna impedance is matched to 50 ohm source (RFID reader) using discrete passive matching network implemented

on a separate PCB, also molded into plastic. Maximum antenna gain of above described antenna is >6dBi and VSWR is better than 1.3 in 865 - 870 MHz band. For the US version (902 – 928 MHz) of antenna has slight different shapes and different components in matching network.

The physically limiting challenge with respect to the performance of a small antenna is to extract high gain from the antenna. Uddin, Reaz, Hasan, Nordin, Ibrahimy & Ali (2010).has shown that to avoid disturbances from operator hand or from other objects, it is best to use a balanced feed antenna. Polarization purity (of the antenna) is often required to enhance the system efficiency; else there can be a performance degradation due to polarization mismatch. In cases where high level of cross polarization is seen, balanced feed can be used to reduce cross polarization fields.. In a balanced feed structure, two symmetrical halves are driven by source and no part of the antenna is used as a ground reference: a simple example is dipole antenna. Balanced antennas are larger than single ended antennas and require using impedance transformer (balun) to be connected to a standard coaxial cable.

A circular polarized antenna for a portable RFID reader operating in the 900MHz frequency band is described by Barbin & Barbin (2010). A high dielectric constant substrate is used for miniaturization. Further size reduction is achieved by cutting slots in the radiating patch. The feeding structure is implemented on another substrate and both circuits are stacked to constitute an electromagnetic coupled antenna.

From the above stated study, it is seen that the usefulness of a universal reader antenna, for both hand-held and fixed readers, is getting recognized. Also, the attempt to reduce the physical size of the antenna (volume), without sacrificing gain and efficiency, is also of prime interest. It is worth noting that in many current applications (e.g. Pharmaceutical industry) UHF RFID deployment is required to operate at close ranges and not the far-field backscatter operation. Under such conditions, a smaller sized antenna design (which in turns will pull the far-field zone closer) without serious degradation of gain is worth investigation.

In the previous sections of this chapter, we have described briefly, the major concerns, constraints and trends in designing UHF RFID reader antennas. It is shown that the need for the hour is development of broadband and small antennas yet maintaining efficiency to the extent possible. The previous discussions lay the foundation for the need to design a new type of universal reader antenna which we are going to describe in details in the subsequent sections.

5. A NEW DESIGN OF UNIVERSAL READER ANTENNA

A new design method for universal reader antennas (for all the geographies at UHF) in compact form and with reasonable gain is presented in this section. It is shown that a modification in the way the ground is placed around the radiating patch can significantly improve performance for a compact antenna.

5.1. Antenna Element Design Methodology

A new compact and broad band antenna design for global UHF RFID band is now described. Some of the characteristics that are needed to be considered (for readers are): frequency band available worldwide, inexpensive, high efficiency, easy tuning, robust and low side-lobe (for reduced interference) (Uddin et. al, 2010) (Nikitin & Rao, 2008). As discussed in early sections, RFID UHF frequency is different in different regions and people are using separate antennas depending on regions (Li., Yang, Gong, Yang, & Liu, 2009). In another scenario, the need is that of a small reader antenna to detect ultra-small tags used in healthcare applications. Keeping those challenges in mind, we started investigation of different antennas and

came up with unique design as shown in Figure1 which can meet all above said requirements. As shown in Figure 1, a planar monopole patch as radiating element is considered. Figure 1 shows top plane view of proposed antenna; the antenna element is placed on a metal plate '9' by using Z-clamp connectors '7' on both sides of the antenna and metal screws '6' are used to fix the antenna element on metal clamps. The volume of the proposed antenna is 150mm×50mm×15mm that is in compact nature while maintaining reasonable gain, compared with other conventional antennas available in present market.

The impedance matching for the proposed antenna is controlled by the feed line position as well as a variable capacitor. The variable capacitor is used in situations, where antenna impedance bandwidth may shift due to environmental conditions; thereby making it impractical to tune by changing antenna dimensions physically. Antenna miniaturization techniques using dielectric or reactive loading are commonly used to increase the antenna's electrical size without a corresponding increase in its physical size (Ali, 2008) (Deshmukh & Kumar, 2005)(Abo-Elnaga, Abdallah & El-Hennawy, 2010b).

In general people use large ground plane to improve the performance of antenna but it is impractical to use large ground plane where compact structure required (Chen, Xianming & Chung, 2009).To change the electrical length of an antenna, one can insert (in series with the antenna) either an inductor (if the antenna is too short for the wavelength), or a capacitor (if is too long). {Note: Below first natural resonance, a small dipole exhibits capacitive reactance at the feed port}

The proposed antenna design brings in a novelty; it involves reactive loading by using z-clamps on both sides of antenna to increase the effective capacitance of antenna structure without increasing total antenna volume. The proximity of these clamps to the radiating patch is critical for obtaining the desired functionality. Also the

Figure 1. Top view of the broad band UHF RFID reader antenna with dimensions (Copyright ©2011, Tata Consultancy Services, Used with permission)

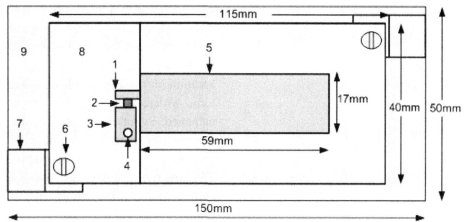

1. *Feed line 2. Variable capacitor 3.Top ground 4.PTH 5.Radiating patch*
6.*Screw 7. Z-clamp connector 8.Bottom ground 9.Metal plate*

antenna has partial ground plane '8' on both sides of the antenna. With the present described antenna setup gives the impedance bandwidth of 840MHz to 960MHz, which will cover all UHF RFID frequency bands globally.

The 3D view of the broad band antenna is shown in Figure 2, where the antenna element is placed above the metal plate with the support of Z-clamped and metal screws on both corners of the metal plate. Figure 3 shows a side view of proposed antenna; here we have used air cavity between antenna element and metal plate to improve the directivity of the antenna. Basically metal plate act as a reflector i.e. reduces the back lobe radiation and improves the directivity of the antenna. Some applications require directional pattern and the antenna need to operate in presence of metal objects. In such scenarios, above described antenna, is suitable. Present design is a directional antenna with linear polarization.

It is seen that this present design brings in additional parameters to optimize. We can reduce the length of the monopole further with appropriate changes in ground plane size; Z-clamp position & dimension vis-à-vis the patch, method of physical connection between the printed ground

(on the substrate) to the metallic ground plane below.. Optimization towards achieving a desired objective can be carried out by full-wave numerical methods (like using commercial solvers) with such variability's put in place. The proposed antenna is small in size and useful in applications where reading range requirement is an overlap of near-field and far-field.

5.2. Results and Discussion

The described antenna structure has been modeled by HFSS (High Frequency Structure Simulator) from M/s Ansys. HFSS works based on FEM (Finite Element Method) which is a very efficient method for 3D antenna designs. The return loss/VSWR is measured using E5062A Network Analyzer (from M/s Agilent Technologies).

The graph as shown in Figure 4 displays the simulated and measured return loss vs. frequency of the broad band antenna. It indicates that the return loss of the proposed antenna is <-10dB throughout the desired frequency band from 840MHz to 960MHz, thus the proposed antenna is tuned for optimum performance in the frequency range from 840MHz to 960MHz band.{Note:

Figure 2. 3D view of the broad band UHF RFID Reader antenna. (Copyright ©2011, Tata Consultancy Services, Used with permission)

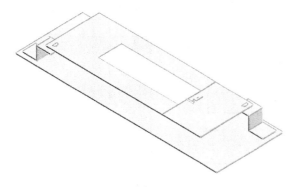

Return loss of 9.54dB corresponds to a VSWR of 2:1; for evaluation of antennas, a VSWR of 2:1 or lower is considered useful} There is slight variation between simulated and measured results due to the mechanical adjustments but due to the broad band nature of the proposed antenna, these effects may be considered negligible.

Figure 5 shows the realized gain (includes miss match losses & efficiency) versus frequency for the proposed antenna. Antenna gain is another fundamental parameter to identify the performance of antenna and is a ratio of the intensity in a given direction to the radiation intensity of isotropic antenna. Figure 5 shows the realized gain is maintained fairly constant throughout the desired frequency band (840-960MHz).

In Figure 6; the radiation patterns at 840, 910 and 955MHz in the elevation plane are shown.

The radiation characteristics of a linear antenna will be modified whenever the antenna is mounted on, or placed in proximity to a ground-plane. Therefore, either by proper design of the ground-plane, wherever feasible (in terms of its size, shape and conductivity), or by choosing an appropriate location on the ground-plane, the radiation patterns with certain desirable properties can be achieved. Symmetrical patterns and wide 3-dB beam width characteristics have been observed from graph. The 3-dB beam width is more than 75°, which is desirable for wide-coverage; basically the metal plate will act as a reflector i.e. it reduces the back lobe radiation and improves the directivity of the antenna.

From above results, it can be concluded that proposed antenna has a bandwidth of >120MHz with stable gain. This is a significant improvement over recent findings in standard literature. It is to be noted that RL >10dB is typically considered a standard notation for reflecting bandwidth of antenna. Being small in size has a major significance in case of a reader antenna.

5.3. Study of Variation of the Design Parameters

This section outlines the effect of reactive loading and air cavity gap on antenna performance. Referring to Figure7, we see the effect of different capacitor values on impedance bandwidth of the proposed antenna. A microstrip antenna can be

Figure 3. Side view of the broad band UHF reader antenna. (Copyright ©2011, Tata Consultancy Services, Used with permission)

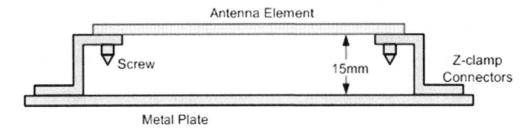

Figure 4. Simulated and measured return loss of broadband antenna (Copyright ©2011, Tata Consultancy Services, Used with permission)

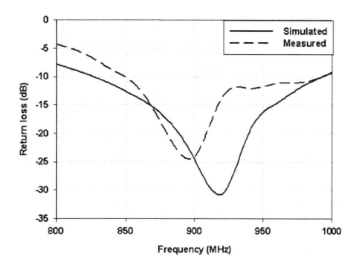

treated as parallel RLC resonator (being leaky cavity). Therefore additional changes in capacitance can vary the resonance frequency of antenna. In some cases, resonance frequency of the antenna may shift depending on object or surface where the antenna is going to be attached. Fine tuning the antenna will improve the additional matching requirement. From the Figure 7, it is clearly identified that the resonance frequency of antenna can be easily fine tuned by varying the capacitor value.

Figure 8 shows the effect of the air cavity on impedance bandwidth. It is clearly identified from

Figure 5. Simulated realized gain vs. frequency of the proposed antenna (Copyright ©2011, Tata Consultancy Services, Used with permission)

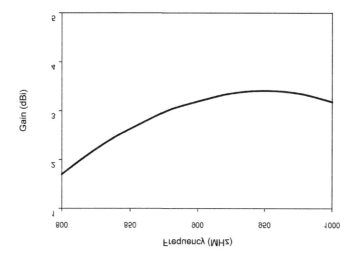

Figure 6. Radiation pattern at 840, 910, and 955 MHz in elevation plane (theta=90deg) (Copyright ©2011, Tata Consultancy Services, Used with permission)

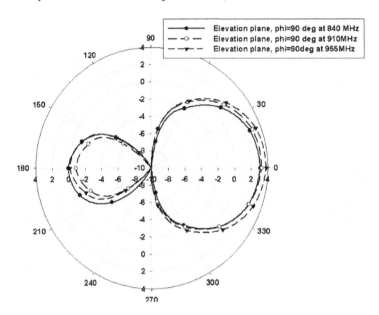

graph that for the described antenna, a minimum of 15mm air gap needs to be maintained for wider impedance bandwidth (840-960MHz). Figure8 shows that if air cavity is less than λ/20, then the impedance bandwidth of proposed antenna is drastically reduced.

The realized gain variance with changing the air cavity is displayed in figure 9. It is clearly identified from graph that 15mm minimum air cavity gap needs to be maintained for better antenna efficiency.

In summary, we present an antenna which is designed to give a broadband performance centered on 900MHz. The design parameters are varied based on the prototype mentioned in figure1. It is observed that, the variations in the design parameters give an indication of tolerance in the frequency response.

6. FUTURE RESEARCH DIRECTIONS

RFID system deployments have seen significant increase in recent times across multiple industry segments. A definite tilt towards UHF and still higher frequencies is being observed. These movements are being driven by considerable advancement in MIC techniques as well as improved ROI (return on investment) for some applications, where Barcode was considered to be only viable option. The main driving force behind rapid growth of RFID is definitely the cost; primarily, consumable cost i.e. the tags. The cost of tag can be pushed to a bottom minimum if the transponder IC is removed i.e. a chip-less RFID tag. Needless to point out that the incorporation of a chip-less RFID tag in a RFID system will require major changes in the reader including reader antenna. Currently, the sensitivity of a system is primarily dependant on the forward link i.e. the reader is interrogating the tag by impinging radio signal. The system sensitivity of the reverse link i.e. passive tag responding to reader is better than the forward link. This will definitely be altered for a chip-less RFID.

The progress, in terms of commercial success, of RFID implementation is seen in terms of frequency of operation that is from HF to UHF.

Figure 7. Effect of reactive loading on impedance bandwidth of broad band antenna (Copyright ©2011, Tata Consultancy Services, Used with permission)

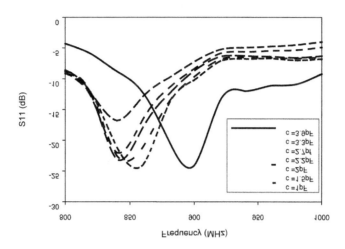

Already, some active tags at 2.4GHz are also being considered for some specific applications like asset tracking in Hospital Management System. As we move towards a still higher frequency and still lower cost, more and more technical challenges are being observed to make future RFID systems, the only Auto-ID method of choice. We try to highlight some of the research areas that can

have significant impact on the future adaptation of RFID

• Designing of orientation insensitive tags that will reduce the complexity in reader system design. Due to advanced IC manufacturing technologies, designing of orientation insensitivity tags are possible i.e. us-

Figure 8. Effect of air cavity gap on impedance bandwidth of broad band antenna (Copyright ©2011, Tata Consultancy Services, Used with permission)

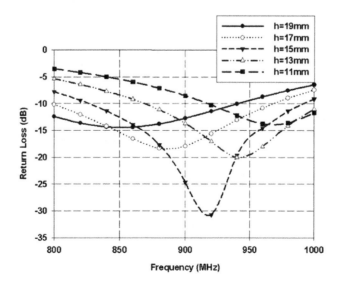

Figure 9. Effect of air cavity gap on gain of broad band antenna (Copyright ©2011, Tata Consultancy Services, Used with permission)

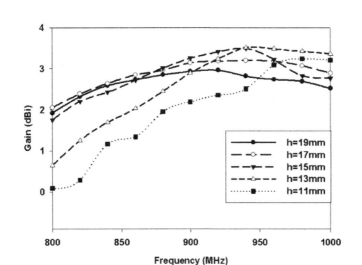

ing dual port IC which can support antenna diversity technique

- Metamaterial based antenna designs; can potentially lead to high gain but compact antennas
- Antenna diversity in readers; Reader design to exhibit antenna diversity by multiplexing their signals between a number of antenna modules mounted orthogonally, or by coordinating multiple readers
- Need to investigate different algorithms to mitigate reader collision system, i.e. reader synchronization
- Smart antenna design; which is broadband antennas with amplitude/delay weights associated with each element for enhanced adaptability in a challenged environment and to identify a weak tag.
- In case of Hand held reader, antenna needs to be compact. However, tolerance of objects including hands near the antenna, radiation away from the handheld display issues need to be addressed

7. CONCLUSION

Through this chapter, an attempt is made to address some of the design challenges envisioned for developing compact antennas as applicable for UHF RFID system design. Such issues have been addressed while designing a new compact antenna for the UHF reader. A compact and broadband patch antenna is described in this chapter for RFID reader operating in global UHF band. The described antenna uses a reactive loading method to increase the electrical length of antenna without increasing corresponding physical dimensions. The lateral size of ground plane is reduced by intelligently tweaking the ground plane by using metal clamps on both sides of antenna and occupies the volume of $150 \times 50 \times 15 mm^3$ which is compact, broadband with moderate gain. Antenna comprises a metallic plate and an antenna element is placed above the metallic plate by using Z-bend shaped clamps, here metal plate act as reflector and the air cavity is used to improve the directional pattern & radiation efficiency of the antenna which is desired in some of RFID applications. Antenna can be connected to RFID reader through a suitable standard coupling, such as a 50Ω coaxial cable with a suitable connector. Antenna input impedance is controlled by feed line position and adjusting capacitor value. From results, it can be concluded that proposed antenna is having bandwidth of >120MHz with stable gain. This may be considered as a major progress in technology advancement.

REFERENCES

Abo-Elnaga, T. G., Abdallah, E. A. F., & El-Hennawy, H. (2010a). *Universal UHF RFID rose reader antenna.* China: PIERS Proceedings.

Abo-Elnaga, T. G., Abdallah, E. A. F., & El-Hennawy, H. (2010b). *UWB circular polarization RFID reader antenna for 2.4GHz Band.* China: PIERS Proceedings.

Ali, J. K. (2008). A new compact size microstrip patch antenna with irregular slots for handheld GPS application. *Engineering & Technology, 26,* 1241–1246.

Balanis, C. A. (Ed.). (2008). *Modern antenna handbook.* New York, NY: John Wiley & Sons. doi:10.1002/9780470294154

Barbin, M. V., & Barbin, S. E. (2010). Antenna design for a portable RFID reader. *PIERS Proceedings,* USA.

Barthel, H. (2009). *Regulatory status for RFID in the UHF spectrum.* Belgium: EPC Global.

Chen, Z. N. (Ed.) (2007). *Antennas for portable devices.* New York, NY: John Wiley & sons

Chen, Z. N., Xianming, Q., & Chung, H. L. (2009). A universal UHF RFID reader antenna. *IEEE Transactions on Microwave Theory and Techniques, 57*(5), 1275–1282. doi:10.1109/TMTT.2009.2017290

Chu, L., J. (1948). Physical limitations of omni-directional antennas. *Journal of Applied Physics, 19,* 1163–1175. doi:10.1063/1.1715038

Collin, R. E., & Rothchild, S. (1964). Evaluation of antenna Q. *IEEE Transactions on Antennas and Propagation, 12,* 23–27. doi:10.1109/TAP.1964.1138151

Deshmukh, A., & Kumar, G. (2005). Compact broadband U-slot loaded rectangular microstrip antennas. *Microwave and Optical Technology Letters, 46*(6), 556–559. doi:10.1002/mop.21049

Dobkin, D. M. (2008). *The RF in RFID: Passive UHF RFID in practice.* Burlington, MA: Elsevier Inc.

Elisabeth, I.-Z., Kemeny, Z., Egri, P., & Monostori, L. (2006). The RFID technology and its current applications. *Proceedings of the Modern Information Technology in Innovation Process of the Industrial Enterprises-MITIP,* (pp. 29-36).

Fante R., L. (1969). Quality factor of general ideal antennas. *IEEE Transactions on Antenna Propagation, 17,* 151-157

Foster, P. R., & Burberry, R. A. (1999). Antenna problems in RFID systems. *IEE Colloquium on RFID Technology: Microwave and Antenna Systems,* (pp. 3/1-3/5).

Harrington, R. F. (1960). Effect of antenna size on gain, bandwidth and efficiency. *Journal of Research of the National Bureau of Standards, 64,* 1–12.

Karmakar, N. C., Zakavi, P., & Kumbukage, M. (2010). Development of a phased array antenna for universal UHF RFID reader. *Proceedings of IEEE 2010 AP-S International Symposium on Antennas and Propagation and 2010 USNC/CNC/URSI Meeting,* (pp. 1-4).

Klaus, F. (2003). *RFID handbook: Fundamentals and applications in contactless smart cards and identification.* New York, NY: John Wiley & Sons.

Li, X., Yang, L., Gong, S.-X., Yang, Y.-J., & Liu, J.-F. (2009). A compact folded printed dipole antenna for UHF RFID reader. *Progress in Electromagnetics Research Letters*, *6*, 47–54. doi:10.2528/PIERL08121303

Nikitin, P. V., & Rao, K. V. S. (2008). Antennas and propagation in UHF RFID Systems. *IEEE RFID Conference Proceedings*, USA

Nikitin, P. V., & Rao, K. V. S. (2010a). Helical antenna for handheld UHF RFID reader. *Proceedings of IEEE RFID Conference*, (pp. 166-172).

Nikitin, P. V., & Rao, K. V. S. (2010b). Compact Yagi antenna for handheld UHF RFID reader. *Proceedings of IEEE APSURSI*, (pp. 1-4).

Uddin, J., Reaz, M. B. I., Hasan, M. H., Nordin, A. N., Ibrahimy, M. I., & Ali, M. A. M. (2010). UHF RFID antenna architectures and applications. *Scientific Research and Essays*, *5*(10), 1033–1051.

Want, R. (2006). An introduction to RFID technology. *IEEE Pervasive Computing / IEEE Computer Society and IEEE Communications Society*, *5*(1), 25–33. doi:10.1109/MPRV.2006.2

Wheeler, H. A. (1947). Fundamental limitations of small antennas. *Proceedings of IRE*, *35*, 1479–1484. doi:10.1109/JRPROC.1947.226199

KEY TERMS AND DEFINITIONS

EIRP: EIRP is the amount of power that a theoretical isotropic antenna (which evenly distributes power in all directions) would emit to produce the peak power density observed in the direction of maximum antenna gain

Quality Factor: Q factor is a measure of the ability of a resonant circuit to retain its energy. A high Q means that a circuit leaks very little energy, while a low Q means that the circuit dissipates a lot of energy, sometimes quality factor is used to characterize the bandwidth of antenna

Return Loss: Return loss is the loss of signal power resulting from the reflection caused due to impedance miss match at a discontinuity in a transmission line

RFID: RFID is a technology is use of an object (typically referred to as an RFID tag) applied to or incorporated into a product, animal, or person for the purpose of identification and tracking using radio waves

VSWR: VSWR is the ratio of the highest voltage anywhere along the transmission line to the lowest

APPENDIX: LIST OF ABBREVIATIONS

AIMS: Automatic Impedance Matching System
EIRP: Equivalent/Effective Isotropic ally Radiated Power
ETSI: European Telecommunications Standards Institute
FDTD: Finite Difference Time Domain
FEM: Finite Element Method
HF: High Frequency
HFSS: High Frequency Structure Simulator
IC: Integrated Circuit
LF: Low Frequency
MIC: Monolithic Integrated circuit
MoM: Method of Moments
PCB: Printed Circuit Board
RADAR: Radio Detection and Ranging
RF: Radio Frequency
RFID: Radio Frequency Identification
ROI: Return on Investment
UHF: Ultra High Frequency
US: United States
VSWR: Voltage Standing Wave Ratio

Chapter 4
Differential Integrated Circuit Package Antenna for RFID Reader

Christina Junjun Wang
Beihang University, China

ABSTRACT

The emphasis of this chapter is to introduce the design of the integrated circuit package antenna (ICPA), which is a compact and cost-effective antenna design method for RFID reader. The concept, the architecture, and the characterizations of the ICPA will be discussed in details. As differential circuitry dominates in RF transceiver integrated circuit design due to its good performance, microstrip antennas can be seen for use in radio systems with differential signal operation. In this chapter, the improved theory of single-ended microstrip antennas based on the cavity model is expanded to analyze the input impedance and radiation characteristics of the differentially-driven microstrip antennas and ICPA. The occurrence of the resonance for the differentially-driven microstrip antennas, which can be tuned by adjusting the ratio of the separation of the dual feeds to the free space wavelength, will be analysed. Furthermore, the frequency band selection capability of the differential ICPA will be presented.

1. INTRODUCTION TO ICPA

Radio Frequency Identification (RFID) procedures, in recent years, have become very popular in many service industries, purchasing and distribution logistics. "An archetypal RFID system consists of an interrogator, more often known as a reader, a transponder or tag, and antennas to mediate between voltages on wires and waves in air" (Dobkin, p.22). Figure 1 shows the overview of the RFID system.

A RFID reader plays a quite important role in the whole RFID system. There is an un-ignorable trend toward more flexible applications and toward higher carrier frequencies for RFID readers. Higher integration density is in need due to the demand for smaller and lighter portable or handheld RFID devices. The ultimate goal would be a single-chip solution. However, the integration of the passives for high frequency bottlenecks the single-chip solution. Until now, no absolutely true single-chip solution is available, a package is still

DOI: 10.4018/978-1-4666-1616-5.ch004

Figure 1. Overview of the RFID system

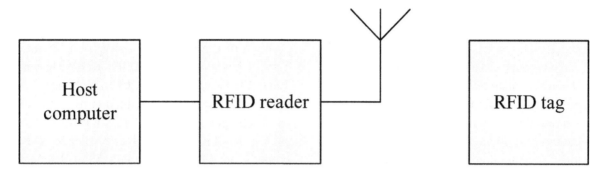

necessary in order to connect the antenna or external filters.

Antenna is another key component of the RFID system. Antennas applied in RFID systems particularly in handheld devices are usually required to have a small size and be easy to be integrated with other components. Compact planar antennas meet the need of high integration density of the RFID devices and become the favorite candidates for the RFID applications because of their clear advantages in terms of weight, manufacturability, compatibility with microwave circuits and cost. Considering good performance of the compact planar antennas, a lot of researchers have paid great efforts to develop novel antenna types that can be applied in RF applications (Hwang, Moon, et al., 2004; Midrio, Boscolo, et al., 2009; Thomas and Sreenivasan, 2010; Avila-Navarro, Cayuelas, et al., 2010). In Hwang, et al. and Avila-Navarro, et al.'s works, printed dipole antennas are presented. The proposed dipole antennas operate in the 2.4 and 5.5 GHz bands with a compact size. Midrio, et al., presented a compact planar dual-band antenna for the wireless application. In this work, a printed dipole antenna and a bowtie antenna are used to get the dual-band characteristic. In Thomas's work, a coplanar waveguide patch antenna is proposed for the 2.4 and 5.5 GHz dual-band application.

Nowadays, differential circuitry dominates in RF transceiver integrated circuit (IC) design because of its good performance. Hence, microstrip antennas can be seen for use in RF systems with

differential signal operation. The differentially-driven microstrip antenna takes full advantages of the standard surface-mounted package; consequently, the system-level board space and the system-level assembly can be reduced and facilitated, respectively. The differentially-driven microstrip antenna introduces a virtual ground; accordingly, there is no need for connecting the ground of the solid-state devices and the ground plane of the antenna, and the parasitic effect caused by interconnect is eliminated. Furthermore, the state of the art single-chip and single-package solutions of RFID readers call for differential antennas to reduce the bill of materials and to improve the receiver noise performance and transmitter power efficiency. Kinds of planar dipoles have got much attention (Hwang, Moon, et al., 2004; Midrio, Boscolo, et al., 2009; Avila-Navarro, Cayuelas, et al., 2010). As a type of balanced antenna, a dipole antenna offers the ability to avoid the disturbances from nearby dielectrics which is quite significant for the handheld RFID devices. However, a dipole antenna always has a large size and this makes it difficult to integrate a dipole antenna and other RF components together with a high integration level. A dipole antenna can be made compact by bending it. Bent dipoles have a much smaller size than conventional straight dipoles and are much easier to be integrated into the RFID devices, but the gain and bandwidth also decrease. One approach to making a high-gain compact antenna is to construct a patch antenna (Dobkin, p.281).

Then the balanced microstrip antenna becomes a more apropos choice for the RIFD application.

To lower the cost and improve the performance of the RFID system, a compact differentially-driven integrated circuit package antenna (ICPA) has been demonstrated. ICPA provides a new and advanced solution for the integration of the RFID reader and the antenna. The antennas realized in this manner enjoy economical advantage of mass production and automatic assembly and what is more important is that they have potential benefits to the system-level board miniaturization and the system-level manufacturing facilitation as they offer the possibility to combine antennas and single-chip RF transceivers into standard surface mounted devices. The potential advantages make the differentially-driven ICPA a quite suitable candidate for the RFID application.

In this chapter, the concept of the ICPA will be introduced firstly, and then the improved theory of single-ended microstrip antennas based on the cavity model will be expanded to analyze the input impedance and radiation characteristics of the differentially-driven microstrip antennas and ICPA. At the end of the chapter, the conclusions will be given.

2. CONCEPT OF ICPA

2.1. Motivation for Integrated Circuit Package Antenna (ICPA)

As mentioned above, there is an important trend toward the higher integration density of the RFID reader. Several researches have been done about system in a package (SIP) (Diels, Vaesen, et al., 2001; Donnay, Pieters, et al., 2000; Lim, Obatoyinbo, et al., 2001; Wambacq, Donnay, et al., 2000). Most of these works are focused on multiple chips and use a thin film multichip module technology (MCM-D) to interconnect these different chips. The MCM technology allows chips manufactured by different technologies (such as silicon and

Gallium Arsenide (GaAs)) to be integrated in a common package. However, partitioning the RF transceiver into multiple chips still faces the problem of cross-talk coupling (Rohit Sharma, Tapas Chakravarty, et al., 2009; Rohit Sharma, Tapas Chakravarty, et al., 2010). In Sharma, et al.'s works, solutions for reduction in cross-talk have been given, however, cross-talk still exists among the chips. And another challenge is the proper grounding of the whole system (Jenshan, 1998). A greater challenge for MCM technology lies in testing and production. This is usually referred to as the known-good-die issue. At the same time, although they said that MCM technology could integrate the full system, few of them have integrated antenna into their RF transceivers. Researchers only demonstrated the possibility of integrating the antenna in MCM technology.

Several researches have integrated the antenna directly on the package (Lim, Obatoyinbo, et al., 2001; Brebels, Ryckaert, et al., 2004; Bhagat, McFiggins, et al., 2002; Fudem, Stenger, et al., 1997). In Lim, et al.'s work, the cavity-backed patch antenna has been developed in the low temperature co-fired ceramic (LTCC) package and fabricated directly on the top of the module. The antenna had advantages over the conventional patch in terms of larger bandwidth, smaller size and less interference with other components. However, they still utilized the multi-chip solution (Brebels, Ryckaert, et al., 2004) or the RF transceivers were for mm-wave frequency band (Fudem, Stenger, et al., 1997).

Several configurations using different layers of the multi-layered organic (MLO) package were studied (Bhagat, McFiggins, et al., 2002). The MLO package consisted of three metal layers of copper separated by two dielectric layers. The most desirable configuration was the antenna and its feed line on the top metal layer with the second metal layer removed and the bottom layer serving as ground. To reduce the coupling of noise to the antenna part, a grounded tuning strip was placed next to the radiating edge of the antenna. But in

MLO package technology, the material has poor loss tangent value (0.02-0.05). And the challenge here is that substrate thickness of each layer of the MLO package is electrically extremely small at high frequency.

To parallel with the single-chip solution trend, Chan, et al. implemented integrated antennas in silicon semiconductor technology on both conventional low-resistivity and proton-implanted high-resistivity silicon substrates (Chan, Chin, et al., 2001). As expected, the integrated antennas on the low-resistivity silicon have rather poor radiation efficiency because of high loss of the silicon substrate, while the integrated antennas on the high-resistivity silicon have acceptable radiation properties. However, the integration of antennas on the high-resistivity silicon substrate has deviated from the mainframe silicon process, which undoubtedly will increase fabrication cost and degrade gate oxide integrity. Furthermore, these integrated antennas are too large to be feasible from the economical viewpoint of very large-scale integrated circuits.

To overcome the problems associated with the integrated antennas and to provide a low-cost antenna solution, miniature surface-mountable chip-scale antennas have received attention recently. Great efforts have been made involving the development of new dielectric materials (Zhang, Lo, et al., 1995; Tanidokoro, Konishi, et al., 1998), the application of novel fabrication technologies (Dakeya, Suesada, et al., 2000; Sim, Kang, 2002), and the optimization of various radiator structures (Choi, Kwon, et al., 2001; Pan, Horng, et al., 2001). These chip antennas are physically small, however, they are still discrete.

Ball grid array (BGA) packaging technology is one of the advanced technologies emerging in response to the challenge for higher integration. As a single-chip package format, BGA package is like chip-scale package with a much smaller package area, which can overcome the know-good-die problem and the packaged circuits can be treated as a single-chip package. To better suit

the innovative development of single-chip solutions of RF transceivers, Zhang has proposed to integrate antennas on chip packages. Microstrip antennas have been integrated on both cavity-up and cavity-down ceramic chip packages for applications at 2 to 5 GHz (Zhang, 2002; Zhang, 2004a; Zhang, 2004b; Zhang, Liu, et al., 2003). A compact ICPA has been demonstrated in a ceramic BGA package format in Zhang's works. The antennas realized in this manner enjoy economical advantage of mass production and automatic assembly and have potential benefits to the system-level board miniaturization and the system-level manufacturing facilitation. This is because they offer the possibility to combine antennas and single-chip RF transceivers into standard surface mounted devices.

2.2. Configuration of the ICPA

Figure 2 shows the configuration of the ICPA in a thin CBGA package format. The ICPA consists of three co-fired laminated ceramic layers, with a bare chip cavity formed at the middle layer. There are two buried layers and one top-layer metallization in the construction. The lower buried layer provides the metallization for the signal traces, while the upper buried layer provides the metallization for the ground plane of the ICPA. The radiating element of the ICPA, which can take any form of a printed circuit antenna, is realized with top-layer metallization. It should be noted that only two of the three laminated layers are used as the effective substrate for the ICPA, which is different from the design of conventional multilayer antennas where all layers are used as the substrates of the antennas (Tentzeris, Li, et al., 2002; Li, Dejean, et al., 2003). There are 48 signal traces in the ICPA. These 48 signal traces follow the current designs of single-chip RF transceivers where 48 input/output pads are often adopted. The outer ends of 48 signal traces are connected to 48 solder balls through 48 vias, while the inner ends of 48 signal traces are connected to the single-chip RF transceiver through 48 bond wires. The single-chip RF transceiver is attached upside down to the ground plane of the ICPA in the chip cavity. The ground plane of the ICPA shields the single-chip RF transceiver from the radiating-element of the ICPA. The feed of the radiating element of the ICPA from the carried single-chip RF transceiver is realized with a bond wire, a signal trace, and a via through an aperture on the ground plane of the ICPA. This feeding technique is compatible with the integrated circuit package technology and can be easily implemented.

2.3. Modeling and Design of the ICPA

Design of the ICPA was done with an in-house CAD code and commercial simulators, for example, high frequency structure simulator (HFSS). The CAD code first calculates the basic physical properties of the rectangular microstrip patch and then outputs its results such as the patch dimensions and the feeding location as initial values to the HFSS engine for fine design. HFSS is the industry-standard simulation tool for 3D full-wave electromagnetic field simulation and is essential for the design of high-frequency and high-speed component design. Users can select the appropriate solver for the type of simulation they are performing. To simulate the ICPA in HFSS and get a correct result, a proper model of the ICPA is quite essential.

The design of the ICPA must consider its fabrication in large quantity, which requires novel manufacturing technologies. LTCC technology that uses noble metals and specific ceramic materials and has the flexibility in realizing an arbitrary number of metallic and ceramic layers is suitable for the mass production of the ICPA. There are a few LTCC material systems in the market. The LTCC material system from Dupont is used in this demonstration. The physical design rules recommended by Dupont are obeyed. The Dupont 951-AX ceramic type has a dielectric constant of 7.8 and a loss tangent of 0.002 at 5 GHz. Two metallization options are available: silver and gold, of which silver metallization is chosen for the ICPA implementation. The design of the ICPA must also consider the package of the single-chip RF transceiver die and the automatic assembly of the ICPA to the printed circuit board. In this regard, the ICPA should be designed as the JEDEC standard compliant. The ICPA has a standard body size 15×15 mm^2 and 48 signal traces and solder balls with a pitch of 0.48 mm. The ICPA holds the stepped cavity that is large enough to accommodate a single-chip RF transceiver die of current size.

Figure 2. View of the ICPA

3. DIFFERENTIALLY-DRIVEN MICROSTRIP ANTENNAS DESIGN

3.1. Theory and Analysis of Differentially-Driven Microstrip Antennas

As introduced before, microstrip antennas have many unique and attractive properties–low in profile, light in weight, compact and conformable in structure, and easy to fabricate and to be integrated with solid-state devices. The differentially-driven microstrip antenna takes full advantages of the standard surface-mounted CBGA package. As a result, differential antennas are called for to reduce the bill of materials and to improve the receiver noise performance and transmitter power efficiency (Behzad, Zhong, et al., 2003).

A microstrip antenna and a coordinate system (Zhang & Wang, 2006) are illustrated in Figure 3.

The microstrip antenna that has dimensions L along x-axis and W along y-axis located on the surface of a grounded dielectric substrate with thickness H, dielectric constant ε_r, and dielectric loss tangent δ is differentially driven at points (x_1, y_1) and (x_2, y_2).

The differentially-driven microstrip antenna can be treated as a two-port network. With refer-ence to the ground plane, the driving point at (x_1, y_1) is defined as port 1 and the driving point at (x_2, y_2) as port 2. Using the Z parameters for ports 1 and 2, one can express the differential voltage V_d as

$$V_d = V_1 - V_2 = \left(Z_{11} - Z_{21}\right)I_1 - \left(Z_{22} - Z_{12}\right)I_2 \tag{1}$$

where V_1 and V_2 are the driving point voltages and I_1 and I_2 are the driving point currents of ports 1 and 2, respectively. Since for the differentially-driven microstrip antenna,

$$I_1 = -I_2 = I \tag{2}$$

Equation (1) simplifies to

$$Z_d = \frac{V_d}{I} = 2\left(Z_{11} - Z_{21}\right) = 2\left(Z_{22} - Z_{12}\right) \tag{3}$$

where Z_d is the input impedance of the differentially-driven microstrip antenna. The Z parameters for the microstrip antenna shown in Figure 3 can be easily calculated with the improved theory based on the cavity model as follows (Richards, Lo, et al., 1981)

$$Z_{11} = j\omega\mu_0 H \sum_{m,n=0}^{\infty} \frac{\phi^2{}_{mn}\left(x_1,y_1\right)j_o^2\left(\dfrac{m\pi d_e}{2L_{eff}}\right)}{k_{mn}^2 - k_e^2} \tag{4}$$

$$Z_{12} = j\omega\mu_0 H \sum_{m,n=0}^{\infty} \frac{\phi_{mn}\left(x_1,y_1\right)\phi_{mn}\left(x_2,y_2\right)j_0^2\left(\dfrac{m\pi d_e}{2L_{eff}}\right)}{k_{mn}^2 - k_e^2} \tag{5}$$

where ω is angular frequency, μ_0 is the permeability of vacuum, d_e is the effective width of a uniform strip of z directed source current of one ampere,

Figure 3. A microstrip antenna and its coordinate system

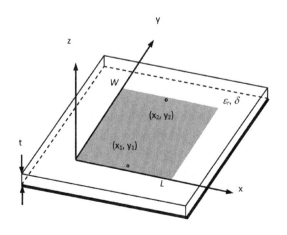

$$k_e^2 = \varepsilon_r \left(1 - j\delta_e\right) k_0^2 \qquad (6)$$

$$k_0 = \omega / v \qquad (7)$$

$$k_{mn}^2 = \left(m\pi/L_{eff}\right)^2 + \left(n\pi/W_{eff}\right)^2 \qquad (8)$$

$$j_0\left(x\right) = \sin\left(x\right)/x \qquad (9)$$

$$\phi_{mn}\left(x,y\right) = \sqrt{\varepsilon_{0m}\varepsilon_{0n}/L_{eff}W_{eff}}\cos\left(m\pi x/L_{eff}\right)\cos\left(n\pi y/W_{eff}\right) \qquad (10)$$

$\varepsilon_{0m} = 1$ for $m = 0$ and 2 for $m \neq 0$, m and n are the mode indices in the x and y directions, respectively, δ_e is the effective loss tangent, v is the speed of light, L_{eff} and W_{eff} are the effective dimensions, taking into account the fringing fields at the edges of the microstrip patch (Abboud, Daniano, et al., 1988). Similarly, Z_{21} and Z_{22} can be computed. L_{eff} may be approximated as follows

$$L_{eff} = L + \left(\frac{W_{eq} - W}{2}\right)\frac{\varepsilon_{eff}\left(W\right) + 0.3}{\varepsilon_{eff}\left(W\right) - 0.258} \qquad (1\text{-}11)$$

where $\varepsilon_{eff}\left(W\right)$ is the effective dielectric constant and W_{eq} is the equivalent width calculated from the planar waveguide model. $\varepsilon_{eff}\left(W\right)$ can be written as

$$\varepsilon_{eff}\left(W\right) = \frac{\varepsilon_r + 1}{2} + \frac{\varepsilon_r - 1}{2}\left(1 + 10\frac{H}{W}\right)^{-1/2} \qquad (12)$$

and W_{eq} as

$$W_{eq} = \frac{120\pi H}{Z_a\left(W\right)\sqrt{\varepsilon_{eff}\left(W\right)}}. \qquad (13)$$

Here $Z_a(W)$, the impedance of a microstrip line of width W and thickness H, is given by

$$Z_a(W) = \frac{60\pi}{\sqrt{\varepsilon_r}}\left[\frac{W}{2H} + 0.441 + 0.082\left[\frac{\varepsilon_r - 1}{\varepsilon_r^2}\right] + \frac{\left(\varepsilon_r + 1\right)}{2\pi\varepsilon_r}\left[1.415 + L_n\left(\frac{W}{2H} + 0.94\right)\right]\right]^{-1} \qquad 14)$$

Similarly, W_{eff} can be calculated by replacing L, W_{eq}, W, $Z_a(W)$ and $\varepsilon_{eff}\left(W\right)$ in equations (12), (13) and (14) with W, L_{eq}, L, $Z_a(L)$ and $\varepsilon_{eff}(L)$, respectively.

Next, the radiation characteristics of the differentially-driven microstrip antenna will be examined. It is known that the radiation of the microstrip antenna originates from two slots and each slot can be thought of as radiating the same field as a magnetic dipole with a magnetic current of

$$M = 2\hat{n} \times \hat{z}E_z \qquad (15)$$

where the factor of 2 comes from the image of the magnetic current in the ground plane, $\hat{n}$ is the outward normal to the magnetic wall, $\hat{z}$ is the unit vector in the z-direction, and E_z is the z component of electric field in the cavity region. For the differentially-driven microstrip antenna, E_z can be written as

$$E_z = j\mu_0 \sum_{m,n=0}^{\infty} I\frac{\phi_{mn}\left(x,y\right)}{k_e^2 - k_{mn}^2}j_0\left(\frac{m\pi d_e}{2L_{eff}}\right)\left[\begin{array}{c}\phi_{mn}\left(x_1,y_1\right)\\-\phi_{mn}\left(x_2,y_2\right)\end{array}\right] \qquad (16)$$

where I is the source current of one ampere in our calculation. Equation (16) reveals that the differentially-driven microstrip antenna introduces the cancellation mechanism, which can be explored to suppress some higher-order modes to reduce the cross polarization radiation.

In the implementation of the improved cavity model, the determination of the effective loss tangent δ_e is critical. It is found that the final value of δ_e is quite sensitive to the value chosen at the start of the iterative process (Lee, Chebolu, et al., 1994). Begin by setting $\delta_e = \delta_0$ instead of the usual δ to shorten the iterative procedure. δ_0 is given (R. Garg, et al., 2001) by

$$\delta_0 = \frac{P_{r0}}{2\omega W_{e0}} + \delta + \frac{s}{H} \qquad (18)$$

where P_{r0} can be expressed as

$$P_{r0} = \frac{V_d^2 A \pi^4}{20340} \left[(1-B)\left(1 - \frac{A}{15} + \frac{A^2}{420}\right) + \frac{B^2}{5}\left(2 - \frac{A}{7} + \frac{A^2}{189}\right) \right] \qquad (19)$$

with $A = \left(\pi L / \lambda_0\right)^2$ and $B = \left(2W / \lambda_0\right)^2$. There appears to be an algebraic error. Working from the approximations indicated in Thouroude, et al.'s work (Thouroude, Himdi, et al., 1990), one obtains 40680 instead of 20340. W_{e0}, the electric energy stored for dominant mode at resonance, can be derived to be

$$W_{e0} = \frac{\varepsilon_0 \varepsilon_r L W V_d^2}{8H}. \qquad (20)$$

3.2. Validation of the Theory for the Differentially-Driven Microstrip Antennas

Theoretical calculations, numerical simulations, and physical experiments were made for a differentially-driven microstrip antenna. The antenna was constructed using Taconic TLY-5 with a relative dielectric constant ε_r of 2.2 ± 0.02 and a loss tangent of approximately 0.0009. It had dimensions $L = 17.2$ mm, $W = 16.7$ mm, and $H = 0.635$ mm. The substrate size is 25×25 mm². The antenna was driven differentially at both radiating edges at $(x_1, y_1) = (0.0054L, 0.5W)$ and $(x_2, y_2) = (0.9946L, 0.5W)$ to excite TM_{10} mode. The fabricated microstrip antenna is driven at both radiat-

ing edges. This differentially-driven scheme was used in Deal and Wei's works (Wang, Zhang, et al., 2004; Deal, Radisic, et al 1999). The calculations were implemented in Matlab, the simulations were executed using the HFSS, and the measurements were conducted with an HP 8510C network analyzer. The measured S parameters were converted to the differential input impedance by

$$Z_d = 2Z_0 \frac{\left(1 - S_{11}^2 + S_{21}^2 - 2S_{21}\right)}{\left(1 - S_{11}\right)^2 - S_{21}^2} \qquad (21)$$

where Z_0 is the reference impedance of $50\,\Omega$.

Figure 4 compares the calculated, simulated, and measured input impedance of the differentially-driven microstrip antenna. The impedance characteristics give insight on how the antenna must be modified to achieve a specified resonant frequency. Here the resonant frequency is defined as where the reactance of the input impedance is equal to zero. According to this definition, one can see that both calculated and simulated resonant frequencies are at 5.77 GHz and the measured resonant frequency is at 5.74 GHz. One also can see that the agreement between the calculated and simulated input impedance is excellent. The agreement between theory and experiment is in general acceptable. The discrepancy is likely due to the fabrication tolerance and the experimental error of our measuring system. It was found that any small phase difference in measured S_{11} and S_{21} would cause a large error in the measured Z_d.

And a comparison was also made between the calculated, simulated, and measured co-polar radiation patterns of the differentially-driven microstrip antenna at 5.75 GHz. The comparison shows that excellent agreement is obtained between the theory, simulation, and experiment in the broadside direction of the forward region.

If we compare the calculated radiation patterns of the microstrip antenna driven for differential and single-ended operations at 5.75 GHz, it is evident

Figure 4. for the differentially-driven microstrip antenna driven at both radiating edges

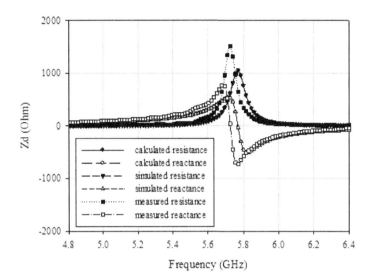

from the results that the cross-polar radiation from the differentially-driven microstrip antenna is weaker than that of the single-ended couterpart.

3.3. Parametric Analysis of the Differentially-Driven Microstrip Antenna

In this section, the differential resonant resistance as a function of ξ/λ_o for the above differentially-driven scheme to excite the TM_{10} mode, is analyzed. ξ is the seperation between the two feeds, and λ_0 is the free space wavelength of the microstrip antenna. The resonant resistance for the single-ended signal operation is included for comparison. For single-ended operation, the microstrip antenna is fed at (x_1, y_1), ξ is calculated with respect to a virtual feed located at (x_2, y_2). Figure 5 shows the resonant resistance for the scheme to excite the TM_{10} mode. It is seen that the resonant resistance increases with ξ/λ_o and the differential resonant resistance is larger than the single-ended one. It is found that when ξ/λ_o is smaller than 0.1, no resonance for differential operation will occur any more, the differential input impedance is inductive. The dependence of

resonance for the differentially-driven microstrip antennas on the separation of the dual feeds can be explained as follows. When the dual feeds are located near to each other $\xi/\lambda_o < 0.1$, the differential signal applied to the dual feeds cancels, hence the resonance does not occur and the feeds themselves make the differential input impedance exhibit inductive.

4. DESIGN AND ANALYSIS OF DIFFERENTIALLY-DRIVEN ICPA

At the present time, single-chip RF transceivers in CMOS technology at 5-GHz bands have been successfully demonstrated. Single-chip RF transceivers have the benefit of cost reduction and system reliability and therefore represent the trend of implementing RFID systems. The antenna, because of the limitation of the size of the chip, is usually left external to the single chip in highly integrated radio systems. The use of the differentially-driven antenna in ICPA will not only translate into the reduction of bill of materials but also the improvement of the receiver noise performance and transmitter power efficiency.

Figure 5. Resonant resistance of the differentially-driven microstip antenna

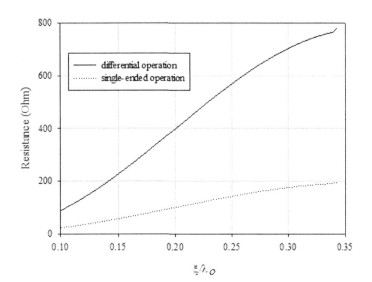

4.1. Integration of the ICPA

The integration of the differentially-driven microstrip antenna on the CBGA package into the differentially-driven ICPA is a challenging task. It requires a versatile numerical tool that can simulate both antenna and package structures. The HFSS from Ansoft can be chosen as the primary design tool. There are many integration issues for the ICPA. This section will be emphasized on such issues as feeding and shielding of the antenna from the transceiver using the packaging techniques. The feeding to the antenna from the RF transceiver is through two identical packaging element networks (shown in Figure 2). A packaging element network includes three bond wires, three signal traces, and three vias in the G-S-G fashion. A bond wire has a loop height of 0.2 mm and a length of 1.3 mm. A signal trace has a width of 0.32 mm and a length of 3 mm. The gap between the two adjacent signal traces is 0.48 mm. A via has a diameter of 0.2 mm. It is of interest to evaluate each of their impact on the overall electrical performance: bond wire, signal trace, and via. The bond wire presents high impedance; it has high inductance and low capacitance.

Keeping the length of the bond wire to a minimum is critical to minimize its disruptive effect on the electrical signal. Using the largest diameter wire possible is also important, and in this work 32.5 μm wire is used. A ground wire is placed on each side of the signal wire. This improves the situation by providing a return path in close proximity to the wire and thus reducing inductance slightly. The signal trace can provide the best electrical performance of the feeding network. It is primarily a coplanar waveguide. The signal integrity is well preserved. However, there is some level of loss due to dielectric material that is surrounding the conductor. As such, the length of this part does have an effect on the overall electrical behaviour but it is much less damaging than the bond wire is. The via is a transition from the signal trace to the microstrip line through an aperture on the ground plane. The diameter of the aperture has the potential to impact the electrical signal and is 0.3 mm.

To model the effect of the single-chip RF transceiver die, a dummy CMOS silicon with dimensions of $4 \times 4 \times 0.4$ mm^3 is attached to the meshed ground in the cavity. The shielding of the antenna from the RF transceiver is realized with the meshed ground. To enhance the shielding, a guard ring with fence vias shorted to the meshed ground is added. The width of the guard ring is 1.5 mm, and the gap between the guard ring and the microstrip patch is 0.4 mm. There are 8 fence vias on each side of the guard ring. The diameter of a fence via is 0.2 mm.

It was compared that the simulated E_z components of the electric field on the middle plane between the microstrip patch and the meshed ground at 5.8 GHz with and without the guard ring. It is evident from the results that the guard ring improves the shielding by 3 dB and there exists a virtual AC ground in the middle of the patch, which is due to the ICPA fed differentially. The virtual ground improves the isolation between the two feeding networks and also can be used to introduce a DC supply without affecting the ICPA operation.

4.2. Measured Results for the ICPA

The designed differentially-driven ICPA was fabricated in Dupont 951-AX LTCC. It measures $15 \times 15 \times 1.6$ mm^3 with the microstrip patch of size 10.04×10.176 mm^2. The ICPA was surface-mounted on the middle of a test board. The test board of size $60 \times 40 \times 0.8$ mm^3 is FR4 substrate (Zhang, Wang, et al., 2008). The ICPA feeds are soldered on two 50 Ω CPW lines for impedance and radiation tests. The differentially-driven ICPA was measured with an HP 8510C network analyzer in the anechoic chamber. The measured S parameters were converted to the measured input impedance of the differentially-driven ICPA by eqn. (1-21). While the measured return loss calculated from Z_d is given by

$$RL = 20 \log 10 \left| \frac{Z_d - Z_o}{Z_d + Z_o} \right| \qquad (22)$$

where Z_0 is of typical value 100Ω, 300Ω or 600Ω. In the study in this section, 100Ω is chosen to calculate the corresponding return loss.

Figure 6 shows the measured input impedance of the differentially-driven ICPA. The impedance characteristics give insight on how the ICPA must be modified to achieve a specified resonant frequency. Here the resonant frequency is again defined as where the reactance of the input impedance is equal to zero. According to this definition, there are two resonant frequencies for the ICPA over the frequency range of interest from 5.6 to 6.2 GHz. It is evident that the impedance characteristics at the frequency of operation exhibit a small peak in the resistance and a gentle swing in the reactance from inductive to capacitive.

Figure 7 illustrates the measured radiation patterns of the ICPA at 5.8 GHz. For the sake of gaining differential signals to feed the differentially-driven ICPA, a 50 Ω balun designed at 5.8 GHz was used. Due to the interaction between the radiation from the ICPA and the feeding cable, the measured radiation patterns show fluctuations particularly in the E-plane, as a result, poor agreement occurs in the lower half plane. The radiation is stronger in the upper hemisphere, i.e., in the direction normal to the ICPA. This feature of the radiation patterns is desirable because it not only helps improve the efficiency of the ICPA but also reduces the interaction of the ICPA with the human body. As shown in Figure 7, the cross-polar components are at least 20 dB lower than the co-polar components. The efficiency of the ICPA was calculated to be 84%. In addition, it should be mentioned that the ICPA has much shorter distance to the RF output of the wireless transceiver than a conventional dielectric chip antenna; this implies a smaller transmission loss, which can be translated as an improvement to the ICPA efficiency by a few percent.

4.3. Band Selection Characterization of the ICPA

In this section, we study the ICPA that features a via shorting scheme, which can be easily shorted to ground or left open at the board-level can achieve frequency band selection. In addition, ICPA of this

Figure 6. Measured Z_d of the differentially-driven ICPA driven at both radiating edges

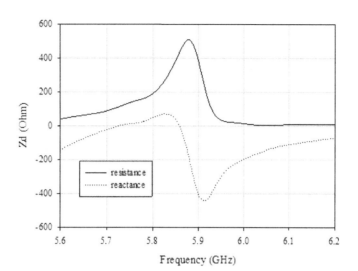

model offers the possibility to achieve shorting using a simple diode circuit in the RFID reader for frequency band selection purpose (Zhang, 2002; Zhang, Liu, et al., 2003). The ability to select different frequency band greatly enhances the ICPA applications in RFID systems (Wang, Xue, et al., 2005). What's more, the ICPA will be treated as a sample to illustrate the antenna measurement equipment and setting up.

The fabricated prototype ICPA was assembled on a test board. The ICPA was fabricated with FERRO A6-M LTCC material system, which has a dielectric constant of 5.9 and a dissipation factor of 0.0012 at 5 GHz. The ICPA was designed in a cavity-down CBGA package structure with nine tape layers. The microstrip patch radiator was printed on the top surface of the ICPA with two vias linked to the microstrip patch radiator. One via was used for antenna excitation and the other via was utilized for frequency band selection. The ground plane of the ICPA was printed on the bottom surface of the third layer. A guard ring was employed, and was shorted to the ground plane to reduce the interference of the antenna to the RFID reader. A multi-tier cavity was formed in the centre of the ICPA, and it can accommodate a 4

$\times 4 \times 0.5$ mm^3 CMOS single-chip RF transceiver die. The ICPA was designed to have 88 I/Os with a JEDEC standard ball pitch of 1.27 mm. The prototype ICPA size is $17 \times 17 \times 2$ mm^3 with the microstrip patch radiator size of 10.5×10.5 mm^2.

Network analyzer measurements were carried out in an anechoic chamber using HP 8510C network analyzer. The return loss characteristic for opening and shorting of the band-select via is shown in Figure 8.

Figure 8 illustrates the measured return loss when the band-select via is shorted and opened on the board-level. The minimum return loss occurs at 5.37 and 5.67 GHz for the short and open cases, respectively. The shorted case has a 6 dB return loss and bandwidth of 80 MHz, while the opened case has a 10 dB return loss and bandwidth of 120 MHz.

5. CONCLUSION

Nowadays, there is an important trend toward more flexible applications and toward higher carrier frequencies for RFID transceivers. And the demand for smaller and lighter portable electronics

Figure 7. Measured radiation patterns driven at both radiating edges

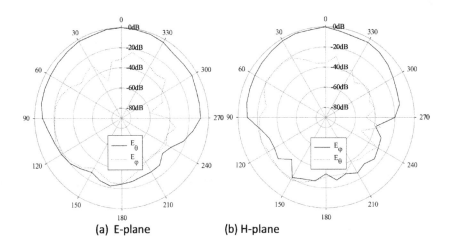

(a) E-plane   (b) H-plane

is driving both the semiconductor and packaging industries to increase integration density.

In this chapter, a concept of integrated circuit package antenna has been further developed, which can be viewed as a kind of integrated antenna solution for RFID devices. A differentially-driven microstrip antenna was used in the ICPA. The input impedance and radiation characteristics of the differentially-driven microstrip antennas and ICPA have been analyzed with the improved theory based on the cavity model of single-ended microstrip antennas. The differentially-driven microstrip antennas were fabricated using Taconic TLY-5. Theoretical and experimental results were found to be in acceptable agreement. It was shown from the analysis that the occurrence of resonance for the differentially-driven microstrip antennas also depends on the ratio of the separation ξ of the dual feeds to the free-space wavelength λ_0. When the dual feeds are located far from each other $\xi/\lambda_o > 0.1$, the resonance occurs,

Figure 8. Measured return loss of the band selection ICPA

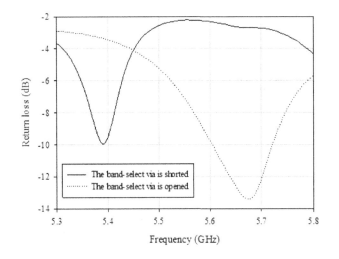

and the input resistance at resonance is rather large. However, when the dual feeds are located near to each other $\xi/\lambda_o < 0.1$, the resonance does not occur, the input resistance is quite small, and the input impedance is inductive. It was found that the differentially-driven microstrip antennas have larger resonant resistance, similar co-polar radiation patterns, and lower cross-polar radiation component compared with the single-ended counterpart. The differentially-driven ICPA fabricated in LTCC technology of Dupont 951-AX was experimentally verified. Results showed that the differentially-driven ICPA of size $15 \times 15 \times 1.6\,mm^3$ achieved impedance bandwidth of 2.2%, radiation efficiency of 84%, and gain of 3.2 dBi at 5.8 GHz.

Furthermore, the band selection function of the ICPA was presented. The measured return loss showed that the minimum return loss occurs at 5.37 and 5.67 GHz for the short and open cases, respectively. The shorted case has a 6 dB return loss and bandwidth of 80 MHz, while the opened case has a 10 dB return loss and bandwidth of 120 MHz.

REFERENCES

Abboud, F., Damiano, J. P., & Papiernik, A. (1988). Simple model for the input impedance of coax-fed rectangular microstrip patch antenna for CAD. *IEE Proceedings. Microwaves, Antennas and Propagation*, *135*(5), 323–326. doi:10.1049/ip-h-2.1988.0066

Avila-Navarro, E., Cayuelas, C., & Reig, C. (2010). Dual-band printed dipole antenna for Wi-Fi 802.11n applications. *Electronics Letters*, *46*(21), 1421–1422. doi:10.1049/el.2010.2000

Behzad, A. R., Shi, Z. M., Anand, S. B., Li, L., Carter, K. A., & Kappes, M. S. (2003). A 5-GHz direct-conversion CMOS transceiver utilizing automatic frequency control for the IEEE 802.11a wireless LAN standard. *IEEE Journal of Solid-state Circuits*, *38*(12), 2209–2220. doi:10.1109/JSSC.2003.819085

Bhagat, M., McFiggins, J., & Venkataraman, J. (2002, October). *Chip package co-design of the RF front end with an integrated antenna on multilayered organic material.* Paper presented at the IEEE Conference on Electrical Performance of Electronic Packaging and Systems, Monterey, Calif.

Brebels, S., Ryckaert, J., Come, B., Donnay, S., Walter De, R., & Beyne, E. (2004). SOP integration and codesign of antennas. *IEEE Transactions on Advanced Packaging*, *27*(2), 341–351. doi:10.1109/TADVP.2004.828822

Chan, K. T., Chin, A., Chen, Y. B., Lin, Y. D., Duh, T. S., & Lin, W. J. (2001, December). *Integrated antennas on Si, proton-implanted Si and Si-on-quartz.* Paper presented at the IEEE International Electron Devices Meeting, Washington, DC.

Choi, W., Kwon, S., & Lee, B. (2001). Ceramic chip antenna using meander conductor lines. *Electronics Letters*, *37*(15), 933–934. doi:10.1049/el:20010645

Dakeya, Y., Suesada, T., Asakura, K., Nakajima, N., & Mandai, H. (2000, June). *Chip multilayer antenna for 2.45 GHz-band application using LTCC technology.* Paper presented at the IEEE MTT-S International Microwave Symposium, Boston, MA.

Deal, W. R., Radisic, V., Qian, Y. X., & Itoh, T. (1999). Integrated-antenna push-pull power amplifiers. *IEEE Transactions on Microwave Theory and Techniques*, *47*(8), 1418–1425. doi:10.1109/22.780389

Diels, W., Vaesen, K., Wambacq, P., Donnay, S., De Raedt, W., & Engels, M. (2001). Single-package integration of RF blocks for a 5 GHz WLAN application. *IEEE Transactions on Advanced Packaging, 24*(3), 384–391. doi:10.1109/6040.938307

Dobkin, D. M. (Ed.). (2008). *The RF in RFID-passive UHF RFID in practice*. Oxford, UK: Elsevier press.

Donnay, S., Pieters, P., Vaesen, K., Diels, W., Wambacq, P., & De Raedt, W. (2000). Chip-package codesign of a low-power 5-GHz RF front end. *Proceedings of the IEEE, 88*(10), 1583–1597. doi:10.1109/5.888997

Fudem, H., Stenger, P., Niehenke, E. C., Sarantos, M., & Schwerdt, C. (1997, June). *A low cost miniature MMIC W-band transceiver with planar antenna*. Paper presented at the IEEE MTT-S International Microwave Symposium, Denver, CO.

Garg, R. (Eds.). (2001). *Microstrip antenna design handbook*. Norwood, MA: Artech House.

Hwang, S. H., Moon, J. I., Kwak, W. I., & Park, S. O. (2004). Printed compact dual band antenna for 2.4 and 5GHz ISM band applications. *Electronics Letters, 40*(25), 1568–1569. doi:10.1049/el:20046579

Jenshan, L. (1998). Chip-package codesign for high-frequency circuits and systems. *IEEE Micro, 18*(4), 24–32. doi:10.1109/40.710868

Kyutae, L., Obatoyinbo, A., Sutono, A., Chakraborty, S., Chang-Ho, L., Gebara, E., et al. (2001, May). *A highly integrated transceiver module for 5.8 GHz OFDM communication system using multi-layer packaging technology*. Paper presented at the IEEE MTT-S International Microwave Symposium, Phoenix, AZ.

Lee, K. F. L., Chebolu, S. R., Chen, W., & Lee, R. Q. (1994). On the role of substrate loss tangent in the cavity model theory of microstrip patch antennas. *IEEE Transactions on Antennas and Propagation, 42*(1), 110–112. doi:10.1109/8.272308

Li, R. L., DeJean, G., Tentzeris, M. M., Laskar, J., & Papapolymerou, J. (2003, June). *LTCC multilayer based CP patch antenna surrounded by a soft-and-hard surface for GPS applications*. Paper presented at the IEEE International Antennas and Propagation Society Symposium and URSI National Radio Science Meeting, Columbus, OH.

Midrio, M., Boscolo, S., Sacchetto, F., Someda, C. G., Capobianco, A. D., & Pigozzo, F. M. (2009). Planar, compact dual-band antenna for wireless LAN applications. *IEEE Antennas and Wireless Propagation Letters, 8*, 1234–1237. doi:10.1109/LAWP.2009.2035647

Pan, C. Y., & Horng, T. S. (2001, December). *Miniaturized dielectric chip antenna in a C-shaped configuration with an array of shorting pins*. Paper presented at the Asia-Pacific Microwave Conference, Taipei, Taiwan.

Sharma, R., Chakravarty, T., & Bhattacharyya, A. B. (2009). Analytical model for optimum signal integrity in PCB interconnects using ground tracks. *IEEE Transactions on Electromagnetic Compatibility, 51*(1), 66–77. doi:10.1109/TEMC.2008.2010054

Sharma, R., Chakravarty, T., & Bhattacharyya, A. B. (2010). Reduction of signal overshoots in high-speed interconnects using adjacent ground tracks. *Journal of Electromagnetic Waves and Applications, 24*, 941–950. doi:10.1163/156939310791285218

Sim, S. H., Kang, C. Y., Yoon, S. J., Yoon, Y. J., & Kim, H. J. (2002). Broadband multilayer ceramic chip antenna for handsets. *Electronics Letters, 38*(5), 205–207. doi:10.1049/el:20020176

Tanidokoro, H., Konishi, N., Hirose, E., Shinohara, Y., Arai, H., & Goto, N. (1998, June). *1-wavelength loop type dielectric chip antennas*. Paper presented at the IEEE Antennas and Propagation Society International Symposium and URSI National Radio Science Meeting, Atlanta, GA.

Tentzeris, E., Li, R. L., Lim, K., Maeng, M., Tsai, E., DeJean, G., et al. (2002, June). *Design of compact stacked-patch antennas on LTCC technology for wireless communication applications.* Paper presented at the IEEE Antennas and Propagation Society International Symposium and URSI National Radio Science Meeting, San Antonio, TX.

Thomas, K. G., & Sreenivasan, M. (2010). Compact CPW-fed dual-band antenna. *Electronics Letters, 46*(1), 13–14. doi:10.1049/el.2010.1729

Thouroude, D., Himdi, M., & Daniel, J. P. (1990). CAD-oriented cavity model for rectangular patches. *Electronics Letters, 26*(13), 842–844. doi:10.1049/el:19900552

Wambacq, P., Donnay, S., Pieters, P., Diels, W., Vaesen, K., De Raedt, W., et al. (2000, February). *Chip-package co-design of a 5 GHz RF front-end for WLAN.* Paper presented at the IEEE International Solid-State Circuits Conference, San Francisco, CA.

Wang, J. J., Xue, Y., & Zhang, Y. P. (2005). Frequency-band selection for an integrated-circuit package antenna using LTCC technology. *Microwave and Optical Technology Letters, 44*(5), 439–441. doi:10.1002/mop.20660

Wang, W., & Zhang, Y. P. (2004). 0.18-um CMOS push-pull power amplifier with antenna in IC package. *IEEE Microwave and Wireless Components Letters, 14*(1), 13–15. doi:10.1109/LMWC.2003.821489

Zhang, C. C., Liu, J. J., & Zhang, Y. P. (2003). ICPA for highly integrated concurrent dual-band wireless receivers. *Electronics Letters, 39*(12), 887–889. doi:10.1049/el:20030582

Zhang, Y. P. (2002). Integration of microstrip antenna on cavity-down ceramic ball grid array package. *Electronics Letters, 38*(22), 1307–1308. doi:10.1049/el:20020937

Zhang, Y. P. (2004a). Finite-difference time-domain analysis of integrated ceramic ball grid array package antenna for highly integrated wireless transceivers. *IEEE Transactions on Antennas and Propagation, 52*(2), 435–442. doi:10.1109/TAP.2004.823889

Zhang, Y. P. (2004b). Integrated circuit ceramic ball grid array package antenna. *IEEE Transactions on Antennas and Propagation, 52*(10), 2538–2544. doi:10.1109/TAP.2004.834427

Zhang, Y. P., Lo, T. K. C., & Hwang, Y. M. (1995, June). *A dielectric-loaded miniature antenna for microcellular and personal communications.* Paper presented at the IEEE Antennas and Propagation Society International Symposium, Newport Beach, Calif.

Zhang, Y. P., & Wang, J. J. (2006). Theory and analysis of differentially-driven microstrip. *IEEE Transactions on Antennas and Propagation, 54*(4), 1092–1099. doi:10.1109/TAP.2006.872597

Zhang, Y. P., Wang, J. J., Li, Q., & Li, X. J. (2008). Antenna-in-package and transmit-receive switch for single-chip radio transceivers of differential architecture. *IEEE Transactions on Circuits and Systems. I, Regular Papers, 55*(11), 3564–3570. doi:10.1109/TCSI.2008.925822

ADDITIONAL READING

Aerts, W., De Mulder, E., Preneel, B., Vandenbosch, G., & Verbauwhede, I. (2008). Dependence of RFID reader antenna design on read out distance. *IEEE Transactions on Antennas and Propagation, 56*(12), 3829–3837. doi:10.1109/TAP.2008.2007378

Ando, A., Kagoshima, K., Kondo, A., & Kubota, S. (2008). Novel microstrip antenna with rotatable patch fed by coaxial line for personal handy-phone system units. *IEEE Transactions on Antennas and Propagation, 56*(8), 2747–2751. doi:10.1109/TAP.2008.927572

Attia, H., Yousefi, L., Bait-Suwailam, M. M., Boybay, M. S., & Ramahi, O. M. (2009). Enhanced-gain microstrip antenna using engineered magnetic superstrates. *IEEE Antennas and Wireless Propagation Letters, 8,* 1198–1201. doi:10.1109/LAWP.2009.2035149

Brzezina, G., Roy, L., & MacEachern, L. (2006). Planar antennas in LTCC technology with transceiver integration capability for ultra-wideband applications. *IEEE Transactions on Microwave Theory and Techniques, 54*(6), 2830–2839. doi:10.1109/TMTT.2006.875448

Chen, Y., & Zhang, Y. P. (2005). A planar antenna in LTCC for single-package ultrawide-band radio. *IEEE Transactions on Antennas and Propagation, 53*(9), 3089–3093. doi:10.1109/TAP.2005.854541

Chen, Z. N., Qing, X. M., & Chung, H. L. (2009). A universal UHF RFID reader antenna. *IEEE Transactions on Microwave Theory and Techniques, 57*(5), 1275–1282. doi:10.1109/TMTT.2009.2017290

Hua, R. C., & Ma, T. G. (2007). A printed dipole antenna for ultra high frequency (UHF) radio frequency identification (RFID) handheld reader. *IEEE Transactions on Antennas and Propagation, 55*(12), 3742–3745. doi:10.1109/TAP.2007.910521

Huang, Y., Wu, K. L., Fang, D. G., & Ehlert, M. (2005). An integrated LTCC millimeter-wave planar array antenna with low-loss feeding network. *IEEE Transactions on Antennas and Propagation, 53*(3), 1232–1234. doi:10.1109/TAP.2004.842588

Kim, J. S., Shin, K. H., Park, S. M., Choi, W. K., & Seong, N. S. (2006). Polarization and space diversity antenna using inverted-f antennas for RFID reader applications. *IEEE Antennas and Wireless Propagation Letters, 5*(1), 265–268. doi:10.1109/LAWP.2006.875892

Lin, Y. F., Chen, H. M., Chu, F. H., & Pan, S. C. (2008). Bidirectional radiated circularly polarised square-ring antenna for portable RFID reader. *Electronics Letters, 44*(24), 1383–1384. doi:10.1049/el:20082579

Lin, Y. F., Chen, H. M., Pan, S. C., Kao, Y. C., & Lin, C. Y. (2009). Adjustable axial ratio of single-layer circularly polarised patch antenna for portable RFID reader. *Electronics Letters, 45*(6), 290–292. doi:10.1049/el.2009.3569

Liu, H. W., Wu, K. H., & Yang, C. F. (2010). UHF reader loop antenna for near-field RFID applications. *Electronics Letters, 46*(1), 10–11. doi:10.1049/el.2010.2868

Lu, A. C. W., Chua, K. M., Wai, L. L., Wong, S. C. K., Wang, J. J., & Zhang, Y. P. (2004, December). *Integrated antenna module for broadband wireless applications.* Paper presented at the Electronics Packaging Technology Conference, Singapore.

Mobashsher, A. T., Islam, M. T., & Misran, N. (2010). A novel high-gain dual-band antenna for RFID reader applications. *IEEE Antennas and Wireless Propagation Letters, 9,* 653–656. doi:10.1109/LAWP.2010.2055818

Ooi, B. L., Zhao, G., Leong, M. S., Chua, K. M., & Lu, A. C. W. (2005). Wideband LTCC CPW-fed two-layered monopole antenna. *Electronics Letters, 41*(16), 9–10. doi:10.1049/el:20051921

Palandoken, M., Grede, A., & Henke, H. (2009). Broadband microstrip antenna with left-handed metamaterials. *IEEE Transactions on Antennas and Propagation, 57*(2), 331–338. doi:10.1109/TAP.2008.2011230

Qing, X. M., & Chen, Z. N. (2007). Proximity effects of metallic environments on high frequency RFID reader antenna: Study and applications. *IEEE Transactions on Antennas and Propagation, 55*(11), 3105–3111. doi:10.1109/TAP.2007.908575

Su, S. W., Wong, K. L., Tang, C. L., & Yeh, S. H. (2006). Wideband monopole antenna integrated within the front-end module package. *IEEE Transactions on Antennas and Propagation, 54*(6), 1888–1891. doi:10.1109/TAP.2006.875929

Sun, M., Zhang, Y. P., & Lu, Y. L. (2006, December). *Ultrawide-band integrated circuit package antenna in LTCC technology.* Paper presented at the Asia-Pacific Microwave Conference, okohama, Yokohama.

Sun, M., & Zhang Yue, P. (2008). A chip antenna in LTCC for UWB radios. *IEEE Transactions on Antennas and Propagation, 56*(4), 1177–1180. doi:10.1109/TAP.2008.919216

Thompson, D., Tentzeris, M., & Papapolymerou, J. (2007). Experimental analysis of the water absorption effects on RF/mm-Wave active/passive circuits packaged in multilayer organic substrates. *IEEE Transactions on Advanced Packaging, 30*(3), 551–557. doi:10.1109/TADVP.2007.898637

Wang, J. J., Zhang, Y. P., Chua, K. M., & Lu, A. C. W. (2005). Circuit model of microstrip patch antenna on ceramic land grid array package for antenna-chip codesign of highly integrated RF transceivers. *IEEE Transactions on Antennas and Propagation, 53*(12), 3877–3883. doi:10.1109/TAP.2005.859907

Yang, X., Yin, Y. Z., Hu, W., & Zhao, G. (2010). Compact printed double-folded inverted-L antenna for long-range RFID handheld reader. *Electronics Letters, 46*(17), 1179–1181. doi:10.1049/el.2010.1031

Ying, C., & Zhang, Y. P. (2004). Integration of ultra-wideband slot antenna on LTCC substrate. *Electronics Letters, 40*(11), 645–646. doi:10.1049/el:20040406

Zhang, Y. P. (2007). Design and experiment on differentially-driven microstrip antennas. *IEEE Transactions on Antennas and Propagation, 55*(10), 2701–2708. doi:10.1109/TAP.2007.905832

Zhang, Y. P., & Liu, D. X. (2009). Antenna-on-chip and antenna-in-package solutions to highly integrated millimeter-wave devices for wireless communications. *IEEE Transactions on Antennas and Propagation, 57*(10), 2830–2841. doi:10.1109/TAP.2009.2029295

Zhang, Y. P., Sun, M., & Lin, W. (2008). Novel antenna-in-package design in LTCC for single-chip RF transceivers. *IEEE Transactions on Antennas and Propagation, 56*(7), 2079–2088. doi:10.1109/TAP.2008.924706

Zhang, Y. Q., Guo, Y. X., & Leong, M. S. (2010). A novel multilayer UWB antenna on LTCC. *IEEE Transactions on Antennas and Propagation, 58*(9), 3013–3019. doi:10.1109/TAP.2010.2052576

KEY TERMS AND DEFINITIONS

Band Selection ICPA: One function of the differentially-driven ICPA which allows the antenna to shift at different bands.

CBGA: The ceramic ball grid array packaging technology.

Differentially-Driven Antenna: Antenna used in differentially-driven scheme which has two feeding points with 180 degree phase difference at both sides of the antenna.

G-S-G: A feeding technology used in ICPA which not only minimizes potential electromagnetic interference but also improves the feeding performance.

ICPA: The integrated circuit package antenna.

LTCC: The low temperature co-fired ceramic technology.

RFID Reader Antenna: The antenna for RFID reader whose function is to transform the energy between the wires and the air.

Chapter 5
Beam Forming Algorithm with Different Power Distribution for RFID Reader

A. K. M. Baki
Independent University, Bangladesh

Uditha Wijethilaka Bandara
Monash University, Australia

Nemai Chandra Karmakar
Monash University, Australia

Emran Md Amin
Monash University, Australia

ABSTRACT

It is possible to achieve higher BE and lower SLL of array antenna by implementing different amplitude or phase distribution technique in the array antenna. The phase errors of the system should also be kept to a minimum in order to maintain lower SLL and higher BE. The phase errors can come from any of the stages: signal detection, MW/RF generation, amplifier/attenuator, phase synchronization, phase shifter, et cetera. The phase error can be reduced by using non-uniform element spacing. In this chapter some methods of SLL reduction and increase of BE by adopting some edge tapering concepts and minimization of phase errors by implementing non-uniform spacing of array elements are discussed. The spectrum below 10 GHz frequency will likely be congested, and the spreading of millimetre wave technology in different emerging wireless applications as well as associated increase in energy consumption will be witnessed in the near future. In this chapter some new and better beam forming techniques for optimization between side lobe levels and beam efficiency are discussed. Different frequency bands of RFID systems are also focused on in this chapter.

1. INTRODUCTION

The antennas used in some applications such as RFID systems, WiMAX systems, and collision avoidance radar, must have very low side lobes. Array antenna technology with higher beam ef-

ficiency (BE) and lower side lobe level (SLL) is required in order to increase the coverage area of RFID systems, transmission bit rate and at the same time decrease the energy consumption and interference levels. The antennas used in some applications such as RFID systems, WiMAX

DOI: 10.4018/978-1-4666-1616-5.ch005

systems, and collision avoidance radar, must have very low side lobes. Array antenna technology with higher beam efficiency (BE) and lower side lobe level (SLL) is required in order to increase the coverage area of RFID systems, transmission bit rate and at the same time decrease the energy consumption and interference levels. It is possible to achieve higher BE and lower SLL of array antenna by implementing different amplitude or phase distribution technique in the array antenna. The phase errors of the system should also be minimum in order to maintain lower SLL and higher BE. The phase errors can come from any of the stages: signal detection, MW/RF generation, amplifier/attenuator, phase synchronization, phase shifter etc. The phase error can be reduced by using non-uniform element spacing. Methods of SLLs reduction and increase of BE by adopting some edge tapering concepts and minimization of phase errors by implementing non-uniform spacing of array elements are presented. The spectrum below 10 GHz frequency will likely be congested and the spreading of millimetre wave technology in different emerging wireless applications as well as associated increase in energy consumption will be witnessed in the near future. A few new beam forming techniques for optimization between SLLs and BE are discussed. Different frequency bands of RFID systems are also focused.

Radio Frequency IDentification (RFID) is a wireless tagging technology that allows a target to be automatically identified at a distance without a direct line-of-sight, by transmitting data through electromagnetic exchange. RFID has gained worldwide popularity in numerous applications such as inventory tracking, animal tagging, security surveillance and authenticity verification(R. Want, 2004). RFID systems are becoming ubiquitous and the growth of RFID uses is rising tremendously. In some sophisticated applications, fast and energy efficient tag reading is desirable, especially when the number of tags is high (D. K. Klair, K. W. Chin, and R. Raad, 2010). In such situations, the RFID cannot read all the tags and most of the transmitted energy from the RFID reader is wasted.

RFID systems can be classified based on their frequency band of operations, such as:

1. **Low frequency (LF):** 125-134 KHz;
2. **High frequency (HF):** 13.56 MHz;
3. **Ultra high frequency (UHF):** 433 MHz, 860-960 MHz;
4. **Microwave (MW):** 2.4 GHz, 5.8 GHz;
5. **Millimetre wave (mm-Wave):** e.g., 60 GHz and 77 GHz (PekkaPursula et al., 2008);

For UHF RFID system, different regions are using different frequency sub-bands:

1. **USA and Canada:** 902-928 MHz;
2. **Australia:** 920 to 926 MHz;
3. **Japan:** 950 MHz;
4. **Europe:** 965-868 MHz;

The main components of a chipless RFID system are a reader antenna and a passive chipless RFID tag. The reader antenna reads the backscattered radio frequency signals from the chipless tag. The block diagram of an RFID reader system is shown in Figure 1.

Adaptive antenna with higher gain and lower side lobe levels (SLL) is required for RFID systems. Adaptive antenna can decrease the energy consumption, interference levels and increase the range of RFID systems. By controlling the phases and amplitudes of antenna elements it is possible to achieve higher beam efficiency (BE) and lower SLLs. Phase errors of antenna elements cause higher SLL and lower gain. In this chapter some phase error reduction technique will be discussed. Edge tapering of array antenna can reduce SLLs and enhance the antenna gain. Although there are some technical demerits of full edge tapering. Some new and better beam forming techniques for different frequency bands by using edge tapering concepts will be discussed in next sections. Design and development process of low

Figure 1. Block diagram of an RFID reader system

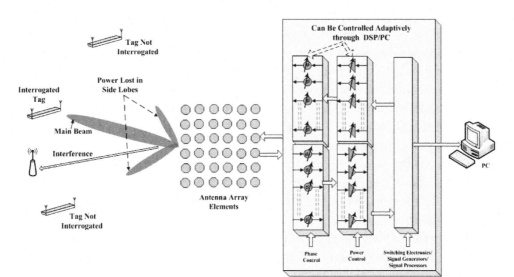

cost beam forming networks for an RFID system will also be discussed in this chapter. An organization chart of the chapter is shown in Figure 2.

2. PERFORMANCE MEASUREMENT OF AN RFID TAG

The performance of an RFID tag can be measured from the following tag characteristics:

1. Tag Range;
2. Backscatter Range;
3. Tag Sensitivity; and
4. Backscatter Efficiency;

The above four terms can be defined in the following way (Pavel V. Nikitin and K. V. Seshagiri, 2009):

The tag range is the maximum distance at which the tag can be read or written in free space. The tag sensitivity is the minimum incident power (signal strength) needed at a tag location either to read or write the tag. Backscatter efficiency is the function of incident power and a measure of the amount of modulated backscattered power by the

tag. Backscatter range is the maximum distance at which the data from the tag can be decoded by the reader with certain receiver sensitivity.

Characteristics (1) and (2) depend also on the reader and environment; maximum tag range is limited by the forward link (reader-to-tag) (Pavel V. Nikitin and K. V. Seshagiri, 2009). Write sensitivity of a tag differs from the read sensitivity by 3 dB, because the RFID IC needs more power for performing write operation; accordingly read and write ranges differ as well by 30% (Pavel V. Nikitin and K. V. Seshagiri, 2009).

2.1. Tag Range and Backscatter Range

The tag range and backscatter range of a passive RFID tag is limited by its ability to provide sufficient power for the tag operation. In order to extend the range, the received power by a passive tag should be increased. The received power by a tag can be expressed by the following Friis transmission Equation (1):

$$P_{tag} = P_T G_T G_{tag} \left(\frac{\lambda}{4\pi r_{tag}}\right)^2 \qquad (1)$$

Figure 2. Organization chart of the chapter

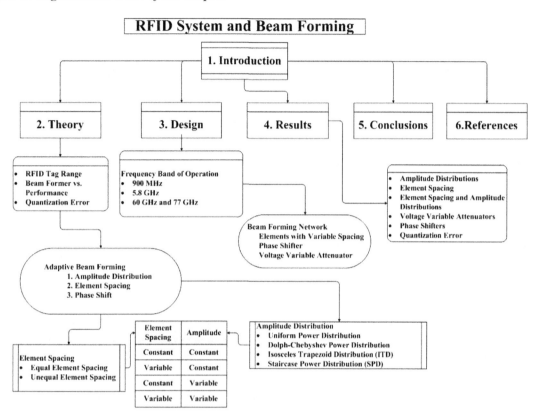

where,

P_T is the transmitted power by the reader system;

G_T is the transmitting antenna gain;

G_{tag} is the tag antenna gain; and

r_{tag} is the distance between the transmitting antenna and the tag;

The transmitted power P_T is limited by the regulation of Federal Communication Comission (FCC). For example in the United States, FCC regulations limit the amount of power, that can be transmitted in the 902-928 MHz frequency range, to 30 dBm (1 watt) 'maximum power output' and to 36 dBm (4 watt) 'maximum Effective Radiated Power (ERP)'.

In (1), the term '$(\frac{\lambda}{4\pi r_{tag}})^2$' is called the free space loss factor which is a function of tag distance and operationg frequency. For certain applications this term is fixed and cannot be changed. For an RFID system of a particular tag sensitivity, the tag range can be increased if the transmitting antenna gain can be increased. Figure 3 shows an example of tag ranges vs. transmitting antenna gains curves for different tag sensitivity. In this case receiving antenna gain is considered as 0 dB.

If the receiving antenna gain of the RFID reader is G_R, then the received power by the reader can be expressed as:

$$P_{reader} = P_T G_T G_R G_{tag} \frac{1}{r_{tag}^2 r_{backscatter}^2} (\frac{\lambda}{4\pi})^4 \xi$$

$$(2)$$

Figure 3. Tag range vs. antenna gain for different tag sensitivity

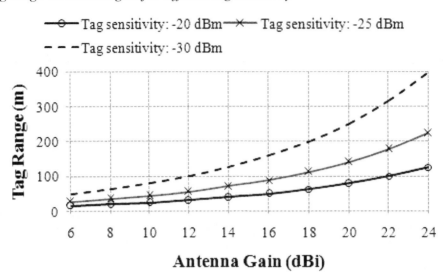

where ξ is the tag backscatter efficiency and $r_{backscatter}$ is the distance between the tag and the receiving antenna of the reader.

The minimum received signal level of the reader is specified (reader sensitivity) for maximum possible range of operation. The other way it can be explained that there is a minimum P_{reader} for which the system is operational. If the minimum reader power (P_{min_reader}) and G_{tag} are fixed, then the ranges $r_{tag}/r_{backscatter}$ can be controlled by controlling G_T and/or G_R.

2.2. Performance Evaluation of Microwave Beam of an RFID Reader Antenna

The longer reading ranges of an RFID reader system can be achieved if the performance of the reader antenna is improved. The link margin and the interferences can be controlled by an efficient antenna. This is particularly important for the system, where the transmitted signal is returned from the tag with a fourth-order reduction in magnitude of the reading distance (R^{-4}) (Karmakar, 2010, pp. 57~82). A little improvement in reader antenna performance, such as, gain and directivity, will play a significant role in improving the reading range, data rates, BER, interference cancelation, collision mitigation, and localization of tag.

The RFID reader antenna must have very low side lobes in its radiation pattern in order to maximize the received power and minimize interferences and noises. Higher side lobes may lead to false alarms and collision may happen in RFID applications. Different RFID reader architectures by using phased array/smart antenna concepts are discussed in details in (Karmakar, 2010, pp. 57~82). RFID reader design, development process and platform are discussed in (Preravodic&Karamakar, 2010, pp. 85~121, Jamali, 2010, pp. 123~137). RFID planar antenna and handheld reader antenna development processes are described in (Roy & Karmakar, 2010, pp. 141~171).Field Programmable Gate Array (FPGA) based phase control method of array antenna for RFID reader has been discussed in (Karmakar et al., 2010, pp. 211~241).

RFID reader antennas can be classified into two categories (N. C. Karmakar, 2010):

1.   Fixed-beam antennas; and
2.   Phased array antennas;

Generally with the fixed-beam antennas it is not possible to achieve higher gain. As a result, localization of tags and anti-collision mitigation are not possible with fixed beam antennas. On the other hand, with a phased array antenna, the direction of the transmitted beam can be controlled by carefully adjusting the amplitudes and phase shift of the antenna elements. With appropriate anti-collision protocol and a phased array antenna, the RFID system can achieve temporal and spatial diversities.

An adaptive beam forming smart antenna is needed for an RFID reader; the antenna should have the gain enhancement and interference cancellation capability. Since the chipless tag will re-transmit the received signal towards the reader antenna, it is most essential to have a smart antenna based transmit/receive system with maximum BE/Beam Collection Efficiency (BCE) as well as minimum SLL. BE is the ratio of power flow within the main beam to the whole transmitted power.

The BE for a two dimensional radiation pattern quantifies the solid angle extent of the main beam relative to that of the entire pattern and can be expressed as (Stutzman& Thiele, 1997):

$$BE_{2D} = \iint\limits_{main\_beam} |E(\theta,\varphi)|^2 \, d\Omega \Big/ \iint\limits_{4\pi} |E(\theta,\varphi)|^2 \, d\Omega \tag{3}$$

where,

$E(\theta,\varphi)$ is the radiated electric field.

BE for one dimensional case can be expressed as:

$$BE_{1D} = \int\limits_{\theta_m} |E(\theta)|^2 \, d\theta \Big/ \int\limits_{\theta_w} |E(\theta)|^2 \, d\theta \tag{4}$$

θ_m is the angle sector due to one dimensional main beam and θ_w is the angle sector of $\pm 90°$.

$E(\theta)$ is the one dimensional radiated electric field.

Similarly BCE is the ratio of power flow that is intercepted by the receiving antenna to the whole transmitted power (Baki et al., 2007, pp. 968~977).

BCE for two dimensional array and rectangular/square receiving antenna can be expressed as (Baki et al., 2007, pp. 968~977):

$$BCE_{2D} = \int\limits_{\theta_{ry}}\int\limits_{\theta_{rx}} |E(\theta_x,\theta_y)|^2 \, d\theta_x d\theta_y \Big/ \int\limits_{\theta_{ty}}\int\limits_{\theta_{tx}} |E(\theta_x,\theta_y)|^2 \, d\theta_x d\theta_y \tag{5}$$

where,

θ_{tx} and θ_{ty} ; ± 90degree angle sector.

θ_{rx} is angle sector due to x dimension of receiving antenna.

θ_{ry} is angle sector due to y dimension of receiving antenna.

$E(\theta_x,\theta_y)$ is the energy of the radiated electric field.

BCE for one dimensional case can be expressed as:

$$BCE = \int\limits_{\theta_r} |E(\theta)|^2 \, d\theta \Big/ \int\limits_{\theta_w} |E(\theta)|^2 \, d\theta \tag{6}$$

θ_r is the angle sector due to one dimensional receiving antenna and θ_w is the angle sector $\pm 90°$.

$E(\theta)$ is the energy of the one dimensional radiated electric field.

BE, BCE, and Maximum SLL (MSLL) are the indices for the performance evaluation of the microwave (MW) beam.

Suppression of Grating Lobes (GL) and Side Lobe Levels (SLL) is necessary for higher BE and to avoid interference to other communication systems. When GLs appear and SLLs increase, the transmitted power is absorbed into these lobes which cause reduction of received power at the sensor tags. Chances of interference also increase when GLs appear.

2.2.1. Adaptive Beam Forming

An RFID reader antenna should have the following properties:

1. Side Lobe Levels as minimum as possible;
2. Beam Efficiency or power transmission/reception efficiency as maximum as possible; and
3. Null steering capabilities;

In adaptive beamforming, array antennas are used to achieve maximum reception/transmission in a desired direction and reject signals in other directions. Maximum reception/transmission is achieved by adjusting the weighting functions (amplitudes and/or phases) of the antenna elements of an array antenna.

An adaptive beam forming antenna will have the beam steering capability to interrogate sensor tags at different locations. The sensor signals can interfere with one another when multiple sensors respond simultaneously to a reader's signal. Because signals from different RFID sensors may occupy the same frequency channel arriving from different directions. It is possible to reduce interferences from sensors of other locations with an adaptive beam forming antenna system since most sensors will be inactive during the interrogation of some particular sensor/sensors group. This feature of adaptive antenna will help collision mitigation and increase the effective Signal to Noise and Interference Ratio (SNIR) both at the reader and sensors sides.

Array antenna elements increase the directivity of the antenna system. If all antenna elements in an array are excited uniformly then the main beam carries only a part of the total energy, since most of the energies are absorbed in higher SLLs.

2.2.1.1. Edge Tapering of Array Antenna

It has been seen that it is possible to increase BE and reduce SLL if edge tapering concept can be adopted in array antennas. Though it is possible to increase the BE of the array antenna by edge tapering but at the expense of technical and thermal complexity and reduced directivity or taper efficiency. The SLLs tend to decrease and the beam width increases as the power amplitude is tapered more toward the edges of the array. The elements at the array centre are excited more strongly than those near the edge. This beamwidth-SLL trade-off can be optimized (Stutzman& Thiele, 1997) by optimizing the phases and amplitudes of the array elements. The thermal complexity is due to the strongest excitation at the array centre and the reduced directivity is due to an increase in beam width.

Some of the amplitude distributions of array antenna are (Baki et al., 2007, pp. 968-977):

1. Uniform Power Distribution.
2. Dolph-Chebyshev Power Distribution.
3. Gaussian Power Distribution.
4. Isosceles Trapezoid Distribution (ITD).

A new method of power distribution for array antenna elements by implementing the staircase concept will be discussed in this chapter. The name of this new power distribution concept is Staircase Power Distribution(SPD).

A chart of the above mentioned amplitude distributions, their respective radiation patterns and some merits/demerits are shown in Figure 4. There are also some other antenna element excitation and SLL reduction techniques, which are out of the scope of this chapter.

Gain factors for different power distributions are discussed in (Mailloux, 1994). Gain factor is the pattern directivity normalized to the maximum directivity of the line source. The gain factor is the highest in uniform distribution (Mailloux, 1994). However, at the same time SLLs are also higher in this case. SLL can be made lower by adapting edge tapering concept at the expense of reduced gain factor or taper efficiency. Taper efficiency is the discrete analogue of the gain factor used for continuous apertures. Taper efficiency for one dimensional phased array antenna can be expressed as (Mailloux, 1994):

$$\varepsilon_T = (1 / N)\left[\left|\sum a_n\right|^2 / \sum \left|a_n\right|^2\right] \qquad (7)$$

where N is the total number of phased array antenna elements and

n: 1, 2, 3, …N.

a_n: Amplitudes of the array elements.

2.2.1.2. Some Merits and Demerits of Edge Tapering

The decrease in beam efficiency of the Dolph-Chebyshev pattern results from the requirement that all SLL are constant. For large arrays, this implies that increasingly more of the energy is in the side lobes region and in the limit of a very large array, maintaining the Dolph-Chebyshev side lobes pattern requires an impractical aperture illumination. One demerit of Dolph-Chebyshev and Taylor distributions is, sometimes the maximum efficiency in these distributions is achieved with a peak in the aperture illumination near the array edge (inverse tapering). This rapid variation in current is difficult to approximate with a discrete array and becomes complex(Mailloux, 1994). One example of a -30 dB Dolph-Chebyshev edge tapering for a 25-element phased array antennas is shown in Figure 5. Here the last antenna element from each side shows the behaviour of inverse tapering.

It is mentioned in (Brown & Eves, 1992, pp. 1239~1250) according to

Figure 4. Chart of different edge tapering of array antenna with radiation pattern

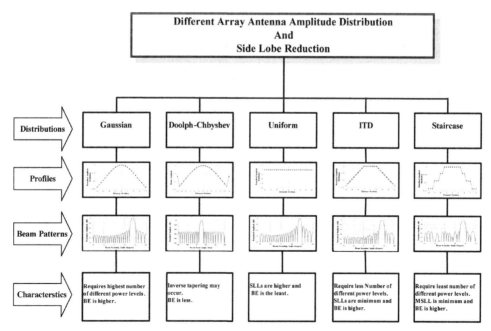

Figure 5. Amplitude distribution of 25 element array antenna with -30 dB Dolph-Chebyshev edge tapering

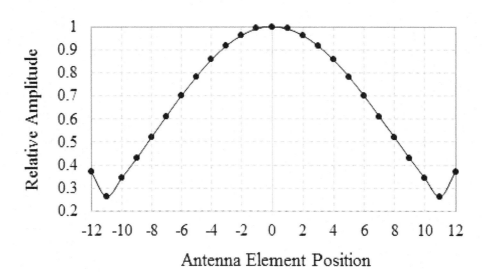

(Goubau&Schwering1961, pp. 248-256), almost 100% efficiency can be achieved by edge tapering which is essentially Gaussian distribution. Figure 6 shows the relative power distribution of -40 dB Gaussian edge tapering for 100% efficiency.

These types of edge tapering and inverse edge tapering near the edges may not be practical for some applications. The effectiveness or technical feasibility of the array antenna also depends on the uniformity of amplitude distribution along the surface, i.e. maximum utilization of antenna surface. Therefore a trade-off of amplitude distribution is needed between uniform and tapered distribution to maintain minimum SLL and maximum taper efficiency or BE.

2.2.1.3. Array Factor with ITD Edge Tapering

A conventional beam forming network for array antenna is shown in Figure 7.

Amplitude distribution for 1D array based on Isosceles Trapezoidal Distribution (ITD) concept is shown in Figure 8. Amplitudes from each side of the array antennas are tapered according to ITD concept and amplitudes of middle elements are kept uniform.

Array Factor (AF) with ITD can be expressed by the following equation (Baki et al., 2007, pp. 968-977):

$$AF = \sum_{n=1}^{N_t} \delta_{tn} e^{jn\psi_1} + \sum_{m=1}^{N} A e^{j(m+N_t)\psi} + \sum_{n=1}^{N_t} \delta_{tn} e^{j(n+N_t+N)\psi_2}$$

(8)

where, N = Number of array element with uniform amplitudes,

N_t = Number of elements tapered from each side,

$$\psi = \beta d(\sin\theta - \sin\theta_0).$$

$$\psi_1 = -\beta d(\sin\theta - \sin\theta_0).$$

$$\psi_2 = [(N-1)d + nd](\sin\theta - \sin\theta_0)\beta.$$

d = inter-element spacing (m).

N_T = Total number of antenna elements= N+ 2N_t.

Figure 6. Relative power distribution along the phased array antenna for 100% efficiency

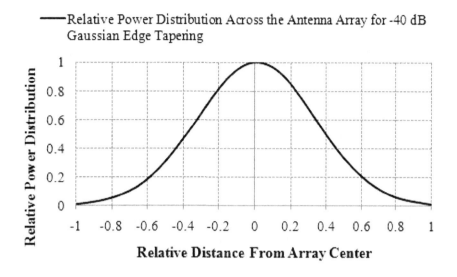

$\beta = 2\pi/\lambda$ = phase constant.

The amplitude distribution of the ITD tapering is:

$$\delta_{tn} = \sum_{n=0}^{N_t} (1 - \delta_{t0})n/N_t + \delta_{t0} \qquad (9)$$

δ_{t0} is the amplitude of each of the end elements.

= Direction of beam maximum along the broad side.

$n = 0, 1, 2...N_t, 0 \leq \delta_{tn} \leq A$.

A= Uniform amplitudes of middle antenna elements.

Figure 7. A conventional array antenna beam forming network

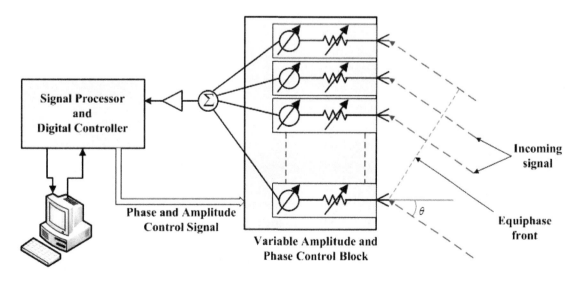

Figure 8. Amplitude distribution of array antenna with ITD concept

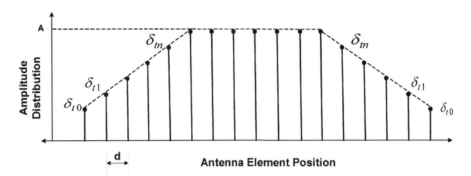

m = 1, 2, 3, ...N.

2.2.1.4. Staircase Power Distribution (SPD)

ITD and ITDU edged tapered concept are discussed in previous sections. A new method of amplitudes distribution of array antenna will be discussed in this section. Instead of using gradual variation of amplitudes distribution of array antenna (as in the cases of ITD, ITDU, Gaussian or Chebyshev etc.), the concept of 'staircase' is implemented for antenna power distribution and named as Staircase Power Distribution (SPD). Fabrication of array antenna, beam forming network and the use of array elements in terms of power distribution with SPD concept are easier and better than those of other kinds of power distributions mentioned above. The number of different amplitudes distribution in SPD is least and stepwise uniform. These aspects of power distribution make the fabrication and construction of array antenna and beam forming network less complex.

Methods of designing transmitter outputs by using on-chip power amplifier (PA) stages in each element need good linearity, high efficiency, high power gain and high output power (Natarajan et al., 2006, Pfeiffer et al., 2007, Van-Hoang Do et al., 2008). A four-stage PA with at least 3-dB gain in each stage, the transistor size doubled in each stage, is discussed in (Natarajan et al., 2006 and Pfeiffer et al., 2007). It is possible to design on-

chip power amplifier stages with variable gains for different antenna elements and this way it would be possible to minimize the SLLs even to lower levels. As a result the BE of the array antenna will increase and interference to other communication systems will decrease. The concept of SPD is shown in Figure 9.

2.2.1.4.1. Array Factor of Staircase Power Distribution

One dimensional Array Factor (AF) with staircase power distribution (SPD) can be expressed as:

$$AF = \sum_{n=-(N-1)/2}^{(N-1)/2} \delta_1 e^{jn\psi} + \sum_{n=[-(N-1)/2]+N_{s1}}^{[(N-1)/2]-N_{s1}} (\delta_2 - \delta_1) e^{jn\psi}$$
$$+ \sum_{n=[-(N-1)/2]+N_{s1}+N_{s2}}^{[(N-1)/2]-N_{s1}-N_{s2}} (\delta_3 - \delta_2) e^{jn\psi} + ...$$
$$+ \sum_{n=[-(N-1)/2]+N_{s1}+N_{s2}...+N_{sl}}^{[(N-1)/2]-N_{s1}-N_{s2}...-N_{sl}} (\delta_l - \delta_{l-1}) e^{jn\psi}$$
$$+ \sum_{n=[-(N-N_s-1)/2]}^{[(N-N_s-1)/2]} (A - \delta_l) e^{jn\psi}$$

(10)

where, N = Total number of antenna elements,

$N_{s1}, N_{s2}, N_{s3...} N_{sl}$ etc. antenna element numbers tapered from each side (starting from edge of the array) for 1st stage, 2nd stage, 3rd stage...last stage.

Here last stage is defined as the stage before the middle antenna elements.

Figure 9. Staircase power distribution (SPD) for array antenna elements

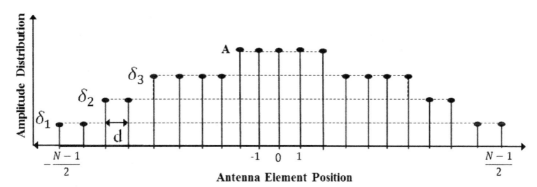

$N_S = N_{s1} + N_{s2} + N_{s3} + ... + N_{sl}$ = Number of elements tapered from each side,

$$\psi = \beta d(\sin \theta - \sin \theta_0).$$

$\delta_1, \delta_2, \delta_3 ....\delta_l$ are the amplitudes of the antenna elements of 1st stage, 2nd stage, 3rd stage...last stage.

d = spacing between elements (m).

$\beta = 2\pi/\lambda$ = phase constant.

A is the amplitude of middle antenna elements.

= Direction of beam maximum along the broad side.

$n = 0, 1, 2...N$.

2.2.1.5. Unequal Element Spacing for ITD

Distribution of elements spacing of array antenna can be classified as two broad categories:

1.   Equal element spacing
2.   Unequal element spacing

Table 1 shows different combination of inter-element spacing and amplitude distributions of array antenna elements.

Radiation pattern of ITD edge tapering can further be controlled and SLL be reduced by implementing unequal spacing of array elements. When both ITD edge tapering and **U**nequal element spacing are applied on array antenna, it is named as 'ITD edge tapering with Unequal (ITDU) element spacing' concept (Baki et al., 2008, pp. 527-535). Calculation process of unequal element spacing for ITD is explained in the following sections.

Let us assume that the uniform element spacing for an antenna array is **d**. And the unequal element spacing for the array antenna is **d**×*m(x)*. Here the variable *x* is the position of antenna elements. The multiplying factor *m(x)* for the determination of unequal element spacing can be expressed as (Baki et al., 2008, pp. 527-535):

$$m(x) = Sinc[A(x)]/\min \langle Sinc[A(x)] \rangle = Sinc\big[A(x)\big]/0.8415 \tag{11}$$

here, *A(x)* is the amplitude distribution of an isosceles trapezoid (and not the power distribution of the ITD edge tapering). If the total number of elements is (2n+1) then the range of *x* is from 1 to 2n. Though several other distributions (such as Gaussian and cosine) were tried to determine the unequal spacing, it was found that the sinc function of *A(x)* best fits for the determination of unequal element spacing in order to reduce MSLL and maintain higher BE/BCE. One example of

Table 1. Different combination of element spacing and amplitude for a phased array antenna

Element Spacing	Amplitude
Constant	Constant
Variable	Constant
Constant	Variable
Variable	Variable

determination process of unequal element spacing is shown in Figure 10. The steps for the determination of unequal element spacing for ITDU of 1D array are described here:

Step 1: Select an Isosceles Trapezoid for the determination of unequal element spacing. In Figure 10, curve 1 [A(x)] is selected for the determination of unequal element spacing. In this case 40 elements from each side of total 101 array elements are selected for unequal spacing. Remaining middle 21 elements are kept for equal spacing.

Step 2: Take the sinc function of A(x) [curve 1] to obtain m(x) [curve 2] by using (11).

Step 3: Determine the value of equal element spacing d. The value of equal element spacing is taken 0.68λ in this example.

Step 4: Multiply d with *m(x)* to obtain the values of unequal element spacing [curve 3].

2.2.2. Beam Forming for 60 Ghz Signal with SPD

2.2.2.1. 60 GHz Signal: Why?

Higher data rates with smaller bandwidths (e.g. WLAN at 2.4 or 5.8GHz), requires a complex modulation scheme in order to squeeze as many bits/Hz as possible. According to Shannon's theorem, channel capacity is proportional to bandwidth and a logarithmic function of SNR. Shannon's theorem for maximum channel capacity (*C*) (in bps) can be expressed as:

$$C = Blog_2(1 + SNR) \qquad (12)$$

where,

B is the bandwidth of in Hz; and

Figure 10. Curves for determination of Unequal element spacing for ITDU

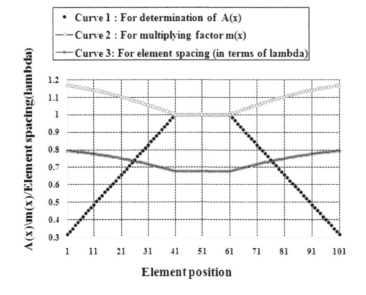

SNR is the ratio of signal power to noise power (not in dB);

It can be seen from (12) that, the channel capacity can be increased by increasing the bandwidth and/or SNR of the signal. At mm-Wave, e.g., 60 GHz, high data rate communication even with gigabit rates is possible (PekkaPursula et al., 2008). There are also some other applications in automatic cruise control (ACC) system, automotive radar for collision detection and mm-Wave imaging for security, if 60 GHz signal can be used.

60 GHz signal experiences higher attenuation. In particular, transmitted wave from mm-Wave 60 GHz signal is rapidly absorbed by atmospheric oxygen molecules over long distances. This high level of atmospheric absorption and resulting range limitations are some of the reasons this band has been relatively unused for long-range communication. 60 GHz signal can be used for short distance communication and the same frequency can be reused in a localized area as shown in Figure 11.

Another benefit of 60 GHz signal is the smaller antenna size. At 60 GHz, a compact antenna array with a larger number of elements can

be integrated even on to an IC. With larger number of antenna elements, pencil like beam with higher gain can be created. In the U.S.A., the maximum limit of power transmission in the 60 GHz band is 40 dBm (10 watt), which is higher than the limit in the UHF band. Since mm-Wave signal with the higher gain and pencil like beam occupies smaller surrounding space, interferences with other communication devices will be minimized if 60 GHz signal is used. As a receiver, mm-Wave antenna will also receive signal through narrower spatial direction, thereby reducing the chances of undesired interferences. 60 GHz RFID system has been proposed by (N.C. Karmakar, 2009, Stefano Pellerano et. al., 2010)

Since large no. of antenna elements can be used in the 60 GHz band, it would be practical to implement SPD concept in this frequency band which will be helpful to overcome the atmospheric attenuation.

Figure 11. Working range vs. frequency

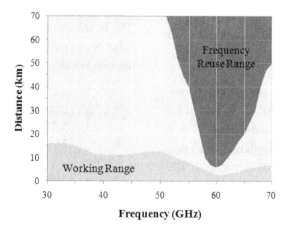

[Adopted from: Federal Communications Commission (FCC), Office of the Communication and Technology, New Technology Development Division, Bulletin no. 70, July 1997. Millimeter Wave Propagation: Spectrum Management Implications.]

2.2.3. Phase Error Due to Quantization and Round off

Phase error of the antenna elements should be minimum in order to maintain higher BE and reduce SLL. The phase error of the antenna is the combined effect of phase errors in every stage: signal detection, MW/RF generation, amplifier/attenuator, phase synchronization, phase shift etc. In this section only the errors due to phase round off in phase shifter is discussed. Phase quantization/round off errors can be minimized by using phase shifter with higher number of bits. However increasing the number of bits causes higher power losses in phase shifters. Also it is expensive to use the phase shifter with higher number of bits. Phase errors as well as phase round off errors cause beam pointing errors. Phase quantization /round off errors are highly correlated and they result in large, well-defined side lobes or grating-lobes-type pattern errors (Mailloux, 1994). Although the array is required to produce a smooth phase taper, an N-bit phase shifter has phase states separated by the least significant bit (Mailloux, 1994):

$$\varphi_0 = \frac{2\pi}{2^N} \tag{13}$$

This discretization allows only a staircase approximation of the continuous progressive shift required for the array and it results in a periodic phase error that can produce the pattern with GL like side lobes and causes the beam pointing error. One example of phase round off errors with 4-bit phase shifter (difference of phase states is 22.5°), which is periodic in nature, is shown in Figure 12. In this case total number of elements is 109 and uniform element spacing is 0.68λ.

It is possible to reduce the peak side lobes or grating lobes by disrupting the total periodicity that leads to the large grating lobes. It is a common engineering practice to randomize the phase round off error in an array steered by the discrete

phase shifters (Mailloux, 1994). Beam pointing error due to phase error will also be decreased due to randomization of phase round off errors. Although the average phase error cannot be reduced, it is possible to reduce the root mean squared (R.M.S) phase error/phase error variance by breaking up the periodicity of the phase quantization/round off error and hence reduce the peak side lobes, beam pointing errors and power loss. For a linear array of $N + 1$ elements, the beam pointing error can be expressed by (Carver et al., 1973, p. 199-202):

$$\delta\theta_{rms} = \frac{2\sqrt{3}\sigma}{\beta d \cos\theta_0 N^{3/2}} \tag{14}$$

Where σ = R.M.S. phase error.

$$\beta = \frac{2\pi}{\lambda} = \text{Phase constant.}$$

d = Spacing between antenna elements.

θ_0 = Main beam pointing angle.

From (14), it can be seen that it is possible to reduce beam pointing error by increasing the element spacing/number of array elements, reducing the beam steering angle or R.M.S phase error. The beam pointing deviation for an array of N elements can be expressed as (Mailloux, 1994):

$$\bar{\Delta}^2 = \bar{\Phi}^2 \frac{\sum w_i^2 x_i^2}{\left(\sum w_i x_i^2\right)^2} \tag{15}$$

where w_i is the amplitude of the ith element excitation; x_i is the element position divided by element spacing d; and $\bar{\Phi}^2$ is the phase error variance.

It can be observed from both (14) and (15) that the beam pointing error can be reduced by reducing the R.M.S. phase error/phase error variance. The effect of phase quantization/round off errors

Figure 12. Phase round off error of 4-bit phase shifter for 109 phased array antenna elements (0.68λ element spacing)

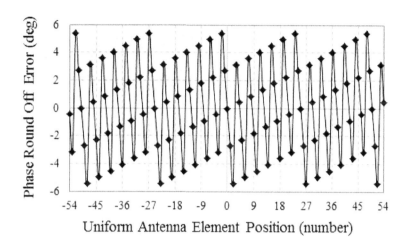

should be minimized to reduce the total R.M.S. phase error/ phase error variance. Since there are practical limitations of using phase shifters with higher number of bits (due to cost, loss of power etc),the other way of reducing the phase quantization/round off errors is the use of unequal element spacing for the phased array antenna. The use of unequal element spacing will disrupt the periodicity of the phase round off error and hence the R.M.S. phase error /phase error variance due to phase round off will decrease.

3. DESIGN OF A LOW COST ADAPTIVE BEAM FORMING NETWORK AT UHF RFID BAND

In order to develop a low cost adaptive beam forming network for an RFID reader system, the following steps can help as a guide line:

1. Development of power divider circuit for providing the signal path to each patch element;
2. Development of phase shifter circuit for controlling the phase of the RF signal for each element.

3. Development of attenuator/controller circuit for controlling the RF signal strength of each patch element;

3.1. Development of Adaptive Beam Forming Network

Figure 13shows a block diagram of an 8×8 elements smart antenna. The operating frequency of the antenna is at 900 MHz with 100 MHz

Figure 13. Top view of the 8x8 elements smart antenna

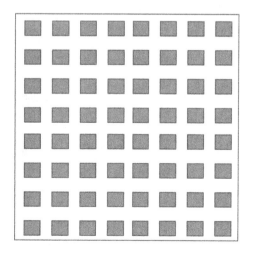

Figure 14. Digital beam former for 8×8 elements smart antenna operating at 900 MHz

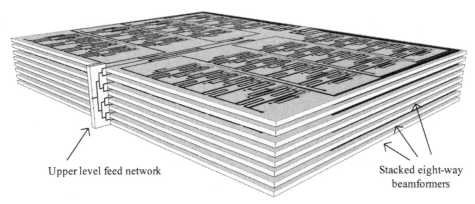

Upper level feed network

Stacked eight-way
beamformers

bandwidth. The shape of each antenna element is L-shaped slot-loaded semicircular patch antennas on air gap layer. (K. P. Ray and D. D. Krishna, 2006) proposed an L-slot loaded semi-circular patch antenna which operates in a wide bandwidth. Air substrate is used in order to reduce the weight and cost of fabrication of the 8x8 elements array. Detailed discussion of the array antenna design is out of the scope of this chapter. Figure 14 shows the digital beam former layout of the smart antenna where eight PCBs are stacked together. Each of the PCBs contains beam former circuits for 8 elements. Figure 15 shows one of

the developed beam former PCB for 8 elements of the 8×8 elements array. The phase of the RF signal for each element can be controlled by a 4-bit phase shifter. The radiation pattern of the antenna can be controlled by the phase shifter and voltage variable attenuator (VVA).

3.1.1. Development of Low Cost Phase Shifter

Phase shifters are the most expensive components in conventional phased array antennas. Relative phases of signals to/from antenna elements are

Figure 15. Digital beam former PCB for 8 elements of the 8×8 elements smart antenna developed by RFID research group of Monash University, Australia (2011)

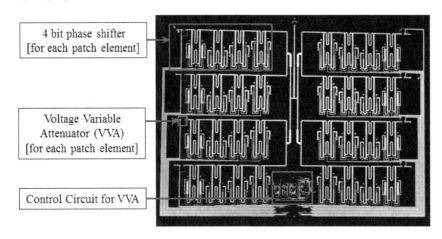

4 bit phase shifter
[for each patch element]

Voltage Variable
Attenuator (VVA)
[for each patch element]

Control Circuit for VVA

controlled by phase shifter. They can be of either analog or digital type. Digital phase shifters are preferred when the array needs digitally control. A digital phase shifter consists of a cascade of 1-bit phase shift elements to form an N-bit phase shifter as illustrated in Figure 16. The microwave/ RF diodes are usually very expensive and they are needed in large quantities in the digital beam forming network. A hybrid coupled reflection-type phase shifter can be used to reduce the number of diodes. The hybrid coupled phase shifter yields larger bandwidth (above 100 MHz) and needs only two diodes per bit. A very low-cost PIN diode can be used to minimize the cost (Karmakar &Zakavi, 2010).In the board shown in Figure 15, a 4-bit phase shifter is used for controlling the phase of each element.

A hybrid coupled reflection-type phase shifter uses a 3 dB hybrid coupler for splitting the power into two equal-amplitude and quadrature phases (L. G. Maloratsky, 2004). These signals are reflected from equal reactive loads placed at the output ports of the coupler and recombine at its decoupled port. The diodes are RF switching elements which include or exclude the reactive elements from the circuit, depending on the state of the diode, hence causing the resultant phase shifts. In order to reduce the space occupied by the phase shifters, the hybrid coupler structure is shrunk (Figure 17) by bending the 50 Ω lines into two 35.5 Ω lines (M. E. Bialkowski& N. C. Karmakar, 1999; N. C. Karmakar& M. E. Bialkowski, 1999). This method provides significant space savings, particularly for applications that employ a large number of phase shifter arrays in tandem.

The 4-bit phase shifters of 180°, 90°, 45° and 22.5° look identical with a variation of reactive loadings. The phase shifters can be driven by ±5 V bias voltages. The DC bias current is generated by V_{CC} and the current limiting resistor R. The DC blocking capacitor C' and the capacitance C of the LC compensation network (M. E. Bialkowski, & N. C. Karmakar,1999, L. G. Maloratsky,2004) force the DC bias current to flow through the diode only. Thus the ON\OFF states of the diode yields the appropriate different phase shift.

3.1.2. Development of Voltage Variable Attenuator (VVA)

A voltage variable attenuator (VVA) is connected with each patch element for controlling the RF amplitude of the signal. The used VVA in the circuit is an 8-lead surface mount IC (HMC346MS8G). The VVA can control the RF signal over a 30 dB amplitude range.

Figure 16. N-bit digital phase shifter

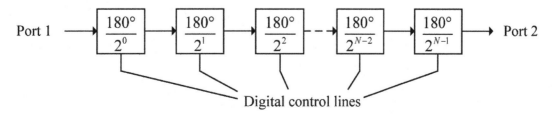

Figure 17. Configuration of compact hybrid coupled phase shifter (1-bit)

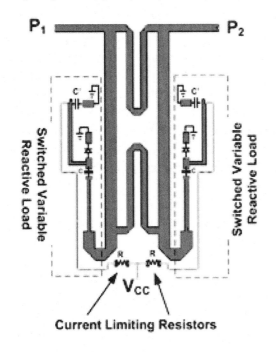

4. PERFORMANCE MEASUREMENTS: EXPERIMENTAL AND SIMULATION RESULTS

4.1. Performance Measurement for 900 MHz Beam Former

4.1.1. Performance Measurement of Voltage Variable Attenuators

Figure 18 shows the measured gain vs. control voltage (V1) of an HMC346MS8G IC at 900 MHz. The IC can operate up to 8 GHz. There is a control circuit on each of the beam former PCBs (Figure 15) which can control the gain of each VVA.

Figure 19 shows the measured insertion losses (S_{21}) between the RF port and the port for a patch element mounted on the PCB board shown in Figure 15. The insertion losses are the functions of different control voltages from the control circuit for VVA. The control voltage (shown in Figure 19) from the control circuit and the control voltage (V1) shown in Figure 18 are not the same. Figure 20 shows the corresponding measured phases which are almost linear for the frequency range 700-900 MHz. Figures 18-20 show that, with the VVA circuit, it is possible to control the gain and to achieve linear phase response of each patch element of the antenna.

Figure 18. Gain vs. control voltage (V1) of the voltage variable attenuator (VVA)

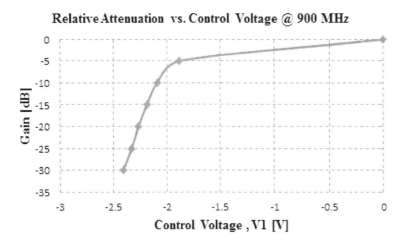

Figure 19. Measured insertion loss (S_{21}) vs. frequency between the RF port and the port for a patch element mounted on the PCB (Figure 24). The curves shown in the figure are the functions of the control voltage from the control circuit of VVA.

Figure 20. Measured phase (S_{21}) vs. frequency between the RF port and the port for a patch element mounted on the PCB (Figure 15). The curves shown in the figure are the functions of the control voltage from the control circuit of VVA.

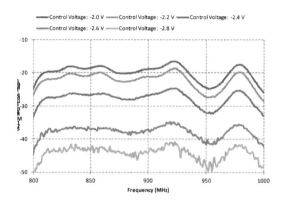

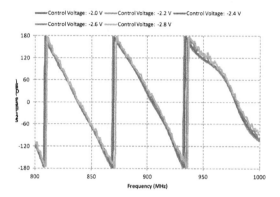

4.1.2. Performance Measurement of Phase Shifters

Figure 21 shows the measured phase shift between the RF port and a port for a patch element mounted on the PCB board shown in Figure 15. The numbers on the right of the graphs show the actual setting of the phase shift values.

4.1.3. Adaptive Beam Forming with 900 MHz Signal

The amplitude of each element of the 8×8 elements smart antenna can be controlled by applying different amplitude control algorithms to the VVA circuit. It is also possible to control the radiation patterns and steer the beam by controlling the phases of each element through 4-bit phase shifter. Staircase Power Distribution (SPD) is not suitable for this 8×8 elements array, since SPD is more suitable for array with larger number of antenna elements. However, radiation pattern of the smart antenna can be controlled by implementing ITD concept. Controlling of radiation patterns of the smart antenna by using ITD concept is explained in the next section.

4.1.3.1. ITD Edge Tapering for the 8×8 Elements Smart Antenna at 900 MHz

One example of controlling the amplitudes of one dimensional 8 elements by using ITD power distribution is shown in Figure 22. The amplitudes of the signal of each element can be controlled by implementing ITD edge tapering concept through the connected VVA circuit. It was mentioned in section X.2.2.1.3 that (Baki et all 2007, 2008)

Figure 21. Measured phase shift vs. frequency between the RF port and the port for a patch element mounted on the PCB (Figure 15). The curves shown in the figure are the functions of the actual phase shift values.

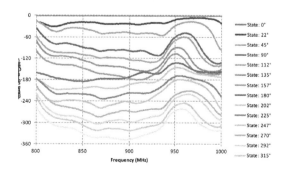

Figure 22. Block diagram of 8 elements smart antenna with beam forming network

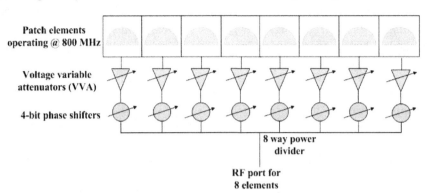

have studied this kind of power distribution for microwave power transmission at 5.8 GHz.

Figure 23 shows the ITD power distributions for two different situations. One is -3 dB ITD with 3 elements tapered from each side; the other is -7 dB ITD with 3 elements tapered from each side. The power distributions shown in the figure can be implemented through the VVA. Figure 24 shows the simulated radiation patterns for these two cases along with one with uniform power distribution. The element spacing for all cases was 0.6λ and the beam steering angle was 25°. Table 2 summarizes different radiation patterns related parameters. The SLLs in case of ITD have decreased significantly; the other way it can be explained that the BE along the direction of main beam are higher in cases of ITD edge tapering. The decrease of these SLL would help discriminating tags along the directions of these SLLs. The main beam width in the cases of ITD has increased. This increase in main beam is suitable for reading multiple tags from the same location. The field trial of the antenna for location finding of RFID tags to various positions is underway and will be reported to future publications. However some results of tag positioning beam steering of a 6-element array antenna can be found in (Karmakar et al, 2010). Table 3 shows some simulation results of power received by a tag at different locations. The values shown in the table are ratios of received power of an RFID tag for

two different amplitude distributions of reader antenna elements. One is the uniform amplitude distribution and the other is the ITD of array antenna. The results shown in Table 3, indicate that with ITD power distribution, it will be possible to increase the tag range.

4.2. Radiation Patterns with Different Amplitude Tapering of Array Antenna at 5.8 GHz

Some new and better methods of amplitude distribution of array antenna elements for microwave power transmission (MPT) by incorporating Isosceles Trapezoidal Distribution (ITD) concept are discussed in (Baki et al., 2007, 2008). The MSLL for an array of uniform power distribution is 13

Figure 23. -3 dB and -7 dBITD edge tapering for UHF band 8 elements smart antennas

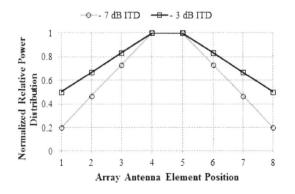

Figure 24. Radiation patterns of 8 elements UHF band smart antenna for different power distributions

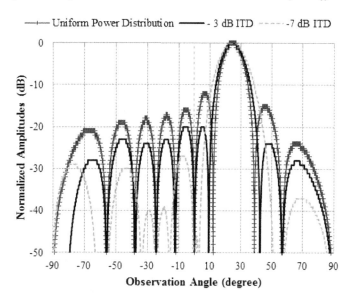

dB less than the main beam peak. The MSLL for -10 dB Gaussian power distribution in 23 dB less than the main beam peak. One example of these three different power distributions is shown in Figure 25. Total 109 elements were considered for each of the uniform, -10 dB Gaussian and -10 dB ITD edge tapering cases. Element spacing was chosen 0.68λ at 5.8 GHz for all three distributions. Calculation process of array factor (AF) for ITD edge tapering will be discussed in next section. Figure 26 shows the radiation patterns for the three cases. Taper efficiency is the highest in uniform distribution, although the BE is the lowest in this case. Taper efficiency in ITD is higher than that of Gaussian distribution. Therefore for the same radiation efficiency, antenna gain is also higher in ITD than that of Gaussian distribution. 33 elements

from each side were tapered in ITD; where as in Gaussian case, all 54 elements from each side were tapered. The SLL for ITD can be reduced further by increasing the number of elements to be amplitude tapered. Table 4 summarizes the results of the three different cases of amplitude distribution of array elements. A detailed experimental procedure based on ITD power distribution is explained in (Baki et al., 2007, pp. 968-977).

4.3. Radiation Patterns with Equal and Unequal Element Spacing

Radiation patterns of array antenna were simulated for two different cases. In one case inter-element spacing and amplitudes of the array antenna elements were kept constant. In another case the

Table 2. Radiation patterns related parameters for different power distribution at 900 MHz

Type of distribution	MSLL (dB)	Location of MSLL (degree)	HPBW (degree)	Beam Steering Angle (degree)
Uniform	-13	4.78	10.87	25
-3 dB ITD	-20	4.51	12.26	25
-7 dB ITD	-27	-8.97	14.35	25

Table 3. Ratio of received powers by an RFID tag through two different amplitude distributions. The ratios are functions of different tag size and tag distance.

		Received power through ITD/Received power through uniform distribution (Ratio)					
Tag distance (m)		1	10	20	30	40	50
Tag Radius (cm)	1	1.65	1.65	1.65	1.65	1.65	1.65
	2	1.65	1.65	1.65	1.65	1.65	1.65
	20	1.65	1.65	1.65	1.65	1.65	1.65

inter-element spacing of the array elements was kept unequal according to the method described in section 2.2.1.5. Total 109 elements were considered in both cases. Equal element spacing was 0.6λ at 5.8 GHz. 40 elements from each side were of unequal spacing. Radiation patterns for the different cases are shown in Figure 27. Table 5 shows some parameters related to the radiation patterns. Radiations results show that the main beam in the case of unequal spacing is narrower than the one in equal element spacing case since the array antenna length became higher in former case. MSLL is also 2.17 dB lower in the case of unequal element spacing.

4.4. Radiation Pattern with ITDU

Figure 28 shows the amplitude distribution versus unequal element spacing graph for 25 antenna elements by using the -20 dB ITDU concept. In this case the amplitude of the edge element is 20 dB less than the maximum amplitude and the unequal element spacing is also calculated by considering another identical ITD. Figure 29 shows the radiation pattern for the case where all the SLLs are below -30 dB. The main beam steering angle is at 10° degree from the array broadside. One demerit with ITDU concept is that grating lobes appear with higher number of antenna elements and with higher beam steering angles. The element spacing

Figure 25. Different array antenna amplitude distributions for 109-element linear array $(d = 0.68\lambda, f = 5.8 GHz$)

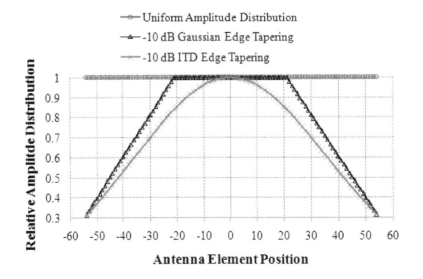

Figure 26. Normalized radiation patterns for uniform amplitude distribution, -10 dB Gaussian, and -10 dB ITD edge taperingfor 109-element linear array $(d = 0.68\lambda, f = 5.8GHz$ $)$

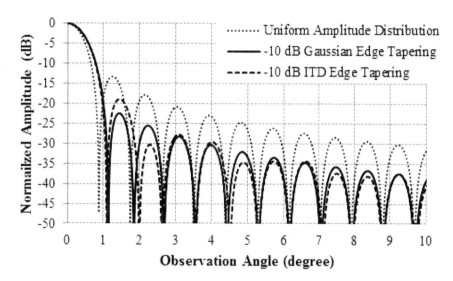

4.5. Experiments with ITDU for 5.8 GHz Signal

A conceptual block diagram of beam forming network by using ITDU concept is shown in Figure 30. The frequency of operation was 5.8 GHz. Element spacing of the 11 elements can be varied from 0.6λ to 1λ. As shown in the block diagram, the amplitudes of side elements can be controlled by variable attenuators. The direction of main beam can be controlled by controlling the phase of each element. The experimentation

was conducted by using the SPORTS (**S**olar **PO**wer **R**adio **T**ransmission **S**ystem) 5.8 GHz "beam forming subsystem" in the **M**icrowave **E**nergy **T**ransmission **LAB**oratory (METLAB) of the Kyoto University. METLAB is an anechoic chamber for MPT experiment. Figure 31 shows the simulated and measured radiation patterns of -10 dB ITDU by using 11 antenna elements.

Figure 31 shows that the simulated MSLL was about -28 dB (2nd SLL in this case) and measured MSLL was about -25 dB (less than those of -10 dB ITD and -10 dB Gaussian edge tapering). The MSLL for -10 dB Gaussian edge tapering is about -23 dB.

Table 4. Summary of radiation patterns related parameters for uniform distribution, -10 dB Gaussian and -10 dB ITD edge tapering for 109-element linear array $(d = 0.68\lambda, f = 5.8GHz$ $)$

Type of distribution	Total no. of array elements	No. elements tapered from each side	BE(%)	MSLL(dB)	Normalized taper efficiency
Uniform	109	0	89.54	-13	1
Gaussian	109	All	98.33	-22.46	0.91
ITD	109	33	98.30	-18.7	0.92

Figure 27. Radiation patterns with equal and unequal spacing of array antenna

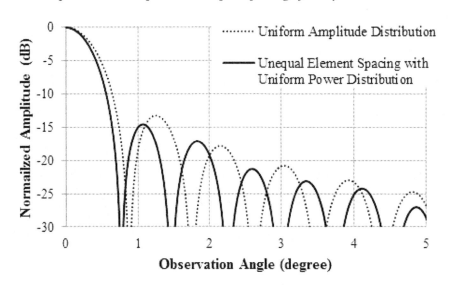

Simulated (required) and measured relative amplitudes for the 11 elements ITDU are shown in Table 6(Baki et al., 2008, pp. 527-535). Simulated (required) and used inter-element spacing for this case is shown in Table 7 (Baki et al., 2008, pp. 527-535).

4.6. Reduction of Phase Quantization/Round off Error by Itdu

Phase round off error of 4-bit phase shifters with a 101-element phased array antenna of-10 dB ITDU is shown in Figure 32. The unequal element spacing was in the range from 0.68λ to 0.79λ. 30 elements from each side were tapered and 40 elements from each side were of unequal spacing in this case. If Figure 12 and Figure 32 are compared then it is evident that the periodicity of the phase round off error is disrupted in Figure 32. The phase error variance/R.M.S. phase errors of only the phase quantization errors are summarized in Table 8. Table 8 shows that the phase error variance/R.M.S. phase errors in ITDU are less than those in uniform element spacing case. These less R.M.S. phase error/phase error vari-

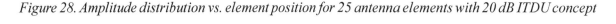

Figure 28. Amplitude distribution vs. element position for 25 antenna elements with 20 dB ITDU concept

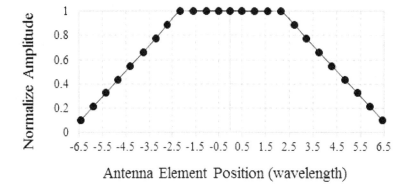

Figure 29. Radiation patterns for -20 dB ITDU edge tapering and for 10° beam steering angle by shifting the phases of the array antenna signals

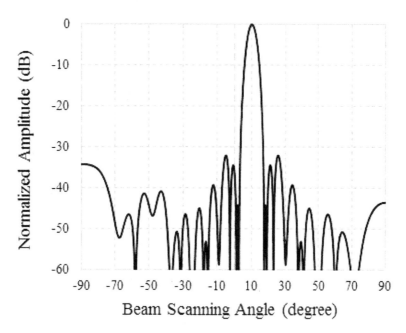

Figure 30. Conceptual block diagram of beam forming network for ITDU

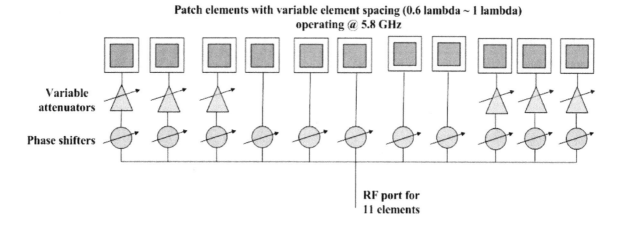

Table 5. Parameters related to the radiation patterns shown in Figure 27

Type of Element Spacing	MSLL (dB)	Location of MSLL (degree)	HPBW (degree)	Beam Steering Angle (degree)
Uniform	-13	1.16	0.76	0
Unequal	-15.17	0.99	0.66	0

Figure 31. Measured and simulated radiation patterns of -10 dB ITDU with 11 phased array antenna elements

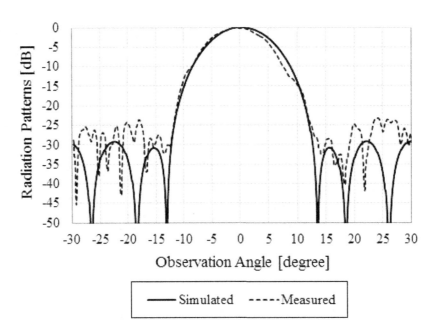

Table 6. Measured and required relative amplitude distributions of -10 dB ITDU of 11 phased array antenna elements

Element Number ⇒	1	2	3	4	5	6	7	8	9	10	11
Required Relative Amplitude Distribution (dB)	-10	-5.90	-2.25	0	0	0	0	0	-2.25	-5.90	-10
Measured Relative Amplitude Distribution (dB)	-10.70	-6.42	-3.65	0	0	0	0	0	-3.62	-6.70	-10.54

Table 7. Simulated and experimental inter-element spacing (cm)of -10 dB ITDU of 11 phased array antenna elements

Element Number ⇒	1-2	2-3	3-4	4-5	5-6	6-7	7-8	8-9	9-10	10-11
Simulated inter-element spacing (cm)	3.63	3.54	3.43	3.3	3.05	3.05	3.3	3.43	3.54	3.63
Experimental inter-element spacing (cm)	3.6	3.5	3.4	3.3	3.0	3.0	3.3	3.4	3.5	3.6

Table 8. Comparison of phase round off error without/with random phase error of the phase shifter

Uniform Element Spacing		Unequal Element Spacing (ITDU)	
R.M.S. phase error	Phase error variance	R.M.S. phase error	Phase error variance
3.24°	0.19°	3.06°	0.17°

Figure 32. Phase round off errors of 4-bit phase shifter with -10 dB ITDU (101 elements of 0.68λ~0.795λ variable spacing was used in the simulation)

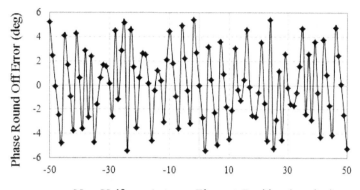

Figure 33. Amplitude distributions of 25 antenna elements for -10 Gaussian, -10 dB ITD and -10 dB SPD

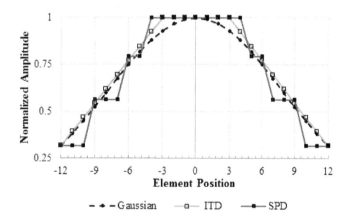

Figure 34. Radiation patterns of 25 antenna elements for -10 Gaussian, -10 dB ITD and -10 dB SPD. The main beam angle is at 35° from array broadside.

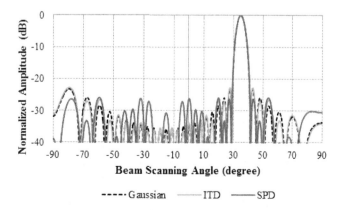

Table 9. MSLL for -10 dB Gaussian, ITD and SPD Edge tapering and for different beam steering angles

Main Beam Angle (deg.)			0	5	10	15	20	25	35
Type of Power Distribution	SPD	MSLL (dB)	-26	-26	-26	-26	-26	-26	-26
	ITD	MSLL (dB)	-23.12	-23.14	-23.05	-22.89	-22.79	-22.79	-22.79
	Gaussian	MSLL (dB)	-23.66	-23.69	-23.58	-23.26	23.25	-23.25	-23.25

ance will decrease the beam pointing error, SLL and loss of power.

4.7. Radiation Pattern with SPD for 60 GHz Signal

Radiation patterns for different amplitude distributions (Gaussian, ITD and SPD) and beam steering angles are compared in this section. Figure 33 shows the amplitude distributions of 25 antenna elements for -10 dB Gaussian, -10 dB ITD and -10 dB SPD tapered cases. Radiation patterns, MSLL and BE for 60 GHz signals for the cases shown in Figure 20 are compared. The analysis was done for different beam steering angles (0° to 35° degree with 5° step). MSLL as well as BE for 0°~35° beam steering angles are shown respectively in Table 9 and Table 10. Figure 34 shows the radiation patterns for 35° beam steering angle. The element spacing was 0.6λ for every case.

Table 9 shows that the MSLL for SPD with every beam steering angle was minimum (-26 dB). Table10 shows that the BE for SPD is also comparable to those of ITD/Gaussian distributions. The radiation patterns of SPD and for different beam steering angle can further be synthe-

sized by varying the number of steps and number of elements in each step.

Maximum power will be contained in the main beam and less power will be in the SLL in SPD since the MSLL is the lowest in this case. Higher SLL causes higher exposure level. With SPD, exposure level to humans and all other living animals/things outside the main beam would be less.

5. CONCLUSION

The RFID reader antenna as well as antennas used in other wireless applications must have very low side lobes in order to maximize power and minimize interferences. Higher side lobes of reader antenna may lead to false alarms and collision also may happen in RFID applications. Array antenna based RFID reader, that has gain enhancement and interference cancellation capability, will be needed for better communication between a reader and RFID tags. It is essential to build an array antenna with minimum SLL, maximum BE and with interference cancellation capabilities in order to achieve the required performance of an

Table 10. Beam Efficiency (BE) for -10 dB Gaussian, ITD and SPD edge tapering and for different beam steering angles

Main Beam Angle (deg.)			0	5	10	15	20	25	35
Type of Power Distribution	SPD	BE (%)	98.05	97.60	97.66	97.49	97.23	97.53	97.17
	ITD	BE (%)	98.72	98.65	98.65	98.58	98.16	97.83	97.02
	Gaussian	BE (%)	98.72	98.67	98.65	98.59	98.31	97.76	97.14

RFID reader. Phase quantization errors of array elements cause higher SLL and reduce the gain of an antenna. Phase quantization errors can be minimized by using unequal spacing of antenna elements. Edge tapering of array elements can reduce the SLL and increase the BE of array antenna. In this chapter some edge tapering concepts and their merits/demerits are discussed. New and better methods for beam forming techniques by implementing Isosceles Trapezoid Distribution (ITD) and Staircase Power Distribution (SPD) concepts of array antenna have been also discussed in this chapter. ITD and SPD edge tapering can reduce the SLLs and enhance the gain of array antenna. Maximum power would be transmitted through the main beam and less power would be in the SLLs in case of ITD and SPD edge tapering; since the SLLs are less in these cases. Development methods of 900 MHz beam forming network for an UHF RFID reader by implementing ITD edge tapering concept has been investigated. SPD power distribution is technically better since less and stepwise uniform power distribution is required in this kind of distribution. It is easier to construct antenna array with SPD and this way it is possible to use the antenna elements in a better way when power distributions are considered. With SPD, the amplitude errors of different antenna elements also would be minimum because least number of different amplitudes would be required. Exposure level to humans and all other living animals outside the main beam would be less due to SPD when it is compared to those of Gaussian distribution or ITD. The RFID systems, automatic cruise control (ACC) system, collision avoidance radar, etc. must have very low side lobes in its radiation pattern in order to maximize the received power and minimize interferences and noises. The applications of ITD or SPD technique would help in achieving these goals.

REFERENCES

Baki, A. K. M., Shinohara, N., Matsumoto, H., Hashimoto, K., & Mitani, T. (2007, April). Study of isosceles trapezoidal edge tapered phased array antenna for solar power station/satellite. *IEICE Transactions in Communication. E (Norwalk, Conn.)*, *90-B*(4), 968–977.

Baki, A. K. M., Shinohara, N., Matsumoto, H., Hashimoto, K., & Mitani, T. (2008, February). Isosceles-trapezoidal-distribution edge tapered array antenna with unequal element spacing for solar power station/satellite. *IEICE Transactions in Communication. E (Norwalk, Conn.)*, *91-B*(2), 527–535.

Bialkowski, M. E., & Karmakar, N. C. (1999). Design of compact L-band 180° phase shifters. *Microwave and Optical Technology Letters*, *22*(2), 144–148. doi:10.1002/(SICI)1098-2760(19990720)22:2<144::AID-MOP19>3.0.CO;2-D

Brown, W. C., & Eugene, E. E. (1992, June). Beamed microwave power transmission and its application to space. *IEEE Transactions on Microwave Theory and Techniques*, *40*(6), 1239–1250. doi:10.1109/22.141357

Do, V. H., Subramanian, V., Keusgen, W., & Boeck, G. (2008, March). A 60 GHz SiGe-HBT power amplifier with 20% PAE at 15 dBm output power. *IEEE Microwave and Wireless Components Letters*, *18*(3), 209–211. doi:10.1109/LMWC.2008.916816

Goubau, G., & Schwering, F. (1961). On the guided propagation of electromagnetic wave beams. *IRE Transactions on Antennas and Propagation*, *9*, 248–256. doi:10.1109/TAP.1961.1144999

Jamali, B. (2010). A development platform for SDR-based RFID reader. In Karmakar, N. C. (Ed.), *Handbook of smart antennas for RFID systems* (pp. 123–137). New Jersey: Wiley Microwave and Optical Engineering Series. doi:10.1002/9780470872178.ch5

Karmakar, N. C. (2008). Smart antennas for automatic radio frequency identification readers. In Sun, C., Cheng, J., & Ohira, T. (Eds.), *Handbook on advancements in smart antenna technologies for wireless networks* (pp. 449–472). Hershey, PA: IGI Global. doi:10.4018/978-1-59904-988-5.ch021

Karmakar, N. C. (2010). Recent paradigm shift in RFID and smart antenna. In Karmakar, N. C. (Ed.), *Handbook of smart antennas for RFID systems* (pp. 57–82). New Jersey: Wiley Microwave and Optical Engineering Series. doi:10.1002/9780470872178.ch3

Karmakar, N. C., & Bialkowski, M. E. (1999). An L-band 90° hybrid-coupled phase shifter using UHF band p-i-n diodes. *Microwave and Optical Technology Letters, 21*(1), 144–148. doi:10.1002/(SICI)1098-2760(19990405)21:1<51::AID-MOP15>3.0.CO;2-H

Karmakar, N. C., Zakavi, P., & Kambukage, M. (2010). FPGA-controlled phased array antenna development for UHF RFID reader. In Karmakar, N. C. (Ed.), *Handbook of smart antennas for RFID systems* (pp. 57–82). New Jersey: Wiley Microwave and Optical Engineering Series. doi:10.1002/9780470872178.ch3

Keith, R. C., Cooper, W. K., & Stutzman, W. L. (1973, March). Beam-pointing errors of planner-phased arrays. *IEEE Transactions on Antennas and Propagation, 21*(2), 199–202. doi:10.1109/TAP.1973.1140434

Klair, D. K., Chin, K. W., & Raad, R. (2010). A survey and tutorial of RFID anti-collision protocols. *IEEE Comm. Surveys & Tutorials, 12*(3), 400–421. doi:10.1109/SURV.2010.031810.00037

Mailloux, R. J. (1994). *Phased array antenna handbook*. Artech House.

Maloratsky, L. G. (2004). *Passive RF & microwave integrated circuits*. USA: Elsevier.

Natarajan, A., Komijani, A., Guan, X., Babakhaniand, A., & Hajimiri, A. (2006, December). A 77-GHz phased-array transceiver with on-chip antennas in silicon: Transmitter and local LO-path phase shifting. *IEEE Journal of Solid-state Circuits, 41*(12), 2807–2819. doi:10.1109/JSSC.2006.884817

Nikitin, P. V., & Seshagiri, K. V. (2009). LabVIEW-based UHF RFID tag test and measurement system. *IEEE Transactions on Industrial Electronics, 56*(7), 2374–2381. doi:10.1109/TIE.2009.2018434

Pellerano, S., Alvarado, J., & Palaskas, Y. (2010). A mm-wave power-harvesting RFID tag in 90 nm CMOS. *IEEE Journal of Solid-state Circuits, 45*(8), 1627–1637. doi:10.1109/JSSC.2010.2049916

Pfeiffer, U. R., & Goren, D. (2007, July). A 20 dBm fully-integrated 60 GHz SiGe power amplifier with automatic level control. *IEEE Journal of Solid-state Circuits, 42*(7), 1455–1463. doi:10.1109/JSSC.2007.899116

Preradovic, S., & Karmakar, N. C. (2010). RFID readers-review and design. In Karmakar, N. C. (Ed.), *Handbook of smart antennas for RFID systems* (pp. 85–121). New Jersey: Wiley Microwave and Optical Engineering Series. doi:10.1002/9780470872178.ch4

Pursula, P., Vaha-Heikkila, T., Muller, A., Neculoiu, D., Konstantinidis, G., Oja, A., & Tuovinen, J. (2008). Millimeter-wave identification—A new short-range radio system for low-power high data-rate applications. *IEEE Transactions on Microwave Theory and Techniques, 56*(10), 2221–2228. doi:10.1109/TMTT.2008.2004252

Ray, K. P., & Krishna, D. D. (2006). Compact dual band suspended semi-circular microstrip antenna with half U-slot. *Microwave and Optical Technology Letters*, 48(10), 2021–2024. doi:10.1002/mop.21844

Roy, S. M., & Karmakar, N. C. (2010). RFID planar antenna-smart design approach at UHF band. In Karmakar, N. C. (Ed.), *Handbook of smart antennas for RFID systems* (pp. 141–171). New Jersey: Wiley Microwave and Optical Engineering Series. doi:10.1002/9780470872178.ch6

Stutzman, W. L., & Thiele, G. A. (1997). *Antenna theory and design* (2nd ed.). New Jersey: John Wiley & Sons, Inc.

Want, R. (2004). The magic of RFID. *ACM Queue; Tomorrow's Computing Today*, 2(7), 40–48. doi:10.1145/1035594.1035619

KEY TERMS AND DEFINITIONS

Beam Collection Efficiency (BCE): The ratio of power flow that is intercepted by the receiving antenna to the whole transmitted power.

Beam Efficiency (BE): The ratio of transmitted power through main beam to the total radiated power.

Channel Capacity: In a data communication channel, **channel capacity** is the measure of data transmission rate in bits per second.

Edge Tapering: In array antenna the edge tapering concept is used to minimize the side lobe levels. Generally, in edge tapering the intensities of side elements are kept minimum and those of middle elements are kept maximum.

Isosceles Trapezoid Distribution (ITD): follows the concept of Isosceles Trapezoid for distribution of values. In array antenna ITD concept is used for amplitude distribution of individual array elements.

Isosceles Trapezoid Distribution with Unequal element spacing (ITDU): In array antenna ITDU follows the concept of Isosceles Trapezoid for amplitude distribution of array elements and it follows some rule for estimation of unequal inter-element spacing.

Microwave Power Transmission (MPT): MPT is a wireless power transmission technique.

Radio Frequency Identification (RFID): RFID is wireless data capturing system for identification purpose.

Staircase Power Distribution (SPD): SPD follows the concept of stair case for amplitude distribution of array antenna elements.

Voltage Variable Attenuator (VVA): VVA can control the amplitude of radio frequency signals of individual elements of array antenna depending on the control voltage applied to a VVA circuit.

Chapter 6
Multi-Input-Multi-Output Antennas for Radio Frequency Identification Systems

Shivali G. Bansal
Deakin University, Australia

Jemal Abawajy
Deakin University, Australia

ABSTRACT

In this chapter the authors discuss the physical insight of the role of wireless communication in RFID systems. In this respect, this chapter gives a brief introduction on the wireless communication model followed by various communication schemes. The chapter also discusses various channel impairments and the statistical modeling of fading channels based on the environment in which the RFID tag and reader may be present. The chapter deals with the fact that the signal attenuations can be dealt with up to some level by using multiple antennas at the reader transmitter and receiver to improve the performance. Thus, this chapter discusses the use of transmit diversity at the reader transmitter to transmit multiple copies of the signal. Following the above, the use of receiver combining techniques are discussed, which shows how the multiple copies of the signal arriving at the reader receiver from the tag are combined to reduce the effects of fading. The chapter then discusses various modulation techniques required to modulate the signal before transmitting over the channel. It then presents a few channel estimation algorithms, according to which, by estimating the channel state information of the channel paths through which transmission takes place, performance of the wireless system can be further increased. Finally, the Antenna selection techniques are presented, which further helps in improving the system performance.

1. INTRODUCTION

The signal transmission in Radio Frequency Identification Systems (RFIDs) takes place through a wireless medium where it tends to deteriorate due to multipath, fading and inter-symbol interfer-

ence (ISI). The RFID technology communicates through radio waves to transfer data between a reader and a tag attached to an unit for the purpose of tracking, sensing, and identifying various targets in wide-range of applications like supply chain, transportation, airline baggage handling,

DOI: 10.4018/978-1-4666-1616-5.ch006

medical and biological industry, homeland security identification, tracking and surveillance and many more. Unlike barcodes, RFID technology offers several other key benefits such as no line of sight (LOS) requirements, robustness, speed, bidirectional communication, reliability in tough environments, bulk detection, superior data capabilities, etc. Because of this, RFID is proving very successful for wide area of applications where traditional identification technologies are inadequate for recent demands. Presently, RFID technology is being wide-spread and applied to real world system. The designing of RFID systems requires a wide range of hardware and software, protocols and algorithms, applications, etc. Signaling through wireless channel is one major technical factor that needs to be carefully understood. The reliability of a wireless channel is closely related to the signal fading conditions occurring during the transmission and reception of signal to and from the tag, respectively.

Wireless communication plays a significant role in understanding and dealing with the effects of channel fading. It has been a topic of study for over half a century now but the past decade has been the most exhaustive period and this research thrust in the past decade has led to a much richer set of perspectives and tools on how to communicate over wireless channels. The high rise in demand for new wireless capacities, low-powered sophisticated signal processing algorithms and coding techniques, the successful second-generation (2G) and third-generation (3G) digital wireless standards surge towards the demands for continued research activities in this area. Adachi (2001) focused on some of the wireless technologies that emerged in the past and also provided an insight into the future wireless technologies. The conventional single-input single-output (SISO) systems failed to meet the growing demands for supporting transmission of images, voice, data and video related services, new wireless multimedia services such as Internet access, and multimedia data transfer, hence, wireless systems with multiple

element antennas (MEAs) were proposed. In the recent years, there has been exploring interests in multiple input multiple output (MIMO) systems because of their ability to greatly enhance the data rates and channel capacity. MIMO technology can be defined as a wireless technology that uses multiple transmitters and receivers to transfer more data at the same time. It takes advantage of the radio-wave phenomenon called multipath where transmitted information bounces off walls, ceilings, and other objects, reaching the receiving antenna multiple times from different angles and at slightly different times due to reflection, refraction and scattering. Due to this multipath phenomenon, an accurate and reliable transmission may not be always possible. To deal with this multipath behavior, communication researchers have thought of many different possibilities to increase the so-called diversity, by using multiple transmitters and receivers. Several different types of diversity modes are used in order to make the communication system more robust like time, frequency, spatial diversity and so on. The higher the diversity is, the lower is the probability of a small channel gain. Besides the advantages of spatial diversity in MIMO systems for improved robustness, multiple antenna technology can also be used to increase data rates by using spatial multiplexing. However, in practice, both diversity and spatial multiplexing can be used, separately or in combination, depending on the channel condition.

The present day invention relates to use of multiple antennas for a RF reader on a RFID system to significantly increase the operating range. RFID systems with single or multiple antenna in reader or in tag have been studied in (Nikitin & Rao, 2008), (Ingram et al., 2001), (Griffin & Durgin, 2008) both theoretically and through simulations. Similar to MIMO systems, spatial diversity can be achieved through the use of multiple antennas at either the reader transmitter or the reader receiver, or at both. It follows the basic principle that by combining the signals from multiple antenna elements, the received signal-to-noise ratio can be significantly increased. One such RF reader performing maximal ratio combining was presented in (Angerer et al., 2009) which used a dual receive antenna in order to deal with the effects of multipath. An analog MIMO frontend for RFID rapid prototyping system allowing for various real time experiments to investigate MIMO techniques as beam-forming, diversity combining, or localization at the reader has been presented in (Langwieser et al., 2010). It also presented measurement based example with one transmitter and two receivers for two different tag positions. Use of multiple antennas offers a wide range of options, for example, a single-input-multiple-output (SIMO) system architecture collects more energy at the receiver and improves the SNR whereas MIMO system architecture opens up multiple data pipes over a communication link. A MIMO antenna system using circular polarization and antenna diversity has been proposed in (He et al., 2008). It shows that by selection of the antenna supported by RFID reader chip sets, the proposed MIMO antenna system can have high receiving signal strength and larger antenna beam coverage area, making it suitable candidate for RFID application especially in warehouse tracking. Different architectures provide different benefits such as array gain, diversity gain, multiplexing gain, and interference reduction. (Foschini (1998), Foschini & Gans (1998), and Gesbert et al., (2003)).

2. WIRELESS COMMUNICATION IN RFID SYSTEMS

In Radio Frequency Identification (RFID) systems, wireless communication takes place between the reader and the tag using radio frequency (RF) signals of very low strength and a wireless link is achieved with antennas built into both tags and readers. When the communication between the reader and the tag occurs, multiple factors like scattering, reflection, refraction, multipath, signal interference, noise, environmental conditions etc.,

distort the signal arriving at the receiver. When combined, all these factors are termed as channel impairments as they occur when the signal from the transmitter to the receiver passes through the wireless propagation channel. As a result, these factors tend to degrade the performance of the readers in detection of these weak signals. For improved reliability and overall throughput, both active and passive RFID tag readers have a need for strong digital signal processing (DSP) techniques (Angerer & Rupp (2009), Engel (2002), Mazurek (2008)). But the architecture and DSP techniques used in commercially available active and passive tags cannot be used with chipless tags (Koswatta & Karmakar, 2010) as chipped tags use high end DSPs and FPGAs and their operating principle is entirely different. This makes the design of the readers quite complex and expensive. On the other hand, a chipless tag reader has a limited amount of processing power as it uses data decoding algorithms and signal processing techniques. To recover the distorted signal at the reader, smart/multiple antennas can be used at both ends.

2.1. Wireless RFID System Model

RFID system is basically based on the principle of wireless communication which uses RF waves which further is a part of the electromagnetic (EM) spectrum. A typical RFID system follows a number of communication sequence and functionalities between the reader and the tag. Figure.1 shows a RFID system model where the communication between the reader and the tag is occurring wirelessly via the wireless channel.

RFID Readers transmits a continuous RF carrier wave towards the tag to communicate with it. The carrier wave can be defined as the wave of a specific frequency which is modulated to carry data. RFID readers can have a single or multiple antennas to transmit radio waves to and receive the signals back from the tags. The complexity of the reader highly depends on the functions to be fulfilled varying from application to

Figure 1. Wireless signal transmission between RFID reader and tag

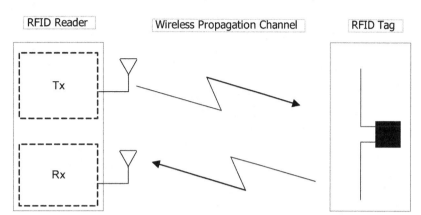

application and based on sophisticated signal conditioning, parity error checking and correction. RFID systems with single tag and reader antennas may require simpler signal processing circuitry while RFID systems with multiple tag and/or reader antennas may require quite sophisticated DSP algorithms. However, in both the case, the function of the reader is to communicate with the tags and facilitate efficient data transfer.

Propagation Channel is the wireless link formed between the reader and the tag in order to communicate. The performance of any wireless system depends highly on the propagation channel properties like path loss and the commonly used channel fading statistics like Rayleigh, Ricean, Nakagami parameters, coherence bandwidth, delay spread, etc.

RFID tags start communicating with the RFID readers as soon as they receive sufficient energy from the RF field as they enter the RF field of the reader. The tags modulate the carrier signal to the data storage on the tags. This modulated carrier signal is resonated to the readers from the tags.

Majority of the signal processing operations in RFID systems are carried out by the readers. The reader initiates the communication with the tags by sending a message in the form of a continuous carrier transmission. This provides energy to the tags to retrieve its information. But, in order

to transmit a message, it needs to be processed in such a way that is suitable for transmission through the wireless channel. The message bits are first encoded as a continuous signal and then modulated. Depending on the number of antennas used and the combination of coding and modulation schemes used, the signals bandwidth, power, and sensitivity to errors while transmission is determined. Figure 2 shows the RFID reader block diagram with an encoder and a modulator.

The main aim of this communication system is to transfer the information from the source to a destination. To do so, the information is modified in appropriate form to allow for its transmission, which in return allows the possibility of errors. The information source is the actual data that needs to be transmitted. In the most fundamental form, the information to be transmitted is either digital or analog. If the data is in analogue format, the source encoder converts it into digital (binary) form. As the information progresses through the communication system, it is modified in both phase and amplitude introducing distortion. Along with, the communication system also introduces noise from various sources that will further degrade the signal. The transmitter accepts the data in form of signal from this information source and processes it into an appropriate format to be transmitted over the channel. The binary

Figure 2. RFID reader block diagram

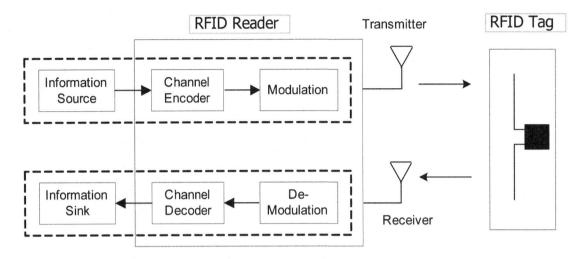

data is then subjected to channel encoder for reliable transmission. The channel encoder adds controlled redundancy to the binary sequence for error detection and correction. The encoded data sequence is then modulated to generate waveforms for transmission over the channel. The channel usually contains disturbances like noise and obstacles like buildings, terrestrial features causing channel impairments. The receiver captures the signal and processes in a form appropriate to trace back the transmitted information. During this whole sequence of data transmission, the signal gets corrupted, degraded and attenuated due to noise and channel impairments. By carefully compensating these channel impairments and system designing, the data rates can be maximized and the signal degradation can be improved.

2.2. Various Communication Schemes

In conventional wireless communication systems, a single antenna is used at both the source and the destination. These systems are one of the simplest in the antenna technology but in some environments these systems are vulnerable to problems caused by multipath effects. In order to minimize

or eliminate problems caused by multipath wave propagation, multiple antennas can be employed at the transmitter and/or the receiver (Elo). The fundamental theoretic research using multiple antennas was initiated by Winters (1987), Telatar (1999), Foschini & Gans (1998), and Foschini (19996). Since then there has been exploring interests in multiple input multiple output (MIMO) systems and it has been proved that these systems greatly enhance the data rates and channel capacity. Recently, there has been an enormous interest for using MEAs in RFID systems. Like conventional wireless communication systems, the transmitter and receiver in the RFID readers can have single or multiple antennas depending on the system requirements. A new signaling scheme based on space-time block coding (STBC) for multiple-antenna RF tags has been recently presented in (Chen & Wang, 2010) while beam steering for a reader with multiple antennas which can be configured to point at discrete directions has been proposed in (Yoon et al., 2008) and (Abbak & Tekin, 2009). Thus, depending on the number of transmit and/or receive antennas deployed at either end there are four possible system models:

1.   Single Input Single Output (SISO) Systems

2. Multiple Input Single Output (MISO) Systems
3. Single Input Multiple Output (SIMO) Systems
4. Multiple Input Multiple Output (MIMO) Systems

- **Single Input Single Output (SISO) Systems:** A Single Input Single Output (SISO) system refers to a wireless communications system in which a single antenna is used both at the transmitter and the receiver as shown in Figure 3a. The reader transmitter transmits the message to the tag using single antenna and the signal from the tag is received over single reader receive antenna. No spatial diversity is exploited here due to use of single antenna. Such systems cause a reduction in data speed and an increase in the number of errors.

- **Single Input Multiple Output (SIMO) Systems:** A Single Input Multiple Output (SIMO) is an antenna technology for wireless communications in which two or more antennas are used at the receiver and a single antenna at the transmitter as shown in Figure 3b. The signals arriving from different directions are combined in various ways at the receiver using receive diversity to minimize errors and optimize data speed.

- **Multiple Input Single Output (MISO) Systems:** A Multiple Input Single Output (MISO) is an antenna technology for wireless communications in which two or more antennas are used at the transmitter and a single antenna at the receiver as shown in Figure 3c. The data stream is converted from serial to parallel by using carious transmit diversity schemes available at the transmitter side and vice-versa at the receiver to minimize errors and optimize data speed.

- **Multiple Input Multiple Output (MIMO) Systems:** A Multiple Input Multiple Output (MIMO) is a wireless communications system in which multiple antennas are used at both the transmitter and the receiver as shown in Figure 3d. The antennas at each end of the communications circuit are combined to minimize errors and optimize data speed. Both transmit and receive diversity is exploited in MIMO technology. Depending on the requirements of the system, different combinations of transmit diversity schemes and receive diversity schemes can be used.

3. WIRELESS FADING CHANNELS

The encoded and modulated signal is now passed through the wireless propagation channel that suffers from various channel impairments.

3.1. Fading Channel Impairments

Signaling between the RF reader and tag occurs wirelessly over radio waves for transmitting information, where the carrier frequency of these radio waves can vary from a few hundred megahertz to several gigahertz depending on the system. The wireless channel can be described as a function of time and space. Signaling over a wireless channel is severely affected by the environmental radio propagation effects as the channel state may change within a very short time span and hence suffers from many channel impairments.

- **Attenuation:** When the wireless transmission takes place where the tag and the reader are separated over a significant distance, there occurs a steady decrease in the power of the transmitted signal at the receiver. This refers to attenuation. The attenuation caused by wireless propagation causes

Figure 3. RFID communication model

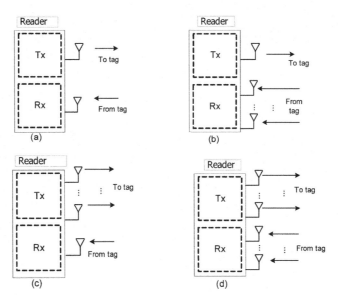

path loss, shadowing loss, and fading loss. Path loss refers to signal attenuation due to distance between the reader and the tag, shadowing loss occurs due to absorption of signals by local physical structures surrounding the tag and the reader, and fading loss occurs due to the constructive and destructive interference of the multiple reflected radio wave paths.

- **Noise:** Noise is present in every communication system in various forms. It can be in a form of thermal noise which occurs due to agitation of electrons, a function of temperature, and is present in all electronic devices and transmission media. Hence it is impossible to eliminate. It can be in terms of Inter-modulation noise which occurs when the signals having different frequencies travel through the same medium. It can also occur due to irregular pulses or noise spikes caused by external EM disturbances, or faults in the communication system and is known as Impulsive noise.

- **Multipath fading:** Multipath refers to the situation where the transmitted signal is reflected by various physical obstacles, thus creating multiple signal paths between the reader and the tag. These multiple paths have randomly distributed amplitudes, phases, and angle of arrivals, which causes fluctuations in the received signal strength causing fading, hence called multipath fading. It also increase the time period required for the transmitted signal to reach the receiver hence causing delay spread. The relative attenuation between the antennas becomes an important issue when dealing with multiple antennas. Channel fading significantly degrades the performance of wireless transmissions.

- **Inter Symbol Interference (ISI):** ISI is one of the most severe limitations encountered in high-speed data transmission systems. Using any practical channel, the effects of filter are unavoidable which causes spreading of individual data symbols passing through the channel and for consecutive symbols this spreading causes part of the symbol energy to overlap with the neighboring symbols. This phenomenon causes ISI and it significantly degrades the system performance.

3.2. Statistical Flat Fading Channel Models

The statistical models are based on measurements made specifically for an intended communication system or spectrum allocation. A significant advantage of the wireless channel models is their flexibility, which means by changing the statistical parameters; the same model can be used to simulate the channel under different conditions. Depending on the nature of the radio propagation environment, there are different models describing the statistical behavior of the multipath-fading envelope. The Rayleigh, Ricean and Nakagami are the most commonly used statistical models to represent small-scale fading phenomenon. In this section, we present different types of fading channels and the mathematical models to describe these channels over which transmission of information takes place or in which information can be stored. These channels can be classified by the environments to which they apply.

Additive White Gaussian Noise (AWGN)

The Additive white Gaussian noise (AWGN) is a transmission channel model in which the only impairment to communication is a linear addition of wideband and is generally used to model an environment which has a large number of additive white noise sources where the white noise has a constant spectral density (expressed as watts per hertz of bandwidth). These additive sources can be modeled as Gaussian random process with a Gaussian distribution of amplitude. The model does not account for fading, frequency selectivity, interference, nonlinearity or dispersion. The statistical model for the AWGN channel with zero mean and variance is given by its probability density function as

$$p\left(x\right) = \frac{1}{\sqrt{2\pi\sigma^2}} exp\left(-\frac{1}{2\sigma^2} x^2\right) \qquad (1)$$

where $\sigma^2 = \frac{N_0}{(2RE_b)}$, R is the coding rate given by the ratio of the number of information bits to the number of transmitted bits, and $\frac{E_b}{N_o}$ is the ratio of the bit energy to noise power spectral density. Thus for an RFID system, when no fading occurs in the propagation channel between the tag and the reader, the channel statistics can be defined using Equation 1.

Rayleigh Fading Channel

AWGN is a hypothetical situation where the transmitter is assumed to be ideal and noiseless. Its merely noise generation at the receiver side assumed to have a constant power spectral density over the entire channel bandwidth and the amplitude following Gaussian distribution. In real-time scenario, when a signal is transmitted it is received as a superposition of many pulses arriving at the receiver due to multi-path fading. Rayleigh fading model is one such model motivated by the fading effects arising due to long distance transmission. In Rayleigh flat fading channel model, it is assumed that the channel induces amplitude, which varies in time according to the Rayleigh distribution. If there is no line-of-sight component and the multiple reflective paths are large in number (as in the case of dense urban environments) then small-scale fading is also termed as Rayleigh fading as the Gaussian distribution has a zero mean and the envelope of the received signal is statistically described by a Rayleigh pdf. The Rayleigh probability distribution function (PDF) of a received complex envelope of a signal $r(t)$ at any time t is given as

$$p(r) = \frac{r}{\sigma^2} e^{\left(\frac{-r^2}{2\sigma^2}\right)} \qquad (r \geq 0) \qquad (2)$$

where σ is the root man square value of the received voltage signal before envelope detection, and is the time-average power of the received signal before envelope detection. It is well known that the envelope of the sum of two quadrature Gaussian noise signals obeys a Rayleigh distribution. This fading distribution could be applied to any scenario where there is no LOS path between transmitter and receiver antennas. Thus, if the reader and the tag in the RFID system are at significant distance such that no LOS take place, the channel can be modeled as Rayleigh faded channel.

Ricean Fading Channel

When there is a dominant stationary (non-fading) signal component present, such as LOS propagation path, the small-scale fading envelope distribution is Ricean. Thus, if the RFID tag and reader are placed such that a LOS path occurs between them, the propagation channel can be modeled as a Ricean faded channel. In such a situation, random multipath components arriving at different angles are superimposed on a stationary dominant signal. At the output of an envelope detector, this has the effect of adding a dc component to the random multipath. The effect of a dominant signal arriving with many weaker multipath signals gives rise to the Ricean distribution. As the dominant signal becomes weaker, the composite signal resembles a noise signal, which has an envelope that is Rayleigh. Thus, the Ricean distribution degenerates to a Rayleigh distribution when the dominant component fades away. With fixed scatterers or signal reflectors in the medium, in addition to randomly moving scatterers, the channel impulse response will have a non-zero mean value and its envelope will have a Rice distribution. This channel is said to be a Ricean

Fading Channel. For a multipath fading channel containing a specular or LOS component, the complex envelope of the received signal can be given by the Ricean distribution

$$p(r) = \begin{cases} \frac{r}{\sigma^2} \, exp\left(-\frac{r^2 + A^2}{2\sigma^2}\right) I_0\left(\frac{Ar}{\sigma^2}\right) & for\,(A \geq 0, r \geq 0) \\ 0 & for\,r < 0 \end{cases}$$

$$(3)$$

where A denotes the peak amplitude of the dominant or LOS signal and $I_0(.)$ is the zeroth order modified Bessel function of the first kind. The Rician distribution is often described in terms of a parameter K called Rician factor, which is defined as the ratio between the deterministic signal power and the variance of the multipath.

K-Factor. The fading signal magnitude follows a Rice distribution, which can be characterized by two parameters: the power P_c of constant channel components and the power from scatter channel components. The ratio of these two (P_c / P_s) is called the Ricean K-factor. The worst-case fading occurs when $P_c = 0$ and the distribution is regarded as Rayleigh distribution $K = 0$. The K-factor is an important parameter in system design since it relates to the probability of a fade of certain depth. Both fixed and mobile communications systems have to be designed for the most severe fading conditions for reliable operation (i.e., Rayleigh fading).

Nakagami-m Fading Channel

The Nakagami distribution is selected to fit empirical data and is known to provide a close match to some experimental data than the Rayleigh distributions. The Nakagami distribution is often used to model multipath fading as it can model fading conditions that are either more or less severe than Rayleigh fading. When $m=1$, Nakagami distribution becomes the Rayleigh distribution, when $m=1/2$ it becomes a one-sided Gaussian distribution and when $m=\infty$ the distribution becomes an impulse (no fading). Even Rice distribution can be closely approximated using Nakagami parameter via the relationship $m=(K+1)^2/(2K+1)$. Thus, in a RFID system where the distance between the RFID tag and reader changes such that the fading conditions during communication between them can change, it is best to model the channel as a Nakagami faded channel.

Considering a receiver with M diversity branches, let the received instantaneous signal at the branch be characterized by the Nakagami distribution. The Nakagami distribution describes the received envelope $z(t)=r(t)$ by a central chi-square distribution with degrees of freedom, i.e.,

$$p(A_k) = \frac{2}{\Gamma m_k}(\frac{m_k}{\Omega_k})^{m_k} r^{2m_k-1} e(-\frac{m_k}{\Omega_k})r^2, k=1,2,...M \tag{4}$$

where $\Gamma(.)$ is the Gamma function, $\Omega_k = \overline{A_k^2}$ is the average power on k^{th} branch, m_k is the fading parameter. For the Rayleigh PDF given by Equation 2 and r denoting the amplitude of the Rayleigh fading, the cumulative distribution function (CDF) of the Rayleigh fading can be written as

$$F_R(r) = 1 - e^{-r^2/2\sigma^2} \tag{5}$$

For the Nakagami-m PDF given by (4), the Nakagami-m CDF can be written as

$$F_N(r) = \int_0^r \frac{2}{\Gamma(m_k)}(\frac{m_k}{\Omega_k})^{m_k} r^{2m_k-1} e(-\frac{m_k}{\Omega_k})^{r^2} dr \tag{6}$$

Then the complex Nakagami-m amplitudes can be calculated as

$$A = F_N^{(-1)}(u) \tag{7}$$

where is the inverse function of Nakagami-m CDF, and is the uniformly distributed variable generated by Rayleigh fading coefficients, $u = F_R(r) = 1 - e^{-r^2/2\sigma^2}$. A is Nakagami-m distributed can be further expressed as

$$A = F_N^{(-1)}(F_R(r)) \tag{8}$$

A general form of the inverse Nakagami-m CDF is obtained as

$$F_N^{-1}(u) = \nu + \frac{a_1\nu + a_2\nu^2 + a_3\nu^3}{1 + b_1\nu + b_2\nu^2} \tag{9}$$

where $v = \left(\sqrt{ln\frac{1}{1-u}}\right)^{(1/m)}$, and m is the Nakagami-m fading parameter, and $0.65 \le m \le 10$.

While these are the most commonly used flat-fading channels, there are still a number of other fading channels available in literature like the Hoyt fading channel model, the Weibull fading channel model, and a few Generalized fading models like the generalized-gamma, η-μ and κ-μ distributions. All these channel models can be used depending on the characteristics of the propagation channel under consideration.

4. DIVERSITY

Diversity is an important way to help mitigate the multipath-induced fading. It is a method that is used to develop information from several signals transmitted over independent fading paths. This means that the diversity method requires that a number of transmission paths be available, all carrying the same message but having independent fading statistics. The mean signal strengths of the paths should also be approximately the same. The basic requirement of the independent fading is that the received signals are uncorrelated. Therefore, the success of diversity schemes depends on the degree to which the signals on the different diversity branches are uncorrelated. To meet the stringent requirements for quality service requirements and spectrally efficient multilevel constellations, antenna diversity is needed to offset penalty on the SNR due to fading and denser signal constellation. By proper combining of the multiple signals severity of fading can be greatly reduced and reliability of transmission can be improved. The use of diversity permits a direct comparison of improvement offered by multiple antenna compared to a single one. The effectiveness of diversity and adaptive combining schemes has been proven in (Kim et al., 2001).

There are several diversity schemes available to choose from which includes time diversity, frequency diversity, polarization diversity, space diversity. The information bearing signal copies are transmitted at different time instants or different frequencies, while using time or frequency diversity, respectively. In frequency diversity, the frequencies have to be separated enough to make sure that the fading coefficients are independent and uncorrelated. In polarization diversity multiple versions of a signal are transmitted and/or received via antennas with different polarization. In space diversity, multiple antennas separated by a few wavelengths are used at the receiver/transmitter to create independent fading channels. Unlike in frequency diversity, no bandwidth efficiency loss occurs in space-diversity. By placing the antennas sufficiently apart, independent channel fading coefficients can be obtained between different antenna pairs and thus independent signal paths can be created. Space diversity can be further classified as transmit diversity and receive diversity where the former uses space-time codes at the transmitter and the later uses combining techniques at the receiver to achieve improved diversity gains. Space-time codes for MIMO systems exploiting both transmit as well as receive diversity schemes is the most popular technique

used in present wireless communication systems, yielding a high quality of reception (Abou-Rjeily & Fawaz (2008), Pawar Et al., (2009)).

4.1. Transmit Diversity

Use of multiple antennas at the transmitter for communication is known as transmit diversity. The effects of fading, outages, and circuit failures can be overcome by using transmit diversity. It uses signals that originate from two or more independent sources that have been modulated with identical information-bearing signals and that may vary in their transmission characteristics at any given instant. Transmit diversity schemes have increasingly grown popular over the last decade as they promise high data rate transmission over wireless fading channels in both the uplink and downlink while putting the diversity burden on the base station. Also it helps to improve the quality and the data rates of the multiple antenna systems in order to obtain diversity orders. The capacity of antenna systems far exceeded by using Multiple Input Multiple Output (MIMO) channels as compared to single-antenna system.

Due to space considerations, it is expected to encompass multiple antennas at the base station while the mobile device having only one (or two) antennas. Also, it can be seen that many of the capabilities alongwith performance gain of receive diversity can be obtained by using multiple antennas at the transmitter to achieve transmit diversity. WLAN and cellular networks have shown special interest in using transmit diversity as performance can be increased without adding extra cost of antennas, system complexity, and power consumption to the mobile devices. Unlike in receive diversity, it is not possible to send same information bearing signal from all elements in antenna array in transmit diversity. The reason being the copies of the transmitted signal which are received at the receiver will add incoherently if all the antennas send the same signal, and thus no diversity gain can be achieved. Therefore, there has to be some sort of transmission plan that can let the signals to combine coherently at the receiver for transmit diversity to work. The most simplest form of transmit diversity at the forward link follows using one receiver antenna at the mobile unit and two transmitter antennas at the base station. Amongst the two one is referred to as the main or common antenna and the other one as the diversity antenna. The transmit diversity design depends on whether the channel gains are known at the transmitter or not. When the channel gains are known to the transmitter, the system is very similar to receive diversity. When the channel is not known, transmit diversity gain requires a combination of space and time diversity like the Alamouti scheme.

Alamouti (1998) devised a two branch transmit diversity scheme by using two transmit and one receive antennas. It provided the same diversity order as with one transmit and two receive antennas. The proposed scheme assumed that there is no ISI. However, ISI plays a major role in signal degradation when signaling with higher order constellations like QAM, 8PSK or 16 QAM or if the channel experiences a non-negligible delay spread. Therefore a new transmit diversity scheme, which can handle the channels with ISI was proposed by Lindskog & Paulraj (2000). The scheme, however, achieved the same diversity benefits as achieved in Alamouti (1998). Alamouti's work was further generalized into orthogonal designs which successfully provided diversity gain but didn't provide optimal rate (Tarokh et al., 1999). In 1998, (Tarokh et al., 1998) introduced Space-time codes (STCs) as a novel means of providing transmit diversity for multiple-antenna fading channel. Since then, space-time codes have attracted considerable attention because of their ability to exploit the capacity increase of MIMO antenna systems capable of maintaining the best trade-off between the fundamental quantities of rate gain, diversity, and receiver complexity. The codes are such designed so as to improve the data and/or reliability of communication over fading channels

using MIMO systems. The presence of multipath rays does not affect the diversity advantage as offered by the original space-time codes' design criteria. Since Space-Time coding is a bandwidth and power efficient method, therefore, one should have a fundamental understanding of the limits of bandwidth efficient delivery of higher bit-rates. Band-limited channels do not accumulate rapid flow of data but by exploiting multiple antennas at the transmitter, higher bit-rates can be achieved. Today, space-time coding has gained enough consideration for implementation over third (and beyond) generation wireless systems.

There are mainly two types of STCs- Space time block codes (STBC) and Space-time trellis codes (STTC). While STBCs operate on a block of input symbols, producing a matrix output whose column represents time and rows represent antennas, STTCs operate on one input symbol at a time, producing a vector symbols whose length represents antennas. While, STBCs do not provide coding gain unless concatenated with some outer code, the key advantage of using STTCs is that they provide coding gain. However, both types of codes provide full diversity gain. STTCs have a disadvantage that they are extremely hard to design and generally require high complexity encoders and decoders while STBCs require simple encoding and decoding at low computational cost. Marked comparisons between these two types of transmit diversity schemes in terms of frame error rate are clearly mentioned in (Sandhu et al., 2001). The main theories of the Space-Time Trellis Code (STTC) (Tarokh et al., 1998) and the Space-Time Block Code (STBC) (Tarokh et al., 1999) have been mentioned in the two key papers for flat independent Rayleigh fading channels. A number of other schemes employing multiple antenna arrays were also developed at about the same time, e.g., the simple and popular Alamouti STBC (Alamouti, 1998), a transmit diversity scheme using pilot symbol-assisted modulation (Guey et al., 1996) and the Bell Labs layered Space-Time (BLAST) multiplexing framework (Foschini, 1996).

Space-Time Trellis Codes (STTC): Encoder Structure and Generator Description

In STTC system, mapping of input binary data to modulation symbols is done by the encoder, where the mapping function is described by the trellis diagram. Considering an M-ary modulated STTC encoder with N_t transmit antennas, the input message stream s given by

$$s = \left(s_0, s_1, s_2, \ldots, s_t, \ldots\right) \tag{10}$$

where s_t is a group of $m = log_2 M$ information bits at time t, then at any time t binary input s_t is fed into the STTC encoder and is given by

$$s_t = s_t^1, s_t^2, \ldots, s_t^m \tag{11}$$

The encoder structure for M-ary modulation is shown in Figure 4.

As can be seen from the figure, the trellis encoder consist of m feed forward shift registers and the m binary input sequences s^1, s^2, ...s^m are

fed into the encoder such that the *i-th* input bit is $s_t^i, i = 1, 2, \ldots, m$ is passed to the *i-th* shift register. It is then subsequently delayed and multiplied by the encoder coefficient set given by Vucetic and Yuan (2005)

$$G^i = \left[\left(g_{0,1}^i, g_{0,2}^i, \ldots, g_{0,N_t}^i\right), \left(g_{1,1}^i, g_{1,2}^i, \ldots, g_{1,N_t}^i\right), \ldots, \left(g_{\nu_i,1}^i, g_{\nu_i,2}^i, \ldots, g_{\nu_i,N_t}^i\right)\right] \tag{12}$$

where $g_{l,k}^i, l = 0, 1, \ldots, \nu_i, k = 1, 2, \ldots, N_t$ is such that it is an element of the M-ary constellation set, and is the memory order of the i-the shift register.

Equation 12 can be further described by the following m multiplication coefficient set sequences

$$g^1 = \left[\left(g_{0,1}^1, g_{0,2}^1, \ldots, g_{0,N_t}^1\right), \left(g_{1,1}^1, g_{1,2}^1, \ldots, g_{1,N_t}^1\right), \ldots, \left(g_{\nu_1,1}^1, g_{\nu_1,2}^1, \ldots, g_{\nu_1,N_t}^1\right)\right]$$
$$g^2 = \left[\left(g_{0,1}^2, g_{0,2}^2, \ldots, g_{0,N_t}^2\right), \left(g_{1,1}^2, g_{1,2}^2, \ldots, g_{1,N_t}^2\right), \ldots, \left(g_{\nu_2,1}^2, g_{\nu_2,2}^2, \ldots, g_{\nu_2,N_t}^2\right)\right]$$
$$g^m = \left[\left(g_{0,1}^m, g_{0,2}^m, \ldots, g_{0,N_t}^m\right), \left(g_{1,1}^m, g_{1,2}^m, \ldots, g_{1,N_t}^m\right), \ldots, \left(g_{\nu_m,1}^m, g_{\nu_m,2}^m, \ldots, g_{\nu_m,N_t}^m\right)\right].$$

$$\tag{13}$$

The *m* multiplication coefficient set sequences are also called the generator sequences as they can

Figure 4. Encoder structure for M-ary modulated STTC encoder

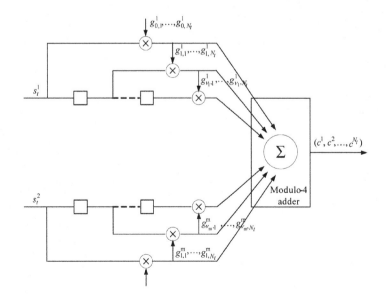

fully describe the encoder structure. The multiplier outputs from all the shift registers are added modulo-M and the encoder output is given by

$$c_t^k = \sum_{i=1}^{m}\sum_{l=0}^{\nu_i} g_{l,k}^i s_{t-l}^i \, modM, \qquad k = 1,2,\dots,N_t$$

(14)

These outputs are mapped into signals from the M-PSK constellation set. The modulated signals from the space-time symbol transmitted at time and are given by

$$c_t = (c_t^1, c_t^2, \dots, c_t^m)^T$$

(15)

These signals are then simultaneously transmitted from the N_t transmit antennas. Thus for an RFID system with N_t reader transmit antennas, Figure 5 shows the STTC encoder structure, where the information is encoded using STTC and the modulated symbols given by Equation 15 are then transmitted through N_t transmit antennas. At the receiver, the symbols are de-modulated and decoded using a ML decoder based on viterbi algo-

rithm. Thus, for $N_t=2$, only two space-time symbols (c_t^1, c_t^2) will be transmitted over the two reader transmit antennas, for $N_t=3$ three space-time symbols (c_t^1, c_t^2, c_t^3) will be transmitted over the three reader transmit antennas, and so on for higher number of reader transmit antennas

The drawback of using STTC is the complexity of the encoder structure as STTC has multiple symbols associated with a single trellis transition. The trellis includes all possible state transitions for every possible input symbol and state combinations due to which the transmitter requires more transmit power. Also at the receiver, use of viterbi algorithm increases the complexity of the system.

Space-Time Block Codes (STBC)

In hope of reducing the exponential decoder complexity of STTC, Alamouti (1998) proposed a simple transmit diversity scheme, which was later expanded by Tarokh et al. (1998) for an arbitrary number of array elements to form the class of Space-Time Block Codes (STBC). It

Figure 5. An RFID system with reader transmit antennas performing STTC

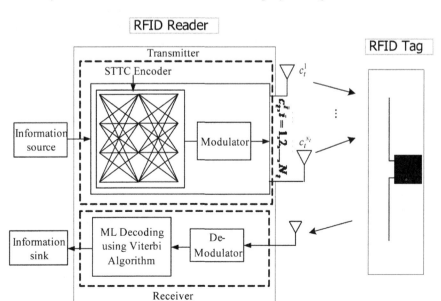

presented a simple two-branch transmit diversity scheme which referred to a full-rate orthogonal design and the orthogonality spread across sequence of symbols transmitted simultaneously from two antennas was exploited at the receiver by the combining rule. STBC can be achieved by a complex orthogonal designed matrix called the encoder (or the transmission) matrix G_{N_t}, where N_t represents the number of transmit antennas, p represents the number of time periods for transmission of one block of symbols. The elements of the encoder matrix are the indeterminants $[\pm x_1, \pm x_2, ..., \pm x_n]$, their conjugates $[\pm x_1^*, \pm x_2^*, ..., \pm x_k^*]$ and the superposition of those. The encoder matrix maps k input m - ary symbols into orthogonal sequence of length p. Thus the rate of $r = k/p$ is achieved.

To construct space-time block codes, the classical mathematical framework of orthogonal designs can be applied. An orthogonal design exists if and only if $N_t = 2, 4\, or\, 8$. That means the STBCs constructed using this method exists only for a few sporadic values of n. These codes have a simple maximum likelihood decoding algorithm based on linear processing and also they exploit full diversity given by transmit and receive antennas and achieve maximum possible transmission rate. However, a generalization of orthogonal designs can provide STBCs for both real and complex constellations for any number of transmit antennas but the maximum possible transmission rate achieved is half for any number of the transmit antennas. STBCs can be classified into STBC with real signals and STBC with complex signals. It is worthwhile to note that full rate orthogonal design exists for arbitrary real signal constellations for N_t number of antenna elements; it does not exist for complex signal constellations like PSK or QAM for more than two antenna elements. For $N_t \geq 2$, the code design aims at constructing high data rate complex transmission matrices G_{N_t} whilst keeping the decoding complexity low and achieving full diversity order. Also the value of p must be minimized to minimize the decoding delay. Alamouti scheme can be regarded as a STBC with complex signal constellations for N_t =2 (Alamouti, 1998). Also it is the only STBC with $N_t \times N_t$ complex encoder matrix that achieves a full rate $r=1$.

There are a few complex orthogonal designs proposed in (Tarokh et al.,1999) for 2, 3 and 4 transmit antennas which accommodates rates as close to one as possible. The STBC encoding algorithm is described by the encoder matrix. For example, for a RFID system with two transmit antennas as shown in Figure 6, encoder matrix G_2 is used which maps a block of $k=2$ bits into N_t=2 sequences each of length $p=2$. G_2^* is given by

$$G_2^* = \begin{pmatrix} x_1 & x_2 \\ -x_2^* & x_1^* \end{pmatrix} \tag{16}$$

where x_1 and x_2 are the set of symbols at the encoder input. The first and second rows of the encoder matrix are transmitted at times t and $t+T$ from the two antennas.

Similarly, complex encoder matrices G_3^* and G_4^* for three and four transmit antennas are orthogonally designed and have the rate $r=1/2$ (Tarokh et al., 1999).

$$G_3^* = \begin{bmatrix} x_1 & -x_2 & -x_3 & -x_4 & x_1^* & -x_2^* & -x_3^* & -x_4^* \\ x_2 & x_1 & x_4 & -x_3 & x_2^* & x_1^* & x_4^* & -x_3^* \\ x_3 & -x_4 & x_1 & x_2 & x_3^* & -x_4^* & x_1^* & x_2^* \end{bmatrix} \tag{17}$$

$$G_4^* = \begin{bmatrix} x_1 & -x_2 & -x_3 & -x_4 & x_1^* & -x_2^* & -x_3^* & -x_4^* \\ x_2 & x_1 & x_4 & -x_3 & x_2^* & x_1^* & x_4^* & -x_3^* \\ x_3 & -x_4 & x_1 & x_2 & x_3^* & -x_4^* & x_1^* & x_2^* \\ x_4 & x_3 & -x_2 & x_1 & x_4^* & x_3^* & -x_2^* & x_1^* \end{bmatrix} \tag{18}$$

Figure 6. RFID system with two transit antennas performing STBC

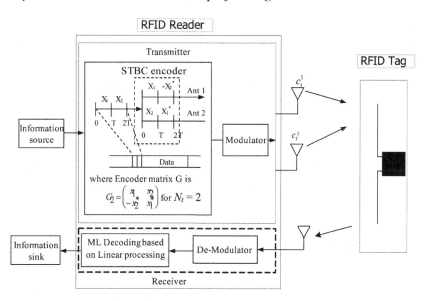

It can be seen that the inner product of any two rows of these matrices is zero which proves the orthogonality of these schemes. Considering the encoder matrix G_3^*, four complex symbols are taken at a time and transmitted by three transmit antennas in eight symbol periods, hence giving the transmission rate $r=1/2$. The encoder matrices H_3 and H_4 are complex generalized orthogonal designs for STBC and achieves the transmission rate of $r=3/4$ (Tarokh et al., 1999).

$$H_3 = \begin{bmatrix} x_1 & -x_2^* & \dfrac{x_3^*}{\sqrt{2}} & \dfrac{x_3^*}{\sqrt{2}} \\ x_2 & x_1^* & \dfrac{x_3^*}{\sqrt{2}} & -\dfrac{x_3^*}{\sqrt{2}} \\ \dfrac{x_3}{\sqrt{2}} & \dfrac{x_3}{\sqrt{2}} & \dfrac{(-x_1 - x_1^* + x_2 - x_2^*)}{2} & \dfrac{(+x_1 - x_1^* + x_2 + x_2^*)}{2} \end{bmatrix}$$

$$(19)$$

$$H_4 = \begin{bmatrix} x_1 & -x_2^* & \dfrac{x_3^*}{\sqrt{2}} & \dfrac{x_3^*}{\sqrt{2}} \\ x_2 & x_1^* & \dfrac{x_3^*}{\sqrt{2}} & -\dfrac{x_3^*}{\sqrt{2}} \\ \dfrac{x_3}{\sqrt{2}} & \dfrac{x_3}{\sqrt{2}} & \dfrac{(-x_1 - x_1^* + x_2 - x_2^*)}{2} & \dfrac{(x_1 - x_1^* + x_2 + x_2^*)}{2} \\ \dfrac{x_3}{\sqrt{2}} & -\dfrac{x_3}{\sqrt{2}} & \dfrac{(x_1 - x_1^* - x_2 - x_2^*)}{2} & -\dfrac{(x_1 + x_1^* + x_2 - x_2^*)}{2} \end{bmatrix}$$

$$(20)$$

The orthogonality of these matrices can be proved since the columns of these matrices fulfill the orthogonality requirements. The orthogonal transmission matrices for different number of antenna elements and input block length for both real and complex signal constellations are summarized in Table 1.

In terms of complexity, STBC is less complex as compared to STTC because of its simpler encoding techniques and the use of linear processing techniques for ML decoding. Thus for an RFID system, where power requirement is a limitation, STBC is preferred over STTC as it requires less processing power.

4.2. Receive Diversity: Combining Techniques

Motivated by the aim to mitigate signal degradation caused by multipath propagation, applying spatial diversity techniques focused on using multiple antennas at the receiver. It led to considerable performance gain in terms of tolerance to co-channel interference (Kim et al.,2001), (Turkmani et al., 1995). Assuming that the paths taken by each of the copies result in statistically independent fading effects, it can be concluded that the signals are unlikely to be in deep fade i.e. highly distorted. Thus an improved signal can be obtained by forming a weighted combination of the received copies. The combining of the received copies can be performed using several different methods in different environments based on the applications.

In RFID systems, the signal after passing through the channel reaches the tag. The tag demodulates and decodes these signals and responds by backscatter modulating the RF carrier, according to the data stored on the tag. The reader detects the modulated response signal from the tag. However, the receiver at the tag reader receives this incoming signal from all directions due to multipath. The receiver needs to collect these independent fading signal branches. This is done by combining these multiple copies of the signal to improve the received SNR. Since the chance of having two deep fades from two uncorrelated signals at any instant is rare, combining them can reduce the effect of the fades. Assuming that the paths taken by each of the copies result in statistically independent fading effects, it can be concluded that the signals are unlikely to be in deep fade i.e. highly distorted. Thus an improved signal can be obtained by forming a weighted combination of the received copies. The combining of the received copies can be performed using several different methods in different environments based on the applications.

The three most commonly used receive diversity techniques when signaling over various statistical fading channels are

1. Selection Combining (SC)
2. Equal Gain Combining (EGC), and
3. Maximal Ratio Combining (MRC).

Selection Diversity

Selection Diversity is the simplest diversity technique where gains of M pre-detection diversity branches are adjusted to provide the same SNR ratio for each branch. The receiver branch having the highest instantaneous SNR will be fed to detector circuit. Selection combining can be used with coherent as well as non-coherent modulations schemes as it does not require the signal phase knowledge of each diversity branch. Figure 7 shows an RFID system with ideal selection combiner at the reader that chooses the signal with the highest instantaneous SNR of all the branches, so the output SNR is equal to that of the best incoming signal and makes it available to the receiver at all times. Multiple branches will improve the probability of having a larger SNR at the receiver.

We assume that the signal received by each diversity branch is statistically independent of the signals in other branches and is Rayleigh distributed with equal mean signal power P_0. The probability density function of the signal envelope, on branch i, is given by

$$p\left(r_i\right) = \frac{r_i}{P_0} exp\left(-r_i^2 \middle/ 2P_0\right) \qquad (21)$$

where $2P_0$ = mean-square signal power per branch and r_i^2 = Instantaneous power in the i-th branch.

Let $\gamma_i = r_i^2 \middle/ 2P_0$ and $\gamma_0 = 2P_0 \middle/ 2N_i$, where N_i is the noise power in the i-th branch, the probability density function for γ_i is given by

$$p\left(\gamma_i\right) = \frac{1}{\gamma_0} exp\left(\gamma_i \middle/ \gamma_0\right) \qquad (22)$$

where γ_0 is the defined threshold for bit energy to noise ratio. Assuming, that the signal in each branch has a constant mean, the probability that

Table 1. Orthogonal encoder matrices and transmission rates for STBC

Orthogonal Design (G_{N_t})	N_t	k (input block length)	p (symbol periods)	Rate $r=k/p$
FOR COMPLEX SIGNAL CONSTELLATIONS				
G_2^*	2	2	2	1
G_3^*	3	4	8	1/2
G_4^*	4	4	8	1/2
H_3	3	3	4	3/4
H_4	4	3	4	3/4
FOR REAL SIGNAL CONSTELLATIONS				
G_2	2	2	2	1
G_4	4	4	4	1
G_8	8	8	8	1
G_3	3	4	4	1
G_5	5	8	8	1
G_6	6	8	8	1
G_7	7	8	8	1

the SNR on any one branch is less than or equal to any given value γ_g is given by

$$P[\gamma_i \leq \gamma_0] = \int_0^{\gamma_g} p(\gamma_i)\, d\gamma_i = 1 - exp\left(\gamma_i \big/ \gamma_0\right) \tag{23}$$

Thus, for a receiver with M branches, the probability that the SNRs in all branches are simultaneously less than or equal to γ_g is given by

$$P_M(\gamma_g) = P[\gamma_1 \leq \gamma_0]\, P[\gamma_2 \leq \gamma_0]\ldots P[\gamma_M \leq \gamma_0] \tag{24}$$

$$= \prod_{i=1}^{M} \gamma_i \leq \gamma_0 = \left[1 - exp\left(\gamma_g \big/ \gamma_0\right)\right]^M \tag{25}$$

Maximal Ratio Combining

In MRC, the signal in each branch is first co-phased and once the phase distortions are cancelled out, the

Figure 7. Selection diversity

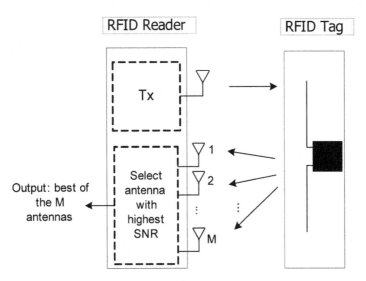

signal in each branch is multiplied by a weighting factor proportional to the signal amplitude. Figure 8 shows a RFID system with multiple antennas at the reader performing MRC where it presents the receiver with a signal-to- noise ratio that is the direct sum of all individual SNRs in the branches.

MRC provides full diversity and also yields the highest SNR but due to channel estimation the complexity is high. An RFID system with dual antenna RFID readers performing MRC was proposed in (Angerer et al., 2009). Experimental evaluation for RFID reader with dual antenna performing MRC has been presented in (Angerer et al., 2010) where a performance comparison of maximum ratio combining and random antenna selection with two receive antennas in a high fading environment was given. Another reader receiver performing maximal ratio combining was presented in (Wang et al., 2007), however, it did not present any realization. Experimental results based on read range increase and orientation insensitivity of RFID tags with multiple RF ports has been presented in (Nikitin et al., 2007). The main drawback of using MRC is that the signal level and noise power at each branch needs to be correctly estimated for all instances in time.

The M signals are weighted proportional to their signal voltage-to-noise power ratios and then summed as

$$r_M = \sum_{i=1}^{M} a_i r_i(t) \tag{26}$$

Since noise in each branch is weighted according to noise power,

$$n_i^2(t) = \sum_{j=1}^{M}\sum_{i=1}^{M} a_i a_j \, n_i(t) n_j(t) \tag{27}$$

The average noise power,

$$N_T = \sum_{i=1}^{M} a_i^2 \, n_i^2(t) = 2\sum_{i=1}^{M} |a_i|^2 \, N_i \tag{28}$$

where $n_i^2(t) = 2N_i$

The probability that $\gamma_M \leq \gamma_g$ is given by:

Figure 8. Maximal ratio combining

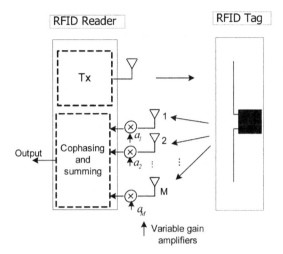

Figure 9. Equal gain combining

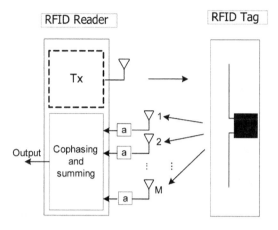

$$P\left(\gamma_M \leq \gamma_0\right) = 1 - exp\left(-\gamma_g\Big/\gamma_0\right) \sum_{K-1}^{M} \frac{\left(-\gamma_g\Big/\gamma_0\right)^{k-1}}{(K-1)!}$$

(29)

Equal Gain Combining

In Equal Gain Combining, each signal branch is weighted with the same factor, irrespective of the signal amplitude. However, the received signal carriers are first co-phased as in the case of MRC to avoid signal cancellation and are then equally weighted by their amplitudes as shown in Figure 9. In other words, the attenuation/amplification factor a_i is the same for all channels i. EGC has performance close to that of MRC but with lower implementation complexity. The exact outage performance for EGC is presented in (Song et al., 2003) for mobile cellular radio systems for Rayleigh fading with limited co-channel interference. EGC receiver is of practical interest because of its reduced complexity relative to optimum maximal ratio combining scheme while achieving near-optimal performance. It is the sum of all the signals received in order to increase the available SNR at the receiver. The gain of all of the branches is set to a particular value that does not change which is in contrast to MRC.

EGC is similar to MRC, but there is no attempt to weight the signal before addition; thus $a_i=1$. The envelope of the output signal with all $a_i=1$. is given by

$$r = \sum_{i=1}^{M} r_i$$

(30)

and the mean output is given as:

$$\gamma_M = \frac{1}{2} \frac{\left[\sum_{i=1}^{M} r_i\right]^2}{\sum_{i=1}^{M} N_i}$$

(31)

For *M=2*, the probability can be written in closed form as:

$$P\left(\gamma_M \leq \gamma_0\right) = 1 - exp\left(-2\gamma_g\Big/\gamma_0\right) - \sqrt{\pi\left(\frac{\gamma_g}{\gamma_0}\right)} exp\left(-2\gamma_g\Big/\gamma_0\right) .erf\sqrt{\left(\frac{\gamma_g}{\gamma_0}\right)}$$

(32)

For *M>2*, the probability can be obtained by numerical integration techniques.

Comparison of the Various Combining Schemes

The use of receive combining techniques in conventional MIMO systems Hourani (2004/2005) as well as in modern day systems (Goel et al., 2010) have shown results that MRC outperforms SC and EGC but also have maximum complexity. The comparison of these combining techniques reveal that in SC, none of the channel state information such as fading amplitudes, phases and any time delays introduced are required. The receiver basically looks at the outputs from each fading channel and selects the one with the highest SNR. The strongest signal is favored, thus signals that have undergone deep fades are unlikely to be picked by the receiver. Since there is also no need for any addition of the fading channel outputs, it further decreases the complexity. Also, it can be used in non-coherent or differentially coherent modulation schemes as no knowledge of the phases is required.

For EGC, the output from each fading channel is rotated by a certain phase angle and these rotated outputs are then summed. This is equivalent to multiplying each output by a different gain factor, which set the outputs to the same amplitude. The summed output is then compared to each of the signals in the constellation alphabet. Whichever signal in the constellation is closest in terms of the Euclidean distance metric to the summed output is declared to be the original transmitted signal. Unlike SC, in EGC the knowledge of the phase shifts that are introduced by the fading channels is required, so only systems that use coherent detection can utilize EGC (Simon & Alouini (2000)). This is a severe limitation, since in general, it is hard to track the phases of the received signals.

MRC is the most commonly used technique over fading channels. Though, it is slightly more complex than SC or EGC, the added benefit in performance, especially in the case of independent fading channels, cannot be unaccounted for, thus, MRC is highly credited for space-time codes. Un-

like EGC, in MRC each fading channel output is not rotated by the same phase angle, instead, each output is multiplied by the complex conjugate of its fading channel response function. The multiplied outputs are then summed to form a hybrid signal. This hybrid signal is then compared to the signals in the input constellation. And based on a simple Euclidean distance metric, the constellation element that is closed to the hybrid signal is selected. An error occurs if this constellation element was not the original transmitted signal. Since MRC depends on perfect channel estimation, special techniques have to be used when noise and time delays are present (Simon & Alouini (2000)). EGC does have an advantage over MRC as the fading amplitudes need not be estimated in former, thus, the implementation complexity for EGC is significantly less than the MRC because of the requirement of correct weighing factors. However, EGC is sub-optimal to MRC for coherent detection. Thus, while selecting a combining technique, a trade-off occurs between complexity and performance.

5. MODULATION

After encoding, the encoded signal is modulated. The process of modulation is to modify a signal so as to carry data over the communication channel. Successful simulation of any wireless communication system depends on the modulation techniques used which is capable enough of capturing all the significant deviations from the ideal behaviour. While designing any wireless communication system, the use of any particular modulation scheme is based on the application as well as on the channel characteristics such as available bandwidth and its susceptibility to fading. When data transmission takes place through a channel, the signal amplitude tends to vary with time due to the variability in the transmission medium. In such situation we need a modulator and demodulator at the transmitter and the receiver side, respectively, in order to recover the signal.

The three basic types of digital modulation schemes most frequently used are amplitude shift keying (ASK), frequency shift keying (FSK) and phase shift keying (PSK). Infact, most of the RFID tag readers with onboard application specific integrated circuits (ASICs) use one of these modulation schemes. During transmission if a binary waveform is superimposed on the carrier then the modulation schemes are called binary digital modulation and any of the amplitude modulation (AM), phase modulation (PM) or frequency modulation (FM) may be used. Similarly, for multiple waveforms, the modulation is called multiple or m-ary modulation schemes like M- ASK, M-PSK, or M-FSK. However, in some cases a combination of AM-PM can be employed such as Quadrature amplitude modulation (QAM). FM shows more resistant to channel impairments and performs better in terms of power efficiency whereas AM and PM outperforms in terms of spectral efficiency.

5.1. ASK Modulation

A binary amplitude-shift keying (BASK) signal can be defined as

$$s(t) = A m(t) \cos(2\pi f_c t) \qquad 0 \leq t \leq T$$
(33)

where A is a constant, $m(t) = 1 \, or \, 0$, f_c is the carrier frequency, and T is the bit duration. It has a power $P = \dfrac{A^2}{2}$ so that $A = \sqrt{2P}$. Thus Equation 33 can be written as

$$s(t) = \sqrt{2P} \cos(2\pi f_c t) \qquad 0 \leq t \leq T$$

$$= \sqrt{PT}\sqrt{\frac{2}{T}} \cos(2\pi f_c t) \qquad 0 \leq t \leq T$$

$$s(t) = \sqrt{E}\sqrt{\frac{2}{T}} \cos(2\pi f_c t) \qquad 0 \leq t \leq T$$
(34)

where $E=PT$ is the energy contained in a bit duration. The effect of multiplication by the carrier signal $A\cos(2\pi f_c t)$ is simply to shift the spectrum of the modulating signal $m(t)$ to f_c. On a similar basis, for an M-ary amplitude-shift keying (M-ASK), the signal can be defined as

$$s(t) = \begin{cases} A_i \cos(2\pi f_c t), & 0 \leq t \leq T \\ 0, & elsewhere \end{cases}$$
(35)

where $A_i = A\big[2i - (M-1)\big]$ for $i = 0,1,\ldots,(M-1)$ and $M \geq 4$. The signal now has the power $P_i = \dfrac{A_i^2}{2}$ so that $A_i = \sqrt{2P_i}$.

5.2. PSK Modulation

If we group together N bits over the time $T_s = NT_b$, where T_b is the bit duration and T_s is the symbol duration, then there are $2^N = M$ possible symbols. Multiple PSK (MPSK) modulation uses these M possible symbols to represent the digital data where each possible symbol encodes $log_2 M$ binary bits to represent a particular symbol. The phase of the carrier signal is shifted by the modulation signal with the phase measured relative to the previous bit interval. The transmitted signals can be identified as

$$x(t) = \sqrt{\frac{2E_b}{T}}\cos(2\pi f_c t + \theta_m)$$
(36)

where E_b is the energy contained per bit, T is the bit duration, f_c is the carrier frequency and θ_m is the phase of the un-modulated carrier given by $(2m + 1)\pi / M$, $m = 0,1,\ldots,M-1$.

Depending on the number of bits being transmitted in a symbol, M can take different values of 2,4,8,16. When $M=2$, each bit is transmitted individually to control the phase of the carrier and is called as binary phase shift keying (BPSK) modulation. During each bit interval, the modulator shifts the carrier to one of the two possible phases which are A radians apart. Thus with two phases 0 and $À$, the binary data is communicated with the following signals:

For binary '0'

$$x_0(t) = \sqrt{\frac{2E_b}{T}} \cos(2\pi f_c t + \pi), \qquad when\,\theta = \pi,\,and \tag{37}$$

For binary '1'

$$x_1(t) = \sqrt{\frac{2E_b}{T}} \cos(2\pi f_c t), \qquad when\,\theta = 0, \tag{38}$$

In the complex form, the modulated signal can be expressed as

$$x(t) = I(t) + jQ(t) \tag{39}$$

where $I(t) = R(t)\cos\theta$, $Q(t) = R(t)\sin\theta$, and

$R(t) = \sqrt{\frac{2E_b}{T}}$ is the bipolar baseband signal.

When $M = 4$, two bits are grouped together to form a symbol and is called quadrature phase shift keying (QPSK) modulation. QPSK modulation uses four constellation points which are equi-spaced around a circle and it is more resilient to noise. With four phases present, QPSK encodes two bits per symbol and there are four possible combinations for each symbol: 00, 01, 10, or 11. Unlike BPSK where the modulator shifts the carrier to 0 or $À$, in QPSK the modulator shifts the carrier to one of the four possible phases corresponding to the four possible input symbols. Thus, a QPSK signal can be defined as

$$x(t) = \sqrt{\frac{2E_b}{T}} \cos(2\pi f_c t + \theta) \tag{40}$$

where θ can take four possible phases of $\frac{\pi}{4}, \frac{3\pi}{4}, \frac{5\pi}{4}, or\,\frac{7\pi}{4}$ as needed. At the receiver the demodulator removes the data from the carrier and the data bits are recovered. In terms of handling noise and distortion, BPSK modulation is the strongest of all the PSKs as it takes the highest noise or distortion level aiding demodulator to reach to a correct decision. But since it is capable of handling only one bit per symbols it is not suitable for band-limited high data-rate applications. Another advantage of using QPSK over BPSK is that in BPSK the data stream is transmitted with T_b bit duration with channel bandwidth of $2f_b$, where $f_b = 1/T_b$, whereas in QPSK the data stream is transmitted with $2T_b$ bit duration using channel bandwidth of f_b, thus using half the available bandwidth.

5.3. FSK Modulation

A binary frequency-shift keying (BFSK) signal can be defined by

$$s(t) = \begin{cases} A\cos(2\pi f_0 t), & 0 \le t \le T \\ A\cos(2\pi f_1 t), & elsewhere \end{cases} \tag{41}$$

where A is a constant, f_0 and f_1 are the transmitted frequencies, and T is the bit duration. The signal has a power $P = \frac{A^2}{2}$ so that $A=2P$. Thus, with $E_b=PT$ Equation 41 can be written as

$$s(t) = \begin{cases} \sqrt{\frac{2E_b}{T}} \cos(2\pi f_0 t), & 0 \le t \le T \\ \sqrt{\frac{2E_b}{T}} \cos(2\pi f_1 t), & elsewhere \end{cases} \tag{42}$$

And for M-FSK, the signal can be defined as

$$s(t) = \begin{cases} A\cos\left(2\pi f_i t + \theta'\right), & 0 \le t \le T \\ 0, & elsewhere \end{cases}$$

(43)

where θ' is the initial phase angle and f_i is the transmitted frequency.

In almost every RFID system, similar modulation techniques are used while communication between RFID reader and tag and vice-versa. However, in some cases, different modulating techniques can be used in each direction (to and from the tags). Thus, each modulation scheme has attributes that favor its use in RFID systems.

6. DE-MODULATION AND DECODING

As the signal is received at the receiver, the demodulator examines the received symbol corrupted by the channel or the receiver using a maximum likelihood (ML) detector. During ML detection, the demodulator selects that point on the constellation diagram as its estimate of actually transmitted symbol which is closest to the received symbol in terms of the minimum Euclidean distance on the constellation diagram. At this point, if the corruption caused the received symbol to shift closer to any other constellation point than the one transmitted, it will incorrectly demodulated the symbol.

Considering x is the encoded and modulated signal transmitted and X be the corrupted signal received at the receiver over the fading channel h, then the received signal y at the receiver can be written as

$$y = \sqrt{E_b}\,Xh + n \tag{44}$$

where the signals are transmitted with energy E_b and n is the added noise during transmission. Equalization is performed by dividing the received signal by the apriori known h

$$\hat{y} = \frac{y}{h} = \frac{Xh + n}{h} = x + \dot{n} \tag{45}$$

where $\dot{n} = \dfrac{n}{h}$ is the additive noise scaled by the channel coefficient h. Maximum Likelihood demodulation of the signal $\hat{y}$ is performed to recover back the actual information as follows

$$\left(\hat{y}\right)_{est} = \left(\hat{y} - MPSK_{map_{set}}\right)'.\left(\hat{y} - MPSK_{map_{set}}\right) \tag{46}$$

where

$$MPSK_{map_{set}} \in \begin{cases} \{0,1\}\,when\,M = 2 \\ \{0,1,2,3\}\,when\,M = 4, \\ and\,so\,on\,for\,higher\,modulations \end{cases}$$

However, depending on the type of modulation used, the demodulation is performed accordingly. The decision is made in favor of the symbol with minimum distance to retrieve the signal with minimum number of errors as

$$\hat{x} = min\left(\hat{y}\right)_{est} \tag{47}$$

For a system model with N_t transmit and N_r receive antennas, assuming that the channel coefficients are constant over the p received symbols and starting with square transmission matrix, let $\in_t$ denote the permutations of the first column to the t-th column and $\in_t(i)$ denote the row position of x_i in the t-th column. Then the ML detection rule is to form the decision variables

$$\tilde{x}_i = \sum_{t=1}^{N_t}\sum_{j=1}^{N_r} y_j^t h_{j,\in_t(i)}^* \delta_t(i) \tag{48}$$

where $\delta_t(i)$ is the sign of x_i in the t-th column and $i=1,2,...N_t$ and then deciding on the constellation symbol x_i that satisfies

$$\hat{x}_i = arg \min_{c \in \mathcal{A}} \left(\left| \tilde{x}_i - c \right|^2 + \left(-1 + \sum_{j=1}^{N_r} \sum_{i=1}^{N_t} \left| h_{i,j} \right|^2 \right) \left| c \right|^2 \right)$$
(49)

with $\mathcal{A}$ the constellation alphabet. For M-PSK signal constellation, $\left(-1 + \sum_{j=1}^{N_r} \sum_{i=1}^{N_t} \left| h_{i,j} \right|^2 \right) \left| c \right|^2, i = 1,2$ are constant for all signal points, given $h_{i,j}$. Therefore, the decision rule in Equation 49 can be further simplified to

$$\hat{x}_i = arg \min_{c \in \mathcal{A}} \left(\left| \tilde{x}_i - c \right|^2 \right)$$
(50)

After the reader demodulates and decodes the low power signal received from the tags in order to retrieve the necessary data from the tag, this information is relayed to the data processing subsystem where more manipulation of the data may be performed, and where relevant information is finally displayed for the user.

A simulation carried out for a BPSK modulated signal over Rayleigh faded channel with different number of transmit and receive antennas showed that the system with no diversity had the worst performance as compared to the systems with applied diversity. The plot for (1×1), (2×1), (1×2), and (2×2) systems is shown in Figure 10. It can be seen that the (2×2) system gave the best performance in terms of bit-error-rates when plotted against the SNR.

7. CHANNEL ESTIMATION AND THE ALGORITHMS

Most of the work on space-time coding and modulation schemes assumed that perfect channel state information (CSI) is available at the receiver. Also,

in some situations, it is better to forego the requirement for channel estimation at the receiver, just to reduce the cost and complexity of the system, or, sometimes fading conditions change so rapidly that channel estimation becomes difficult. In such cases, the coherence time can be too small to estimate the channel accurately or alternately it may require quite a large number of pilot symbols to achieve the desired accuracy. While dealing with space-time modulation, this problem increase as the number of transmit antenna increases as now the receiver has to estimate the path gain from each transmit antenna to each receive antenna.

Similar to conventional MIMO systems, when the data is transmitted from RFID tag to RFID reader, the transmission suffers from ISI and other channel impairments. Apart from using various space-time coding and modulation techniques, estimation of the channel state between the tag and the reader will further help in enhancing the performance on multipath environments. Since the estimator at the receiver has practically no access to the transmitted signal that enters the channel due to siginificant distances between the transmitter and the receiver, the CSI can be obtained through the use of channel estimation algorithms. These identification algorithms can be categorized as

- Blind Algorithms, and
- Training-based Algorithms.

Blind algorithms do not rely on the knowledge of the transmitted signal rather they depend on the demodulated and detected sequence at the receiver to reconstruct the transmitted signal. This reconstructed signal is then used as if it was the actual transmitted signal. Decision-directed or decision feedback algorithms are amongst the popular class of Blind algorithms. However, this method has an obvious drawback that the decision or the bit error rate at the receiver will result in the construction of an incorrect transmitted signal. This decision

Figure 10. Performance comparisons of Rayleigh, 1X2 MRC, 2X1 Alamouti, and 2X2 STBC for a BPSK modulated communication system

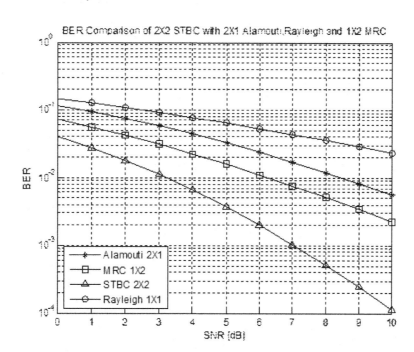

error will introduce a bias while estimating the channel thus making it less accurate.

In training-based method, the receiver uses the training sequence known and sent by the transmitter to reconstruct the transmitted waveform irrespective of the fact that the receiver might not have direct access to the transmitted signal. The channel is identified at the receiver by exploiting this known training sequence. In comparison to blind techniques where the channel is continually updated, training-based techniques produces the most accurate estimates of the channel during the training interval but the estimates become out of control between these intervals. The channel estimates are obtained using different criteria like the Least Square (LS) Criteria, the Minimum Mean Square Estimation (MMSE) Method, Feedback based Auto-Regressive (AR) Models, and Adaptive Methods for estimating the time-varying channels like Modified MMSE. A least squares (LS) equalization method was proposed in Lunglmayr

& Huemer (2010) for counteracting the ISI where it discussed the application of LS equalization for tag to reader communication by load modulation.

7.1. Least Square (LS) Estimation Algorithm

The LS channel coefficients can be represented by

$$\hat{h}_{LS} = arg \max_{h} \left(y - xh\right)^{H} \left(y - xh\right) \tag{51}$$

Differentiating (51) with respect to the channel coefficients and equating to zero, the LS channel estimate can be represented as

$$\hat{h}_{LS} = x^{\#}y \tag{52}$$

where $x^{\#} = \left(x^{H}x\right)^{-1} x^{H}$ denotes the pseudo-inverse of x. Substituting Equation 44 in Equation

52, the following expression can be obtained for the LS channel estimate

$$\hat{h}_{LS} = h + \left(x^H x\right)^{-1} x^H n \qquad (53)$$

7.2 Minimum Mean Square Estimation (MMSE) Algorithm

An MMSE estimator treats the channel coefficients as random variables. Also this approach is independent of the channel coefficients; however, it depends on the moments of the channel coefficients. The MMSE channel estimator is defined as

$$\hat{h}_{MMSE} = arg \min_{h} \left[E\left\{ \left(h - \hat{h}\right)^H \left(h - \hat{h}\right) \right\} \right] \qquad (54)$$

Solving for Equation 54, the MMSE solution can be expressed as

$$\hat{h}_{MMSE} = R_{hh} \times c_p^* \times G_{mmse} \times y_p \qquad (55)$$

where $G_{mmse} = \left(c_p R_{hh} c_p^* + \sigma_n^2 I\right)^{-1}$, R_{hh} is the auto-covariance matrix of h and is assumed to be known, and σ_n^2 is the noise variance. Similar to Maximum a-posteriori Probability (MAP) algorithm, MMSE also requires prior information of the channel coefficients. Also MMSE is similar to MAP under the assumption that the channel coefficients are Gaussian. Hence, MMSE estimates the channel $\hat{h}_{MMSE}$ for the Gaussian channel $h(k)$.

7.3. Auto-Regressive (AR) Model

AR process can be used for modelling time-varying channels. It can be very well described by a hidden Markov Model and can be modelled by a multi-channel auto-regressive (AR) process of order P, generally defined as

$$h(t) = \sum_{l=0}^{L-1} ℎl(t)\delta(t - \tau_l(t)) \qquad (56)$$

where $\tau_l(t)$ is the delay, and $ℎl(t)$ is the complex amplitude of the $l - th$ multipath tap. In order to keep the complexity of the receiver at a reasonable level, we have preferred to use only the first order AR process to model the time-varying channel. The performance of this time-varying nature of the fading channel can be analyzed by assuming that the channel matrix varies according to the following first-order auto-regressive model

$$H_t = \sqrt{\alpha} H_{t-1} + \sqrt{1 - \alpha^2} w_t \qquad (57)$$

where H_t is the time-varying channel, w_t is the $N_t \times N_r$ matrix containing independent zero-mean Gaussian noise processes with variance ½. The coefficient α is related to the doppler spread f_d according to the first order approximation of Jakes channel model as $\alpha = J_0(2\pi f_d T_s)$ where $f_d T_s$ is used as the measure for fading rate. Typically its value falls in the range of (0.01-0.1).

8. ANTENNA SELECTION TECHNIQUES

Communication systems exploiting multiple antennas at the transmitter and/or the receiver are able to provide both data rates (capacity) and performance (BER). These two main advantages of the MIMO systems can be achieved in two different ways namely diversity methods and spatial multiplexing. Apart from the benefits achieved from using MIMO systems, they impose a few drawbacks as well. These include poor link reliability, little advantage of antenna diversity, and the need of sub-optimum detection interfaces at the receiver when large number of antennas is used. Another major problem that arises while using MIMO systems is the increased complexity, leading to increased cost, due to the need of multiple $N_t \left(N_r \right)$ RF chains. While deploying multiple antennas on RFID readers such high degree of hardware complexity becomes totally detrimental. Also, the ever rising desire of owning smaller and lighter RFID units without significant performance loss forces to devolve more processing burden on the transmit side. Therefore, considerable efforts have been put in exploring new MIMO systems that can significantly reduce this complexity but continue to provide similar capacity and performance improvements.

A promising technique to achieve this goal is based on selecting antennas at the transmitter and/or receiver. Antenna selection procedure may involve selecting a single antenna or multiple antennas and can be at the transmitter and/or receiver. The selection procedure involves selecting a subset of the transmit and/or receive antennas based on selection criteria including maximization of the received SNR and the channel capacity, similar to MIMO systems. Achieving a certain diversity order may be opted as the system design criteria. On the other hand, processing power of the receiver can also be accounted for while determining the number of active antennas that a system can support. Antenna Selection can be performed at either transmitter or receiver end or both depending on the cost involved and the required performance levels.

8.1. Receive Antenna Selection

When selecting single antenna at the receiver, by keeping track of the received power periodically, the signal of the receive antenna observing the largest instantaneous SNR can be selected and

fed to the RF chain for processing. Due to availability of only one RF chain, the problem of knowing the entire branch SNR arises (which is required for optimal selection). This problem can be solved by using preamble at the start of every transmitted frame and known pilot symbols be sent with this preamble. The receiver can exploit this preamble while scanning the antennas, finds the antenna with the highest channel gain and selects it for receiving the next data burst. For multiple antenna selection at the receiver where there are multiple RF chains available (but always less than the number of receive antennas) subset selection criteria can be followed where L_r branches with the highest SNR are chosen. The signals of the selected subset are combined.

8.2. Transmit Antenna Selection

For transmit antenna selection, a feedback path is required from the receiver to the transmitter. Single antenna selection at the transmitter is approximately similar to the single receive antenna selection i.e. the antenna with highest equivalent receive SNR is selected. For selecting multiple antennas at the transmitter, the most suitable L_t antennas out of N_t transmit antennas are chosen assuming that there is only one receive antenna and that there are L_t RF chains and N_t transmit antennas, and $L_t < N_t$. The phase and amplitude of the transmit signals have to be such that their superposition at the receiver provides the maximal receive SNR. In doing so, the L_t transmit antennas with highest channel gain are chosen. This is equivalent to beamforming over selected antennas and is known as hybrid maximal ratio transmission (MRT). Hybrid MRT requires that apart from knowing the L_t most suitable transmit antennas, the relative complex valued channel gains from each transmit antenna to the receiver should also be known, which in return requires more feedback as compared to selecting single transmit antenna.

8.3. Joint Transmit and Receive Antenna Selection

Antenna selection can be performed at both transmit and receive end simultaneously also. In this case selection diversity is applied at both ends. Assuming that there are N_t transmit and N_r receiver antennas, the transmit and receive ends have L_t and L_r RF chains. Thus space-time codes are used to provide diversity at the transmitter to transmit L_t parallel data streams. Joint transmit/receive antenna selection is a quite complex task as it requires to choose a subset of the rows and columns of H, where H is the $N_t \times N_r$ overall channel matrix, such as to maximize the sum of the squared magnitudes of transmit-receive channel gains. The problem that occurs is selecting the best receivers and then best transmitters may not necessarily result in the overall best possible choice. Thus efficient joint transmit/receive antenna selection requires systematic solution apart from exhaustive search for antenna selection.

Although Channel Estimation and Antenna Selection techniques helps in reliable data transfer and improve system performance and fight channel vulnerabilities, their application in RFID systems may still be limited due to many factors like the size of the RFID readers and tags, more complex DSP unit required at the tag readers for channel estimation and the feedback path involved for performing the task of antenna selection. However, a trade-off can be maintained between the cost involved and the BERs achieved depending on the sensitivity of the system towards the surroundings involved and the need for reliable information retrieval.

9. CONCLUSION

In this chapter, we provided an insight into the current state of technology in the field of wireless communication in RFID systems. By using smart antennas, channel impairments can be controlled for reliable information retrieval. We also conclude that there are various ways by adopting which the performance of any wireless system can be significantly improved. By combining various modulation and coding techniques, by efficient channel estimation and by considering the amount of diversity involved, new communication systems can be modelled. Antenna Selection is yet another technology that helps in improving the system's performance.

REFERENCES

Abbak, M., & Tekin, I. (2009). RFID coverage extension using microstrip patch antenna array. *IEEE Antennas and Propagation Magazine, 51*(1), 185–191. doi:10.1109/MAP.2009.4939065

Abou-Rjeily, C., & Fawaz, W. (2008). Space-time codes for MIMO ultra-wideband communications and MIMO free-space optical communications with PPM. *IEEE Journal on Selected Areas in Communications, 26*(6). doi:10.1109/JSAC.2008.080810

Adachi, F. (2001). Wireless past and future – Evolving mobile communications systems. *IEICE Transactions in Fundamentals. E (Norwalk, Conn.), 84-A*(1).

Alamouti, S. M. (1998). A simple transmit diversity technique for wireless communications. *IEEE Journal on Selected Areas in Communications, 16*, 1451–1458. doi:10.1109/49.730453

Angerer, C., Langwieser, R., Maier, G., & Rupp, M. (2009). Maximal ratio combining receivers for dual antenna RFID readers. In *Proceedings of International Microwave Workshop on Wireless Sensing, Local Positioning, and RFID.*

Angerer, C., Langwieser, R., & Rupp, M. (2010). Experimental performance evaluation of dual antenna diversity receivers for RFID readers. *Proceedings of the 3rd International EURASIP Workshop on RFID Technology.*

Angerer, C., & Markus, R. (2009). Advanced synchronisation and decoding in RFID reader receivers. *IEEE Radio and Wireless Symposium, RWS '09,* (pp. 59-62).

Baro, S., Bauch, G., & Hansmann, A. (2000). Improved codes for space-time trellis coded modulation. *IEEE Communications Letters, 4*(1), 20–22. doi:10.1109/4234.823537

Chen, H., & Wang, Z. J. (2010). Gains by a space-time-code based signaling scheme for multiple-antenna RFID tags. *23rd Canadian Conference on Electrical and Computer Engineering (CCECE),* (pp. 1-4).

Chen, Z., Yuan, J., & Vucetic, B. (2001). *An improved space-time trellis coded modulation scheme on slow Rayleigh fading channels.* IEEE.

Elo, M. (2007). *SISO to MIMO: Moving communications from single-input single-output to multiple-input multiple-output.* White Paper, Keithley Instruments.

Engel, J. (2002). DSP for RFID, circuits and systems. *45th Midwest Symposium on MWS-CAS-2002,* Vol. 2, (pp. II-227- II-230).

Foschini, G. J. (1996). Layered space-time architecture for wireless communication in a fading environment when using multi-element antennas. *Bell Labs Technical Journal, 1*(2), 41–59. doi:10.1002/bltj.2015

Foschini, G. J., & Gans, M. J. (1998). On limits of wireless communications in a fading environment when using multiple antennas. *Wireless Personal Communications, 6*(3), 311–335. doi:10.1023/A:1008889222784

Gesbert, D., Shafi, M., Shiu, D. S., Smith, P., & Naquib, A. (2003). From theory to practice: An overview of MIMO space-time codes. *IEEE Journal on Selected Areas in Communications, 21*(3), 281–302. doi:10.1109/JSAC.2003.809458

Goel, S., Abawajy, J. H., & Kim, T. (2010). Performance analysis of receive diversity in wireless sensor networks over GBSBE models. *Sensors (Basel, Switzerland), 10,* 11021–11037. doi:10.3390/s101211021

Griffin, J. D., & Durgin, G. D. (2008). Gains for RF tags using multiple antennas. *IEEE Transactions on Antennas and Propagation, 56*(2), 563–570. doi:10.1109/TAP.2007.915423

Guey, J. C., Fitz, M. P., Bell, M. R., & Kuo, W. Y. (1996). Signal design for transmitter diversity wireless communication systems over Rayleigh fading channels. *IEEE Vehicular Technology Conference,* (pp. 136–140).

He, W., Huang, Y., Wang, Z., & Zhao, Y. (2008). A circular polarization MIMO antenna system applied for RFID management. *8th International Symposium on Antennas, Propagation and EM Theory, ISAPE 2008,* (pp. 225-228).

Hourani, H. (2004/2005). *An overview of diversity techniques in wireless communication systems. Postgraduate Course in Radio Communications.* Helsinki University of Technology.

Ingram, M. A., Demirkol, M. F., & Kim, D. (2001). *Transmit diversity and spatial multiplexing for RF links using modulated backscatter.* International Symposium on Signals, Systems, and Electronics (ISSSE'01).

Kim, S. W., Ha, D. S., & Kim, J. H. (2001). *Performance gain of smart antennas with diversity combining at handsets for the 3GPP WCDMA system*. 13th International Conference on Wireless Communications (Wireless 2001), Calgary, Alberta, Canada.

Koswatta, R., & Karmakar, N. C. (2010). Moving average filtering technique for signal processing in digital section of UWB chipless RFID reader. *Microwave Conference Proceedings (APMC), Asia-Pacific,* (pp. 1304-1307).

Langwieser, R., Angerer, C., & Scholtz, A. L. (2010). A UHF frontend for MIMO applications in RFID. *Radio and Wireless Symposium (RWS), IEEE,* (pp. 124-127).

Lindskog, E., & Paulraj, A. (2000). A transmit diversity scheme for channels with intersymbol interference. In *Proceedings of International Conference on Communications,* (p. 1).

Lunglmayr, M., & Huemer, M. (2010). Least squares equalization for RFID. *2nd International Workshop on Near Field Communication,* (pp. 90-94).

Mazurek, G. (2008). Experimental RFID system with active tags. *International Conference on Signals and Electronic Systems, ICSES '08,* (pp. 507-510).

Nikitin, P. V., & Rao, K. V. S. (2007). Performance of RFID tags with multiple RF ports. In *Proceedings of the IEEE Antennas and Propagation Society International Symposium,* (pp. 5459–5462).

Nikitin, P. V., & Rao, K. V. S. (2008). *Antennas and propagation in UHF RFID systems*. IEEE International Conference on RFID, Las Vegas, USA, April.

Pawar, S. A., Raj Kumar, K., Elia, P., Vijay Kumar, P., & Sethuraman, B. A. (2009). Space–time codes achieving the DMD tradeoff of the MIMO-ARQ channel. *IEEE Transactions on Information Theory, 55*(7). doi:10.1109/TIT.2009.2021332

Sandhu, S., Heath, R., & Paulraj, A. (2001). *Space-time block codes versus space-time trellis codes*. IEEE.

Simon, M. K., & Alouini, M. S. (2000). *Digital communication over fading channels: A unified approach to performance analysis*. New York, NY: Wiley-Interscience Series in Telecommunications and Signal Processing. doi:10.1002/0471200697

Song, Y., Blostein, S. D., & Cheng, J. (2003). Exact outage probability for equal gain combining with cochannel interference in Rayleigh fading. *IEEE Transactions on Wireless Communications, 2,* 865–870. doi:10.1109/TWC.2003.816796

Tarokh, V., Jafarkhani, H., & Calderbank, A. R. (1999). Space-time block codes from orthogonal designs. *IEEE Transactions on Information Theory, 45,* 1456–1467. doi:10.1109/18.771146

Tarokh, V., Seshadri, N., & Calderbank, A. R. (1998). Space-time codes for high data rate wireless communication: Performance criteria and code construction. *IEEE Transactions on Information Theory, 44*(2), 744–764. doi:10.1109/18.661517

Telatar, I. E. (1999). Capacity of multi-antenna Gaussian channels. *European Transactions in Telecommunications, 10*(6), 585–595. doi:10.1002/ett.4460100604

Turkmani, A. M. D., Arowogolu, A. A., Jefford, P. A., & Kellent, C. J. (1995). An experimental evaluation of performance of two-branch space and polarization diversity schemes at 1800 MHz. *IEEE Transactions on Vehicular Technology, 44*(2), 318–326. doi:10.1109/25.385925

Vucetic, B., & Yuan, J. (2005). Space-time trellis codes. In *Space-time coding*. Chichester, UK: John Wiley & Sons, Ltd.

Wang, J. J. M., Winters, J., & Warner, R. (2007). *RFID system with an adaptive array antenna*. US Patent, No. 7212116.

Winters, J. H. (1987). On the capacity of radio communication systems with diversity in a Rayleigh fading environment. *IEEE Journal on Selected Areas in Communications*, *5*, 871–878. doi:10.1109/JSAC.1987.1146600

Yoon, C. S., Jeon, K. Y., & Cho, S. H. (2008). The performance enhancement of UHF RFID reader in multi-path fading environment using antenna diversity. In *Proceedings of the 23rd International Technical Conference on Circuits/Systems, Computers and Communications*, (pp. 1749–1752).

KEY TERMS AND DEFINITIONS

Antenna Selection: A low-cost low-complexity alternative to capture many of the advantages of MIMO systems.

Channel Estimation: Estimation for the instantaneous channel on a short-term basis as the conditions vary.

Diversity Combining: The technique applied to combine the multiple received signals of a diversity reception device into a single improved signal.

MIMO Systems: A system that use of multiple antennas at both the transmitter and receiver to improve communication performance.

Modulation: The process of varying one or more properties of a high-frequency periodic waveform, called the carrier signal, with a modulating signal which typically contains information to be transmitted.

Space-time Coding: A method employed to improve the reliability of data transmission in wireless communication systems using multiple transmit antennas.

Wireless Fading Channels: The deviation of the attenuation that a carrier-modulated telecommunication signal experiences over certain propagation media channels.

Chapter 7
Design of High Efficiency Power Amplifier for RFID Readers

Liming Gu
Nanjing University of Science and Technology, People's Republic of China

Yang Yang
Monash University, Clayton Campus, Australia

Shichang Chen
City University of Hong Kong, Hong Kong

Nemai Chandra Karmakar
Monash University, Clayton Campus, Australia

ABSTRACT

In RFID reader systems, power amplifier plays a critical rule for efficiency enhancement. A high efficiency power amplifier may not only increase the life expectancy of portable RFID devices but also reduce the reliance on heat sinks. Heat sinks usually occupy plenty of space and lead to packing difficulties. A well-designed power amplifier with high efficiency and output power may also increase the reading range of RFID and system reliability, especially for the applications requiring long reading range (e.g. vehicle tagging in complicated traffics) or in a lossy environment (e.g. in sensing in rainy weather). This chapter systematically introduces the typical power amplifiers classified as Class A, AB, B, E, and F. The principles of Class F are emphasized due to its outstanding performance in efficiency enhancement. A practical design example is also presented, and also some recent typical techniques for improving the performances of Class F power amplifier are summarized.

1. INTRODUCTION

In modern wireless communication systems, efficiency enhancement has become one of the most popular topics, which significantly attracts much attention of researchers. For radio frequency identification (RFID) systems (especially for the portable readers and active tags), high efficiency power amplifiers play a crucial rule due to the limited capacity of the portable battery. Moreover, it is usually expensive and difficult to improve the capacity of a battery, so the best way to prolong the lifetime of a portable RFID device is efficiency enhancement.

DOI: 10.4018/978-1-4666-1616-5.ch007

RFID is a contactless data-capturing technique, which uses radio frequency (RF) waves for automatic identification of objects. In terms of implementations, the applications of RFID can be categorized into five different groups (Glover & Bhatt, 2006): smart shelf, access control, tag and ship and track/trace. In different implementations, the RFID tags and reader systems have to be specially designed based on the conditions of the environments. A general RFID system usually consists of a tag system and a reader system. The tag system with active components and on-tag battery is known as active RFID, while the system with only passive circuits is known as chipless RFID. Active and chipless RFID are two classifications of RFID system distinguished by the criterion that whether the tag system is using a battery or not. For the RFID readers, both active and chipless systems have the same topologies of transmitter and receiver parts (Preradovic & Karmakar, 2007; Karmakar, 2010). The transmitter part is composed of a transmitting antenna, a power amplifier (PA), a bandpass filter and an oscillator, while the receiver part is composed of a receiving antenna, a bandpass filter, a low noise amplifier (LNA), a demodulator and base band processing units.

As a key component in RFID reader system, a power amplifier plays a significant rule of transmitting RF signals to the tag system. For an active tag system or a portable reader system, power added efficiency (PAE) of the power amplifier affects the lifetime of a battery. High PAE means an endurable performance of the system and that the system is more applicable for the long time sensing situations (for example of equipments tracking and patients monitoring in hospital [Cangialosi, Monaly, & Yang, 2007; Hakim, Renouf, & Enderle, 2006]). Moreover, high PAE can effectively reduce the requirements of heat sinks, which will significantly reduce the packaging complexity and cost.

The active device in a power amplifier can be a field-effect transistor (FET) or a bipolar-junction transistor (BJT). For convenience, this chapter only utilizes FET models. This chapter is suitable for those readers, who have certain elementary knowledge of power amplifiers and RFIDs. A summarized theory of the classical high efficiency power amplifiers - Class F is provided, including a practical design example. The readers will be benefit from the example which can be further applied during the development of future RFID system.

This chapter is organized as following: section 2 presents the background of the conventional RFID reader system and the general knowledge of power amplifier classification. In section 3, as a classical high efficiency power amplifier, Class F power amplifiers is introduced, including the principal theories, design methodologies and the latest development. In last section, a summary of the chapter is presented.

2. BACKGROUND

2.1 RFID Readers

RFID is an automated contact-less data capturing technique which can be found in more and more applications in numerous fields. The RFID system mainly includes two basic parts: RFID tag and RFID reader. An RFID reader is a device that is applied to interrogate an RFID tag. The reader emits radio waves via reader antenna to the tag system and the tag responds by sending back a unique identification data. Figure 1 shows the block diagram of a reader system. In a conventional RFID reader system, HF (high frequency) interface is used to enable RF signal to be transmitted and received. HF interfaces consists of two paths with signal flows from and to the antenna — one is transmitting path and the other is receiving path. In the transmitting path, the oscillator circuits generate a HF signal. However, the power of the generated signal is relatively low and the generated signal cannot be transmitted over a long

Figure 1. Block diagram of a RFID reader

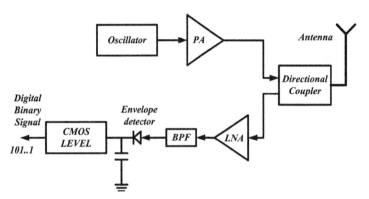

distance in the atmosphere. To ensure a reliable transmission link, the output power needs to be boosted to an adequate level. Therefore, we may see that the power amplifier is indispensable in RFID reader transmitting path.

RFID readers can be classified on various bases such as power supply, communication interface, mobility, tag interrogation, frequency response, supporting protocols. The RFID reader based on their power supply can be classified into two types of readers (Preradovic & Karmakar, 2007b): *a.* readers with power supply from the power network; and *b.* battery powered (BP) readers. For both cases, the efficiency is very important due to the consideration of heat dissipation and battery capacity, while the efficiency of power amplifier always determine the total efficiency of the reader system, so high efficiency power amplifier circuits are especially important in RFID reader system.

2.2 Power Amplifiers for RFID Readers

RFID systems have been widely used in various applications such as logistics (Gao & Yuen, 2011), management of library bookshelf (Lau, Yung, & Yung, 2010), secure banknote application (Preradovic & Karmakar, 2009), wireless health monitoring (Keinhorst et al., 2007), disease diagnosing (Occhiuzzi & Marrocco, 2010) and hospital

facility management (Jiao, Zhen, & Jiao, 2008). Along with more and more extensive applications of RFID systems, there is an increasing demand in efficiency enhancement of RFID readers due to the marketing requirements of low cost, low power consumption and low profile (especially for portable device). In RFID reader systems, power amplifiers consume the highest power in the transmitter architecture; therefore high output power amplifiers with high efficiency are highly desired and become a challenging task in most of RFID systems. Advanced integrated circuit technology based on very cheap and mature CMOS technology enables the RFID to be more and more popular (Kuo & Lusignan, 2001; Han et al., 2006; 2011; Gao et al., 2008; Khannur et al., 2008; Shim, Han, & Hong, 2008; Joo et al., 2010). Gao et al. (2008) reported a fully integrated on-chip CMOS power amplifier for RFID readers. By using Class E topology, this work enhanced the PAE of the power amplifier to 32.1% with an output power of 20 dBm. In another recently published work, Khannur et al. (Khannur et al., 2008) proposed a universal UHF RFID reader IC using CMOS technology in 2008. This system successfully realized the main RFID reader architecture into a 6×6 mm² chip including both transmitter and receiver parts. However, the power amplifier used in this system only presented a power added efficiency (PAE) of 16% which significantly reduced the overall performance of the chip based system.

During the same period, Shim et al. (Shim, Han, & Hong, 2008) presented another switch-mode power amplifier based on 0.25 μm CMOS process achieving a better PAE of 35% in UHF mobile RFID reader using binary-ASK. This work successfully realized RFID reader system on a chip with an enhanced PAE in comparison to (Khannur et al., 2008). However, still 65% of the power was dissipated due to the non-ideal switch operation of the common gate transistors. In 2010, a significant achievement of 56% PAE for a power amplifier with 0.18 μm CMOS process in UHF RFID readers was reported by Joo et al. (Joo et al., 2010). In this work, the Class E topology using fast transistor switching operation was combined with quasi four pair structure and integrated passive device (IPD) transformers. However, power amplifiers with higher PAE have already been realized using other process in different frequency bands for various applications (Warr et al., 2009; Wakejima et al., 2005; Lee & Jeong, 2007; Kim et al., 2011; Gao et al., 2006; Lee, Lee, & Jeong, 2008; Kim, Derickson, & Sun, 2007). Therefore, there is still a promising perspective for RFID engineers to implement these power amplifier technologies into RFID reader systems.

2.3 Classification of Power Amplifiers

Traditionally, power amplifier can be classified into four types according to the quiescent operation point or conduction angle; they are Class A, Class AB, Class B and Class C respectively. Class A is the most linear type compared with the other three types; where the conduction angle is 2π in radian. However, Class A power amplifier can only reach a maximum PAE of 50% (Cripps, 1999) with the optimal load impedance in an ideal situation. Class AB, B and C belong to the reduced conduction angle amplifier modes. The efficiency of these classes can be improved by reducing the conduction angle with less DC supply power. For example —the efficiency of

Class B can be improved to 78.5% (Cripps, 1999) maximally by decreasing the DC supply power with a factor of $\pi/2$ but with the same output as class A. For conduction angle less than π, corresponding to Class C condition, it has been demonstrated (Cripps, 1999) that the maximum theoretical efficiency of 100% can be achieved in a limiting deep Class C condition, where, current conduction angle dropping to zero. This operation in turn implies that no power is transferred to the external load, thus it is not a feasible solution in practice (Colantonio, Giannini, & Limiti, 2009). Other than the classifications mentioned above, Class D, Class E and Class F (or Class F^{-1}) are classical and well known high efficiency power amplifiers. By reducing the overlap between current and voltage waveforms in drain/collector of the transistor, the power dissipation in transistor will be significantly reduced, and accordingly the efficiency will be improved. Among these four classes, Class D and Class E are switch modes due to the switching operation of transistors, while Class F (or Class F^{-1}) belongs to harmonic control mode owing to the fact that the current waveform and voltage waveform are shaped, by harmonics, to half sinusoidal (square) and square (half sinusoidal) without any overlap between them in ideal situation. So the efficiency of them can reach theoretically up to 100% (Colantonio, Giannini, & Limiti, 2009). However, Class E amplifier requires fast switching driver signal which is not easy to realize in practice. Moreover, Class E amplifiers usually have scale problems with the trend toward lower-power technology with lower breakdown voltage (Colantonio, Giannini, & Limiti, 2009), especially for CMOS technology due to its low breakdown voltage (Lee, 1999). In contrast, the problems suffered by Class E is not applied to Class F. Due to these reasons, more researches have been recently addressed into Class F power amplifier development especially at higher frequency. Considering the difficulties of Class E power amplifiers in RFID reader designs, this chapter only focuses on the Class F power amplifiers.

The following section presents a detailed technical description of Class F power amplifiers, design techniques and a recently published design example (Chen & Xue, 2011) using compact microstrip resonant cell (CMRC) (Xue, Shum, & Chan, 2000) with high PAE.

3. CLASS F POWER AMPLIFIERS

3.1 Basic Theory

As stated above, RF power amplifiers are commonly categorized as Class A, AB, B, C, D, E, and F (F^{-1}). The classes of operation differ in the quiescent operation point, drain efficiency, linearity, and output power capability. One reason for the popularity of Class F as opposed to Class B or Class C is that Class F power amplifier can achieve higher efficiency while delivering higher output power (Colantonio, Giannini, & Limiti, 2009). It has gradually become the most popular research topics during the recent years (Negra et al., 2008; Sheikh et al., 2009; Kuroda, Ishikawa, & Honjo, 2010; Cipriani et al., 2010; Ishikawa & Hondo, 2011; Woo, Yang, & Kim, 2006) owing to its obvious advantages and feasibility in high frequency band. The benefits of flattening the bottom of the plate-voltage waveform were known in 1919 (Gao, 2006). After that, the early implementation was described by Tyler in 1958 (Tyler, 1958). A few years later, the first application of the Class F technique to UHF power amplifiers was proposed by Snider in 1967 (Snider, 1967). Raab demonstrated the achievable benefits in terms of output power and efficiency by using a limited number of harmonics with the aim to synthesize maximally flat voltage and current waveforms; maximum efficiency and output power (Raab, 1996; 1997; 2001). The detailed analysis procedure and results have become the classic theoretical basics for designing Class F power amplifiers. In (Colantonio, Giannini, & Limiti, 2009), a detailed analysis method of Class F power amplifier has been demonstrated.

In Figure 2, the ideal output voltage (v_{DS}) and current (i_D) waveforms at the drain of a Class F power amplifier are shown. With the condition of $V_{knee} = 0$V, the drain current and voltage waveforms can be described as (Colantonio, Giannini, & Limiti, 2009):

$$i_D(\theta) = \begin{cases} I_{Max} \times \cos(\theta); & when -\pi/2 \leq \theta \leq \pi/2 \\ 0; & otherwise \end{cases} \quad (1)$$

$$v_{DS}(\theta) = \begin{cases} 0; & when -\pi/2 \leq \theta \leq \pi/2 \\ 2V_{DD}; & otherwise \end{cases} \quad (2)$$

As can be seen from Figure 2, there is no overlap between $i_D(\theta)$ and $v_{DS}(\theta)$, therefore $P_{diss} = i_D(\theta) \times v_{DS}(\theta)$ equals to zero, indicating that there is no power dissipation in Class F power amplifier.

Expending equations (1) and (2) with Fourier series, the drain current and drain to source voltage can be expressed respectively as (Colantonio, Giannini, & Limiti, 2009):

$$i_D(\theta) = \sum_{n=0}^{\infty} I_n \cdot \cos(n\theta) \quad (3)$$

where,

$$I_n = \begin{cases} I_{Max}/\pi; & n = 0 \\ I_{Max}/2; & n = 1 \\ \frac{2 \cdot I_{Max}}{\pi} \frac{(-1)^{\frac{n}{2}+1}}{n^2-1}; & n = even \\ 0; & n = odd \end{cases} \quad (4)$$

and

$$v_{DS}(\theta) = \sum_{n=0}^{\infty} V_n \cdot \cos(n\theta) \quad (5)$$

Figure 2. Ideal output voltage (v_{DS}) and current (i_D) waveforms at the drain of a Class F power amplifier

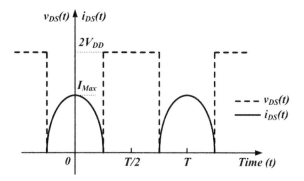

where,

$$V_n = \begin{cases} V_{DD}; & n = 0 \\ -4 \cdot V_{DD} / \pi; & n = 1 \\ 0; & n = even \\ \dfrac{4 \cdot V_{DD}}{\pi} \dfrac{(-1)^{\frac{n+1}{2}}}{n}; & n = odd \end{cases} \quad (6)$$

To the described ideal waveforms, typical harmonic impedance should be provided; these values can be calculated as the ratio between the respective Fourier components V_n and I_n (Colantonio, Giannini, & Limiti, 2009).

$$Z_n = \frac{V_n}{I_n} = \begin{cases} \dfrac{8}{\pi} \cdot \dfrac{V_{DD}}{I_{Max}}; & n = 1 \\ 0; & n = even \\ \infty; & n = odd \end{cases} \quad (7)$$

At the fundamental frequency, the load impedance is purely resistive, since it results from the perfectly phased waveforms, as anticipated, it results a short and open circuit for even and odd harmonics respectively.

Based on the analysis in (Colantonio, Giannini, & Limiti, 2009), the output power at fundamental frequency, DC power and drain efficiency of the Class F power amplifiers can be evaluated by

the following equations (Colantonio, Giannini, & Limiti, 2009):

$$P_{out,f} = \frac{I_{Max} \cdot \left(V_{DD} - V_{knee} \right)}{\pi^2} \cdot \frac{\Phi - \sin \Phi}{1 - \cos \dfrac{\Phi}{2}} \quad (8)$$

$$P_{DC} = V_{DD} \cdot \frac{I_{Max}}{2\pi} \cdot \frac{2 \sin \dfrac{\Phi}{2} - \Phi \cdot \cos \Phi}{1 - \cos \dfrac{\Phi}{2}} \quad (9)$$

$$\eta = \frac{2}{\pi} \cdot \frac{\Phi - \sin \Phi}{2 \sin \dfrac{\Phi}{2} - \Phi \cdot \cos \Phi} \quad (10)$$

where, V_{knee} is the knee voltage; Φ is the conduction angle.

In order to achieve 100% efficiency, infinite numbers of harmonics should be included. However, it is impossible to control all the harmonics simultaneously. It is required to realize the full amplitude at each harmonic frequency and this may lead to a more complicated circuit for harmonics control. Moreover, it may result in additional losses at the output and therefore degrades the efficiency of the circuit. The maximum efficiency and output power capability of Class F power amplifiers are shown in Table 1 (Raab, 2001). It is shown that the number of harmonics determines the maximum attainable efficiency. However, the

Table 1. Maximum efficiency and output power capability of class F power amplifiers (Raab, 2001)

m	η			
	n = 1	n = 3	n = 5	n = ∞
1	0.500	0.577	0.603	0.637
2	0.707	**0.817**	0.853	0.900
4	0.750	0.866	0.905	0.955
∞	0.785	0.907	0.948	1.000
P_{max}				
	n = 1	n = 3	n = 5	n = ∞
	0.125	0.144	0.151	0.159

m, n denote maximum order of harmonics in drain current and voltage, respectively.

harmonic impedances determine the output power capabilities. When m = 2, n = 3, the efficiency and output power increase significantly compared with the cases of m ≤ 2 and n ≤ 3 (except the case of m=2 and n=3). In contrast, when m ≥ 2 and n ≥ 3 (except the case of m=2 and n=3), the efficiency and output power do not increase significantly and, at the same time, the amount of increased efficiency and output power are counteracted by the loss of the additional higher order harmonics control circuits. Therefore, the case of m = 2, n = 3 is the most reasonable trade-off between high efficiency and output network complexity since both the size and power consumption are important factors in RFID reader systems. Thus in the practical designs of Class F power amplifiers, the considerations are mostly focused on the second and third harmonics (Schmelzer & Long, 2007; Kim, Oh, & Kim, 2007; Wu & Boumaiza, 2009; Lee & Park, 2009).

Noting that, not any FET device is suitable for Class F operation. To terminate the third harmonic impedance into a perfect open circuit, the shunt capacitance of the transistor itself must be very tiny. One of the keys to achieve high efficiency of Class F power amplifier is the ability to manipulate the required harmonics to shape the voltage and current waveforms at the drain. For this reason, devices with low unity gain, such

as Laterally Diffused Metal Oxide Semiconductor (LDMOS) are only suitable for low frequency Class F power amplifier design, since higher harmonics will be shorted to the ground and the ability to shape the waveforms at the drain will be lost (Wu & Boumaiza, 2009). More recent technology with both high unity gain and high device breakdown voltage, such as GaN, is a promising candidate for effective design of high frequency Class F power amplifiers (Schmelzer & Long, 2007; Wu & Boumaiza, 2009; Ramadan et al., 2010; 2011; Saad et al., 2009).

The gate bias point which dictates the output linearity will also have an impact on waveform shaping at the drain. For Class F operation, the transistor should be biased at deep Class AB or Class B mode depending on the amount of harmonic components desired (Wu & Boumaiza, 2009).

The basic frame of a Class F power amplifier is shown in Figure 3. The optimum load impedance Z_{opt} at the drain can be calculated from equation (7) in theory or from load pull simulation or measure in practical, and then the passive output matching and input matching circuits can be determined. While the load impedance at 2nd and 3rd harmonics should be terminated as shorted and open by harmonics control circuits. Generally, the typical harmonics control circuits (Woo, Yang, & Kim, 2006; Gao, 2006) are shown as in Figure 4.

In Figure 4 (a), the harmonics control circuit is realized by microstrip, which can be easily fabricated on the PCB, or in the chip at high frequency. In Figure 4 (b), only the 3rd harmonics are controlled at the drain of the transistor, other odd harmonics and all even harmonics are shorted to the ground by a parallel resonant circuit resonating at fundamental frequency f_0. In Figure 4 (c), at f_0, the drain sees a pure resistive load of Z_{opt}, and the quarter-wave transformer transforms the 50 Ohm load to Z_{opt} at the drain. Z_{opt} can be determined from load-pull measurement or simulation. Due to the quarter-wave transformer, the drain sees a short circuit at even harmonics;

Figure 3. The basic frame of Class F power amplifier

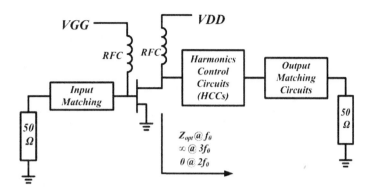

Figure 4. Typical harmonics control circuits in Class F power amplifiers

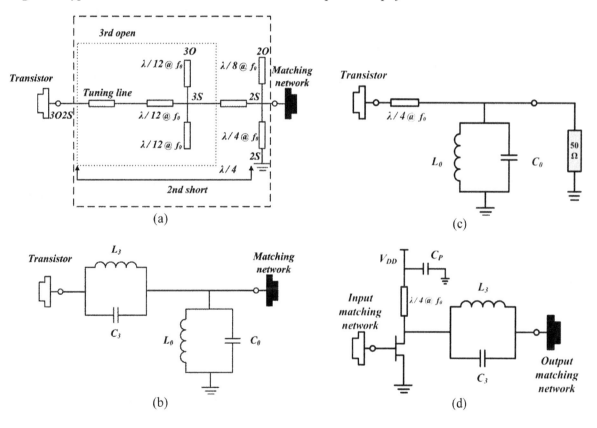

conversely, at odd harmonics, the drain sees an open circuit. In Figure 4 (d), at the drain, the DC feed is seen to be an open circuit at f_0, while the even harmonics are shorted to the ground by the DC feed. The parallel resonant circuit resonating at $3f_0$ forms a bandstop filter to rejects the 3rd harmonics to the drain.

3.2 Design Procedure and Example

A. Design Procedure

Based on the detailed analysis of Class F power amplifier in section 3.1, a design procedure of a

Figure 5. The design procedure of a basic Class F power amplifier

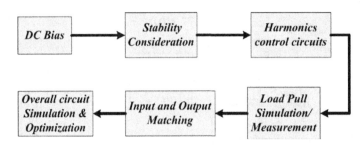

basic Class F power amplifier can be summarized in Figure 5.

All the steps as shown in Figure 5 can be stated as:

- *Step 1*: DC bias selection. At the beginning of the power amplifier design, the operation point of the FET (DC supplies, V_{DD} and V_{GG}) has to be determined according to its simulated or measured DC features. Considering the output power and the desired amount of the harmonic components, deep Class AB and Class B are usually selected as the best DC bias mode.
- *Step 2*: Stability consideration. To eliminating the oscillation of the transistor after DC provided, the stabilizing components should be added into the circuit to make the stability factor be larger than 1 (Gonzalez, 1996).
- *Step 3*: Design of harmonics control circuits. The basic frames of harmonics control circuits have been listed in Figure 4.
- *Step 4*: Load pull simulation. In practical design, the optimum load impedance for maximum output power or maximum PAE at the fundamental frequency is usually complex impedance, which can be obtained by the load pull simulation or measurement.
- *Step 5*: Input and output matching. The power matching should be adopted as the output matching method to maximize the output power. The standard impedance (50

Ohm) at the output port should be matched to the optimum impedance which can be obtained in *step 4*; while at the input port; the conjugated matching should be adopted to minimize the reflection of the input signal power.
- *Step 6*: Overall circuit simulation and optimization.

B. Design Example

In this section, a succinct Class F power amplifier with compact microstrip resonant cell (CMRC) is introduced (Chen & Xue, 2011). The CMRC is a section of a microstrip transmission line etched with incorporated band gap structure (Xue, Shum, & Chan, 2000). It exhibits typical band-stop characteristic which can be utilized to reject undesired frequencies. In this particular design, the CMRC is exploited as harmonics control circuit to reflect the 3rd harmonic. In addition, its intrinsic slow-wave effect leads to a size reduction of the circuit.

Figure 6 shows the schematic of the whole Class F power amplifier. The CMRC is inserted between the transistor and the output fundamental matching network. As the operation frequency f_0 of the Class F power amplifier is set to 2.4GHz, so the stop band of CMRC should be centered at the 3rd harmonic frequency 7.2GHz by carefully tuning the dimensions of CMRC. The detailed structure and performances of the CMRC are shown in Figure 7. From the measured results, the insertion loss at 2.4GHz is less than 0.5dB, then the CMRC acts as a transmission line with

Figure 6. Schematic of the succinct Class F power amplifier with CMRC

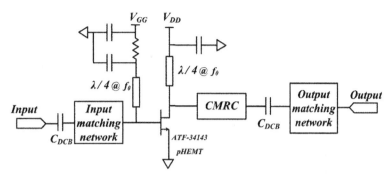

characteristic impedance of 50Ω; meanwhile, the rejection at 7.2GHz is about -20dB, the 3rd harmonic can be reflected and by controlling the phase, the drain thus sees an open circuit. Besides, the size of the harmonics control circuit can be greatly reduced, the physical length is only 29° in this particular design, but its electrical length is 90° due to slow-wave effect. A quarter-wavelength line at the fundamental frequency is employed to bias the drain of the transistor, which also behaves as a short circuit for the 2nd harmonic, and open circuit for third harmonic respectively. Figure 8 shows the simulated voltage and current waveforms at the drain of the transistor. The voltage and current look like a square and half sinusoid.

In this design, the transistor is biased at Class B mode. Figure 9 shows the photograph of the fabricated Class F power amplifier with CMRC for 3rd harmonic suppression. The measured performances are shown in Figure 9; the maximum PAE value can reach up to 74% near the 1 dB compression point.

3.3 Recent Development of the Class F Power Amplifiers

The growing demand of communication devices speed up the development of Class-F power amplifier. Recently, many techniques have been proposed for improving the performances of Class F power amplifiers with hybrid integrated circuit (HIC) process (Lee, Lee, & Jeong, 2008; Chen & Xue, 2011; Ramadan et al., 2010; 2011;

Saad et al., 2009; Ingruber et al., 1998; Nemati et al., 2009; 2011) and RF CMOS process (Kuo & Lusignan, 2001; Fortes & do Rosario, 2001; Zhe, A'ain, & Kordesch, 2007; Javidan, Torkzadeh, & Atarodi, 2008; Bozani & Sinha, 2009; Liao et al., 2009; Carls et al., 2009).

A. Class F Power Amplifiers with HIC Process

As we all know, the parasitic inductor and capacitor exist in either the commercial transistors or the bare transistor dies, as shown in Figure 11 (a) (Lee, Lee, & Jeong, 2008). The capacitor conducts most of the high order harmonics to the ground, while the parasitic inductor and capacitor at the drain output make it nearly impossible to make even harmonics short-circuited and odd harmonics open-circuited. In practical designs, in order to gain higher PAE and output power, the parasitic inductor and capacitor must be taken into consideration in the design of Class F power amplifiers. In (Lee, Lee, & Jeong, 2008), Lee, et al. used a simple and effective compensation circuit consisting of a series capacitor and a shunt inductor (as shown in Figure 11 (b)) to compensate the internal parasitic components. Hence the drain efficiency was improved by 20%. In the basic Class F power amplifiers, the output harmonics at the drain are always concerned, while, if there is 2nd harmonics injection at the gate port of transistors, as shown in Figure 12, the efficiency will be improved further more (however, the efficiency

Figure 7. (a) The detailed structure of the CMRC; (b) The simulated and measured performances of the CMRC

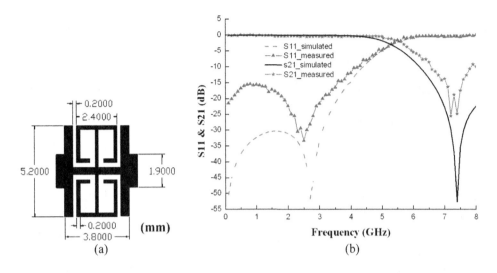

(a)   (b)

Figure 8. Simulated waveforms of the voltage and current at the drain

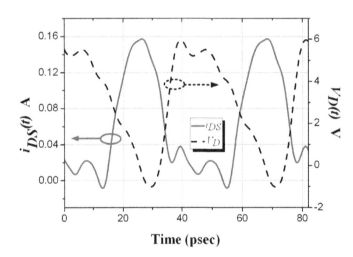

Figure 9. Photograph of the fabricated Class F power amplifier with CMRC for 3rd harmonic suppression

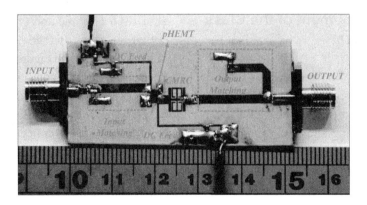

Figure 10. The measured performances of the Class F power amplifier

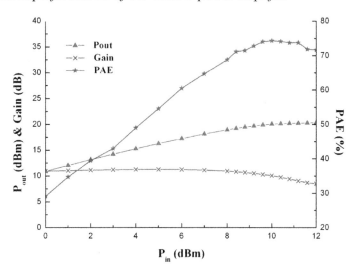

will be decreased by the 2nd harmonics injection in Class F[-1] condition) (Ramadan et al., 2010). The principle of this method is that the injection of 2nd harmonics at the gate will shape the input waveform to be a quasi-half sine wave, thus it will reduce the conduction angle and efficiently improve the PAE. The quasi half sine wave can be obtained using appropriate proportions of the fundamental component and the second harmonic, as introduced in (Ramadan et al., 2010). However, the problem is the complexity of its design, which may also introduce some losses. An alternative way of shaping the input waveform is to use a Class F[-1] power amplifier as the 1st stage and a Class F power amplifier as the 2nd stage, as presented in Figure 13. The Class F[-1] power amplifier can produce the output voltage as a half sinusoidal waveform, which is fed directly into the input of the 2nd stage in Class F situation (Ramadan et al., 2011). In order to achieve higher PAE of Class F, the researchers have built some dedicated and precise large signal models of the transistor (Saad et al., 2009; Ingruber et al., 1998; Nemati et al., 2009), while they have also paid more attention on the bare-die mounting techniques and considered the optimum load impedance in higher order (2nd and 3rd) harmonics. Meanwhile, a succinct output matching circuit was adopted to meet the imped-

ance matching at fundamental, 2nd harmonic and 3rd harmonic frequencies respectively. A peak PAE of 80% was measured at 3.5 GHz with an output power of 38.7 dBm and power gain of 15.5 dB (Nemati et al., 2011).

B. Class F Power Amplifiers with RF CMOS Process

CMOS is preferred with respect to the low costs in mass fabrication and the ability for digital baseband integration. Due to the requirement of high breakdown voltage to the devices using Class E power amplifiers, Class F power amplifiers are more suitable for low breakdown voltage CMOS technology (Kuo & Lusignan, 2001). Based on a 0.6 μm CMOS standard technology (Forbes & do Rosario, 2001), an integrated Class-F power amplifier with a maximum efficiency of 42% and 200 mW output power at 1.9 GHz and a 3-V power supply has been reported. In 2007 (Zhe, A'ain, & Kordesch, 2007), an integrated two-stage Class-F power amplifier with standard 0.18 μm RF CMOS process was proposed. This work successfully achieved an output power of 18.9 dBm, a maximum PAE of 42% and a maximum gain of 19.7 dB measured with DC supply of 3V. In order to alleviate the disadvantage of low breakdown

Figure 11. (a) The internal parasitic circuit in transistor; (b) The external compensation circuit

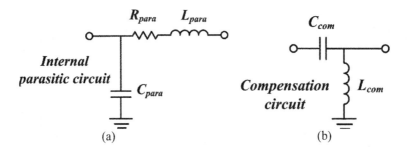

Figure 12. Input and output harmonics control scheme

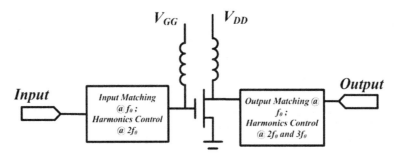

Figure 13. A two-stage cascaded power amplifier

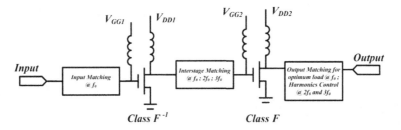

voltage and poor features of passive device, a thick oxide cascode CMOS PA with modified matching network configuration was presented in (Javidan, Torkzadeh, & Atarodi, 2008). The proposed Class F power amplifier with 0.18 µm CMOS process exhibited a PAE of 56% at 900MHz and maximum output power of 29.8 dBm. In (Bozani & Sinha, 2009), the design flow for a Class F CMOS power amplifier was presented including a spiral inductor search algorithm. This algorithm can be used to generate layouts of inductors with Q-factors optimized at a desired frequency. As is well known, at low operation frequency, the passive harmonic control circuit is difficult to be integrated into

a chip. Due to this reason, a recent work (Liao et al., 2009) implemented the harmonic control circuit and output matching circuit on an off-chip PCB. The measured PAE of the presented 2.6 GHz Class F power amplifier using 0.18 µm RF CMOS process was more than 24.4% with 1 dB compression point P_{1dB} = 20.2 dBm and power gain G_p = 13.1 dB at a low drain voltage of 1.8 V. The adjacent channel power ratio (ACPR) was below -33 dBc at 5 dBm input power level. Similar to (Bozani & Sinha, 2009), a fully integrated 5–6 GHz Class-F power amplifier in 0.18 µm CMOS was presented in (Carls et al., 2009). At a V_{dd} of 1.5 V, a PAE up to 42.7% and a P_{1dB} of 16.2

dBm were measured. By tuning V_{dd} to 1.9 V, the performances of P_{1dB} =18.4 dBm, PAE = 42.8% and gain = 8.8 dB were achieved.

4. CONCLUSION

In this chapter, a typical structure of the traditional RFID reader system and its basic principle of operation have been presented in the introduction and background sections. At the same time, the importance of power amplifier in RFID has been emphasized in terms of high efficiency and output power. Also, a comprehensive study of various power amplifiers in class A, AB, B, C, E and F has been given theoretically. Then, a classical type (Class F) of high efficiency power amplifiers was introduced in section 3. A practical example for Class F power amplifier with CMRC was then given. Finally, the latest developments of high efficiency Class F power amplifier with HIC and CMOS process are discussed respectively. This chapter is useful for RFID engineers who are interested in employing high efficiency power amplifiers for RFID applications. By using the high output power amplifiers, the RFID reading range can be effectively extended. The high efficiency power amplifier can significantly reduce the reliance on battery for portable applications (e.g. portable RFID reader designs). Moreover, by enhancing the efficiency of power amplifiers, the dependence on heat sinks can be markedly alleviated so that the system complexity is accordingly reduced. The techniques of enhancing efficiency and output power in power amplifier can be accordingly applied in different RFID applications.

REFERENCES

Božani, M., & Sinha, S. (2009). Design flow for CMOS based Class-E and Class-F power amplifiers. *IEEE AFRICON, 2009*, 1–6. doi:10.1109/ AFRCON.2009.5308187

Cangialosi, A., Monaly, J. E., & Yang, S. C. (2007). Leveraging RFID in hospitals: Patient life cycle and mobility perspectives. *IEEE Communications Magazine, 45*, 18–23. doi:10.1109/ MCOM.2007.4342874

Carls, J., Ellinger, F., Joerges, U., & Krcmar, M. (2009). Highly-efficient CMOS C-band Class-F power amplifier for low supply voltages. *Electronics Letters, 45*, 1240–1241. doi:10.1049/ el.2009.1252

Chen, S., & Xue, Q. (2011). A class-F power amplifier with CMRC. *IEEE Microwave and Wireless Components Letters, 21*, 31–33. doi:10.1109/ LMWC.2010.2091265

Cipriani, E., Colantonio, P., Giannini, F., & Giofre, R. (2010). Theoretical and experimental comparison of Class F vs. Class F^{-1} PAs. *Microwave Integrated Circuits Conference* (pp. 428-431).

Colantonio, P., Giannini, F., & Limiti, E. (2009). *High efficiency RF and microwave solid state power amplifiers.* John Wiley & Sons, Publication

Cripps, S. C. (1999). *RF power amplifiers for wireless communication.* Artech House.

Fortes, F., & do Rosário, M. J. (2001). A second harmonic Class-F power amplifier in standard CMOS technology. *IEEE Transactions on Microwave Theory and Techniques, 49*, 1216–1220. doi:10.1109/22.925529

Gao, B., & Yuen, M. F. (2011). Passive UHF RFID packaging with electromagnetic band gap (EBG) material for metallic objects tracking. *IEEE Transactions on Components . Packaging and Manufacturing Technology, 1*, 1140–1146. doi:10.1109/TCPMT.2011.2157150

Gao, S. (2006). High-efficiency class F RF/microwave power amplifiers. *IEEE Microwave Magazine, 7*, 40–48. doi:10.1109/MMW.2006.1614233

Gao, S., Butterworth, P., Sambell, A., Sanabria, C., Xu, H., Heikman, S., et al. (2006). Microwave Class-F and inverse Class-F power amplifiers designs using GaN technology and GaAs pHEMT. *36th European Microwave Conference* (pp. 1719-1722).

Gao, T., Chi, B., Zhang, C., & Wang, Z. (2008). Design and analysis of a highly integrated CMOS power amplifier for RFID reader. *11th IEEE Singapore International Conference on Communication Systems* (pp. 1480-1483).

Glover, B., & Bhatt, H. (2006). *RFID essentials.* O'Reilly Media, Inc.

Gonzalez, G. (1996). *Microwave transistor amplifiers: Analysis and design* (2nd ed.). Prentice Hall.

Hakim, H., Renouf, R., & Enderle, J. (2006). Passive RFID asset monitoring system in hospital environments. *IEEE 32nd Annual Northeast Bioengineering Conference,* (pp. 217-218).

Han, J., Kim, Y., Park, C., Lee, D., & Hong, S. (2006). *A fully-integrated 900-MHz CMOS power amplifier for mobile RFID reader applications.* IEEE Radio Frequency Integrated Circuit Symposium.

Han, S., Kim, M., Kim, H., & Yang, Y. (2011). A CMOS power amplifier for UHF RFID reader systems. *13th International Conference on Advanced Communication Technology* (pp. 57-60).

Ingruber, B., Pritzl, W., Smely, D., Wachutka, M., & Magerl, G. (1998). High efficiency harmonic-controlled amplifier. *IEEE Transactions on Microwave Theory and Techniques, 46,* 857–863. doi:10.1109/22.681213

Ishikawa, R., & Honjo, K. (2011). Distributed class-F/inverse class-F circuit considering up to arbitrary harmonics with parasitics compensation. *IEEE MTT-S International Microwave Workshop Series on Innovative Wireless Power Transmission: Technologies, Systems, and Applications* (pp. 29-32).

Javidan, J., Torkzadeh, P., & Atarodi, M. (2008). A 1W, 900MHz Class-F RF power amplifier with 56% PAE in 0.18-um CMOS technology. *2008 International Conference on Microelectronics* (pp. 90-93).

Jiao, Y.-Y., Zhen, Y.-W., & Jiao, R. J. (2008). Hospital linens inventory control re-engineering based on RFID. *IEEE Conference on Cybernetics and Intelligent Systems* (pp. 612-617).

Joo, T., Lee, H., Shim, S., & Hong, S. (2010). CMOS RF power amplifier for UHF stationary RFID reader. *IEEE Microwave and Wireless Components Letters, 20,* 106–108. doi:10.1109/LMWC.2009.2038552

Karmakar, N. C. (2010). *Handbook of smart antennas for RFID systems* (pp. 13–52). doi:10.1002/9780470872178

Keinhorst, T., Tsigkourakos, P., Yaqoob, M. A., van Driel, W. D., & Zhang, G. Q. (2007). Test and health monitoring of microelectronics using RFID. *The 8th International Conference on Electronic Packaging Technology* (pp. 1-6).

Khannur, P. B., Chen, X., Yan, D. L., Shen, D., Zhao, B., & Raja, M. K. (2008). A universal UHF RFID reader IC in 0.18-μm CMOS technology. *IEEE Journal of Solid-state Circuits, 43,* 1146–1155. doi:10.1109/JSSC.2008.920355

Kim, B., Derickson, D., & Sun, C. (2007). A high power, high efficiency amplifier using GaN HEMT. *Asia-Pacific Microwave Conference* (pp. 1-4).

Kim, J. H., Jo, G. D., Oh, J. H., Kim, Y. H., Lee, K. C., & Jung, J. H. (2011). Modeling and design methodology of high-efficiency Class-F and Class-F^{-1} power amplifiers. *IEEE Transactions on Microwave Theory and Techniques, 59,* 153–165. doi:10.1109/TMTT.2010.2090167

Kim, J.-Y., Oh, D.-S., & Kim, J.-H. (2007). Design of a harmonically tuned class-F power amplifier. *Asia-Pacific Microwave Conference* (pp. 1 – 4).

Kuo, T.-C., & Lusignan, B. (2001). *A 1.5W Class-F RF power amplifier in 0.2 μm CMOS technology.* IEEE International Solid-State Circuits Conference.

Kuo, T.-C., & Lusignan, B. (2001). *A 1.5W Class-F RF power amplifier in 0.2 μm CMOS technology.* IEEE International Solid-State Circuits Conference.

Kuroda, K., Ishikawa, R., & Honjo, K. (2010). Parasitic compensation design technique for a C-band GaN HEMT Class-F amplifier. *IEEE Transactions on Microwave Theory and Techniques, 58,* 2741–2750. doi:10.1109/TMTT.2010.2077951

Lau, P.-Y., Yung, K. K.-O., & Yung, E. K.-N. (2010). A low-cost printed CP patch antenna for RFID smart bookself in library. *IEEE Transactions on Industrial Electronics, 57,* 1583–1589. doi:10.1109/TIE.2009.2035992

Lee, C., & Park, Y. (2009). Design of compact-sized Class-F PA for wireless handset applications. *IEEE MTT-S International Microwave Symposium Digest* (pp. 405 – 408).

Lee, T. H. (1999). *The design of CMOS radio-frequency integrated circuits.* Cambridge.

Lee, Y.-S., & Jeong, Y.-H. (2007). A high-efficiency Class-E GaN HEMT power amplifier for WCDMA applications. *IEEE Microwave and Wireless Components Letters, 17,* 622–624. doi:10.1109/LMWC.2007.901803

Lee, Y.-S., Lee, M.-W., & Jeong, Y.-H. (2008). High-efficiency Class-F GaN HEMT amplifier with simple parasitic-compensation circuit. *IEEE Microwave and Wireless Components Letters, 18,* 55–57. doi:10.1109/LMWC.2007.912023

Liao, H.-Y., Chen, J.-H., Chiou, H.-K., & Wang, S.-M. (2009). Harmonic control network for 2.6 GHz CMOS Class-F power amplifier. *IEEE International Symposium on Circuits and Systems* (pp.1321-1324).

Negra, R., Sadeve, A., Bensmida, S., & Ghannouchi, F. M. (2008). Concurrent dual-band Class-F load coupling network for applications at 1.7 and 2.14 GHz. *IEEE Transactions on Circuits and Systems II, 55,* 259–263. doi:10.1109/TCSII.2008.918993

Nemati, H. M., Fager, C., Thorsell, M., & Herbert, Z. (2009). High-efficiency LDMOS power-amplifier design at 1 GHz using an optimized transistor model. *IEEE Transactions on Microwave Theory and Techniques, 5,* 1647–1654. doi:10.1109/TMTT.2009.2022590

Nemati, H. M., Saad, P., Fager, C., & Andersson, K. (2011). High-efficiency power amplifier. *IEEE Microwave Magazine, 12,* 81–84. doi:10.1109/MMM.2010.939314

Occhiuzzi, C., & Marrocco, G. (2010). The RFID technology for neurosciences: feasibility of limbs' monitoring in sleep diseases. *IEEE Transactions on Information Technology in Biomedicine, 14,* 37–43. doi:10.1109/TITB.2009.2028081

Preradovic, S., & Karmakar, N. C. (2007). *Modern RFID reader.* Retrieved on 6th September, 2011, from http://mwjournal.com/Article/Modern_RFID_Readers/AR_4830/

Preradovic, S., & Karmakar, N. C. (2007). Modern RFID readers – A review. *The 4th Conference on Electrical and Computer Engineering,* (pp. 100-103).

Preradovic, S., & Karmakar, N. C. (2009). Design of fully printable chipless RFID tag on flexible substrate for secure banknote applications. *The 3rd International Conference on Anti-counterfeiting, Security, and Identification in Communication* (pp. 206-210).

Raab, F. H. (1996). An introduction to class-F power amplifiers. *RFID Design, 19,* 79–84.

Raab, F. H. (1997). Class-F amplifiers with maximally flat waveforms. *IEEE Transactions on Microwave Theory and Techniques, 45,* 2007–2012. doi:10.1109/22.644215

Raab, F. H. (2001). Maximum efficiency and output of Class-F power amplifiers. *IEEE Transactions on Microwave Theory and Techniques, 49,* 1162–1166. doi:10.1109/22.925511

Ramadan, A., Reveyrand, T., Martin, A., Nebus, J. M., Bouysse, P., & Lapierre, L. (2010). Experimental study on effect of second-harmonic injection at input of classes F and F-1 GaN power amplifiers. *Electronics Letters, 46,* 570–572. doi:10.1049/el.2010.0392

Ramadan, A., Reveyrand, T., Martin, A., Nebus, J.-M., Bouysse, P., & Lapierre, L. (2011). Two-stage GaN HEMT amplifier with gate–source voltage shaping for efficiency versus bandwidth enhancements. *IEEE Transactions on Microwave Theory and Techniques, 59,* 699–706. doi:10.1109/TMTT.2010.2095033

Saad, P., Nemati, H. M., Thorsell, M., Andersson, K., & Fager, C. (2009). An inverse class-F GaN HEMT power amplifier with 78% PAE at 3.5 GHz. *EuMC 2009 European* (pp. 496-499).

Schmelzer, D., & Long, S. I. (2007). A GaN HEMT Class F amplifier at 2 GHz with > 80% PAE. *IEEE Journal of Solid-state Circuits, 42,* 2130–2136. doi:10.1109/JSSC.2007.904317

Sheikh, A., Roff, C., Benedikt, J., Tasker, P. J., Noori, B., Wood, J., & Aaen, P. H. (2009). Peak Class F and inverse Class F drain efficiencies using Si LDMOS in a limited bandwidth design. *IEEE Microwave and Wireless Components Letters, 19,* 473–475. doi:10.1109/LMWC.2009.2022138

Shim, S., Han, J., & Hong, S. (2008). A CMOS RF polar transmitter of a UHF mobile RFID reader for high power efficiency. *IEEE Microwave and Wireless Components Letters, 18,* 635–637. doi:10.1109/LMWC.2008.2002490

Snider, D. M. (1967). A theoretical analysis and experimental confirmation of the optimally loaded and over-driven RF power amplifier. *IEEE Transactions on Electron Devices, 14,* 851–857. doi:10.1109/T-ED.1967.16120

Tyler, V. J. (1958). A new high-efficiency high power amplifier. *Marconi Review, 2,* 96–109.

Wakejima, A., Asano, T., Hirano, T., Funabashi, M., & Matsunaga, K. (2005). C-band GaAs FET power amplifiers with 70-W output power and 50% PAE for satellite communication use. *IEEE Journal of State Circuits, 40,* 2054–2060. doi:10.1109/JSSC.2005.854596

Warr, P. A., Morris, K. A., Watkins, G. T., Horseman, T. R., Takasuka, K., & Ueda, Y. (2009). A 60% PAE WCDMA handset transmitter amplifier. *IEEE Transactions on Microwave Theory and Techniques, 57,* 2368–2377. doi:10.1109/TMTT.2009.2029021

Woo, Y. Y., Yang, Y., & Kim, B. (2006). Analysis and experiment for high-efficiency class-f and inverse class-f power amplifiers. *IEEE Transactions on Microwave Theory and Techniques, 54,* 1969–1974. doi:10.1109/TMTT.2006.872805

Wu, D. Y.-T., & Boumaiza, S. (2009). *10W GaN inverse Class F PA with input/output harmonic termination for high efficiency WiMAX transmitter.* 10th Annual IEEE Wireless and Microwave Technology Conference.

Xue, Q., Shum, K. M., & Chan, C. H. (2000). Novel 1-D microstrip PBG cells. *IEEE Microwave Wireless Components Letters, 10,* 403–405.

Zhe, H. M., A'ain, A. K. B., & Kordesch, A. V. (2007). Two stage integrated Class-F RF power amplifier. *International Symposium on Integrated Circuits* (pp. 108-110).

Section 3
Chipless RFID Tags

Chapter 8
Mastering the Electromagnetic Signature of Chipless RFID Tags

Smail Tedjini
Grenoble-INP/LCIS, France

Etienne Perret
Grenoble-INP/LCIS, France

Arnaud Vena
Grenoble-INP/LCIS, France

Darine Kaddour
Grenoble-INP/LCIS, France

ABSTRACT

The rapid development in wireless identification devices and subsequent applications is at the origin of intensive investigations in order to fulfill various constraints that can exist when implementing applications in practice. Chipless technologies have many advantages. They are fundamentally wireless and powerless devices, and can be all passive components, which potentially means infinite lifetime. However, chipless technology is still in its infancy age, even if it is the most effective for cost reduction. One of the most important features of chipless is coding capacity and ways to imprint it into the device. This chapter will review and discuss various coding techniques. It will address a comparison of the most relevant coding techniques. For sake of clarity some global parameters that can be used as figure of merit will be introduced and applied to compare different practical chipless tags.

1. INTRODUCTION

The history of Radio Frequency IDentification (RFID), birth and development, are described in several relevant publications (Landt, 2001; Finkenzeller, 2004). The principle of RFID communication was clearly explained in an IRE publication by H. Stockman in 1948 (Stockman, 1948). Even if it is generally said that the first application of RFID is the Identify Friend or Foe (IFF) system introduced by Watson-Watt, the first real device that we can consider as the ancestor of modern tags is the "thing" designed by Leon Theremin (Glinsky, 2005). The later is

DOI: 10.4018/978-1-4666-1616-5.ch008

a passive spying device that was embedded in a craved wooden plaque of the US Great Seal and offered by Russians to U.S. Ambassador as a « gesture of friendship » at the end of World War II. The "thing" is composed of an electromagnetic resonant cavity coupled to a monopole antenna. One of the walls of the electromagnetic cavity is transformed into a membrane acting as a microphone. Figure 1 briefly describes the "thing" and its main components.

The communication principle of the "thing" is straightforward. Under a continuous wave (CW) electromagnetic illumination (actually @ 330 MHz) and when the microphone detects some sounds, the resonance frequency of the cavity will change. Consequently, the reflected signal by the monopole antenna will be amplitude modulated by the detected sounds around the "thing".

It is quite amazing to remark that the first RFID device, ever developed, was in the same time a sensor (microphone) and a chipless tag (no IC, neither digital communication protocol). Wireless sensing and chipless tags are currently topics of great interest under consideration in numerous applications where powerless and robustness features are highly desired.

Last decades, roughly since the mid of 1960's, very effective IC based tags have been developed and implemented in numerous applications, from the most simple like the Electronic Surveillance Article (EAS) to the more sophisticated for Internet Of Things application purposes (Giusto, Iera, Morabito & Atzori, 2010). Nowadays the vast majority of RFID tags (or transponders) are usually comprised of an IC chip and an antenna. They are used or under consideration in a large variety of domains and thousands of applications. However, despite the success of IC based tags, there are some drawbacks inherent to their manufacturing process, economic cost, security and privacy of data, electromagnetic and mechanic robustness in specific applications.. On the other hand, IC based tags are often wrongly compared to barcode, which is very challenging when com-

Figure 1. The «Thing», first RFID chipless tag. operatingo at 330 MHz, it was used as a sound sensor

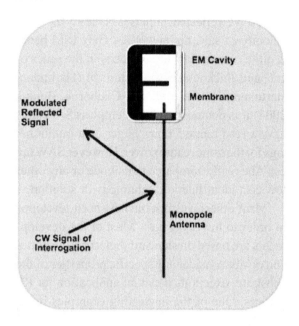

paring the costs of both identification solutions. Therefore, several research projects have been developed towards the concept of chipless RFID tags with no ICs, known also as RF barcode by some authors (Preradovic, Karmakar & Balbin, 2010). Modern chipless tags have been developed since the mid of 1990's. They are usually passive and exhibit very low cost that is comparable to barcode. On the other hand, contrary to the traditional chip-based tags where the read range is limited by the sensitivity of the tag (i.e. minimum power for chip activation), the read range of the chipless tags is limited by the reader sensitivity leading to larger read range

Perhaps the most known chipless example was introduced by RFSAW® (Hartmann, 2002). Based on the Surface Acoustic Wave (SAW) properties, this tag exploits the properties of the propagation of an acoustic wave on piezoelectric substrate on which reflectors are printed. The implemented reflectors will generate a specific sequence of reflected signals when excited by a short impulse.

The identification code is given by the geometry of the reflectors as the distance between two reflectors is converted into time delay between the successive reflected impulses. Due to the high velocity of saw, the tag @ 2.4 GHz ISM band, is quite compact as the surface is in the order of cm^2 and thickness is some mm. In (Hartmann, Hartmann, Brown, Bellamy, Claiborne, Bonne, 2004) it is demonstrated that chipless SAW tags have a read range 5 times larger than traditional tags for the same reader power. However SAW tag can't be really compared to barcode or any other low cost identification technology or solution.

Many other investigations have been developed in order to lower tag cost. Most of the developments are based on standard dielectric substrates and do not consider any specific properties of the substrate, except its potential application for RF devices. One of the interesting examples is the technique known as SARcode (Surface Aperture Radar) introduced by Inksure®. This is mostly an imaging approach that exploits radar techniques. When designed at millimeter wave (60 GHz), SARcode has potentially an identification capacity exceeding hundred bits, which makes it a serious competitor of optical barcode (Taylor, 2009). This potential high coding capacity is due to the use of very efficient signal processing routines, because when two symbols are placed too near each to other, reflected waves will interfere complicating then the recognition of the symbols.

Still now, chipless technology is in its infancy age, and many developments are found in the recent literature (Preradovic, Karmakar & Balbin, 2010). Chipless tags differ in the ways, how data are coded, into the physical structure of the tag. Figure 2 is a tentative of classification of chipless configurations regarding coding techniques for implementing the data into the tag.

As we can notice there are several criteria for possible classification:

- Coding could be made on surface or on volume. Even if coding in the volume is possible, it remains not really considered in the literature. There are only two examples published in literature, both are studied and proposed for THz domain. A multilayer configuration where the coding is due to the defects implemented into the Terahertz Band Gap (TBG) structure has been proposed (Tedjini, Perret, Deepu & Bernier, 2009). Cumming & Drysdale (2003) have reported an authentication technique using THz signals and an imaging technique for identification. The two previous coding techniques are very hard to clone. Thus, they are quite interesting for secure identification and possibly authentication.
- When considering the surface coding, there are two major concepts. In both cases coding is obtained thanks to the metallic

Figure 2. Possible classification of chipless tags and coding methods

printed structure on the surface of the substrate. However the printed structure could be composed of only a set of radiating elements or antennas and an RF passive processing circuit. Both cases are available in the literature, but the tags made with antennas only are roughly more compact, meanwhile the tags composed of antenna and RF passive circuit exhibit larger density of coding. Configurations based on an RF processing circuit could be of two kinds: monostatic with only one antenna or bistatic with two antennas: one antenna for the reception of the interrogation signal and one antenna for the backscattered signal.

- When considering frequency band of operation, there are two main categories. The first category uses a limited number of frequencies, which are mostly part of ISM bands at 2.4 GHz and 5.8 GHz. The second category is based on the possibility to exploit the UWB band as defined by regulations (-41.3 dBm/MHz). Potentially the later category is more interesting in term of coding capacity. Even if UWB communication distances are roughly limited to some tens of centimeters, higher read range distances can be practically achieved with chipless RFID. However, as UWB regulations are not the same worldwide, this could generate more restrictions.

- The last criterion regards the method on which the information is encoded on the interrogation signal. Different approaches can be considered. They consist in mastering the modulation techniques and the relevant signal processing methods that are necessary for signal generation by the reader and analysis of the reflected signal by the tag. As a rule of thumb, two complimentary techniques can be implemented: time domain approach and frequency domain. However in both cases, modulation

methods of the reader signal and processing methods of thebackscattered signal can take various formats, as it will be discussed in the next paragraphs of this chapter.

This chapter will consider the different techniques to implement a given code, usually under a binary format, into the physical structure of the tag. We will restrict our discussion to chipless tags based on isotropic substrate. The coding will be entirely implemented by the printed metallic structure on the surface of the substrate. The organization of the chapter is as follows:

Part 2 will focus on the principle of chipless tags. It will introduce the concept of communication by means of reflected wave and how this principle has been applied with a great success in radar technology. The modern chipless approach will be introduced and the main aspects of the signal generation and signature implementation will be presented. Some examples of real chipless tags will be presented.

Part 3 will introduce the coding approach based on time domain. Some examples from the literature will be presented and discussed. A special attention will be given to SAW technology and how it is used in chipless solution. Finally, ways to implement anticollision capability for chipless RFID will be discussed.

Part 4 will concern the coding in the frequency domain. Again many examples from the literature will be presented and discussed. These methods are mainly amplitude modulation schemes. This part will discuss possible strategies for high capacity coding.

Part 5 will consider the coding of the information by exploiting the phase of the backscattered signal. As in advanced modulation schemes the phase of the signal could be used to code the information. Thanks to mastering the phase of signals, such as the control of resonances, group delay and some other parameters, various chipless tags are described and simulation examples are reported and discussed.

Part 6 will introduce the concept of hybrid coding. Such method exploits several coding techniques in the same time. This approach can be considered to increase and improve the coding efficiency of chipless tags. As for communication systems, modulation techniques will play a major role. Constellation diagrams will be used in order to define the efficiency of coding.

Part 7 will underling a set of parameters in order to evaluate the performance of chipless tags. A discussion on the impact of geometrical and physical parameters on the coding capacity of chipless technology is carried out.

Part 8 will introduce the concepts of coding Density per Surface (DPS) and coding Density per Frequency (DPF) as an Indicator of Merit.M Performance of several chipless tags will be evaluated and compared thanks to DPS and DPF parameters.

Part 9 is dedicated to experimental set-up for chipless characterization. Two measurement techniques are briefly discussed.

Finally, part 10 concludes the chapter and addresses some important remarks and perspectives.

2. OPERATING PRINCIPLE OF THE CHIPLESS TAGS

Such as traditional RFID, the family of RFID chipless tags described here is associated with a specific RF reader. This reader interrogates the tag and retrieves its ID. The operating principle of the reader is based on the emission of a specific EM signal in the direction of the tag, and the capture of the backscattering signal (see Figure 3). The processing of the received signal allows recognizing and tracing the tag ID.

However, chipless tags are fundamentally different from conventional RFID tags. Indeed, in the case of classical RFID tags, the reader sends a signal containing a query to the tag. To generate this signal, the reader will use a predefined standard modulation and communication protocol.

The tag demodulates this signal, processes the query, and returns a response by modulating its load. On the contrary, chipless RFID tags do not use any communication protocol. They can be seen as radar targets having a particular frequency or time signature. Thus, reading the identifier of a chipless tag consists in analyzing the radar signature of that tag.

3. DESCRIPTION OF THE PRINCIPLE OF THE CHIPLESS RFID

Figure 4 shows a basic example of chipless tag. This structure is composed of short-circuited dipoles; it is one of the very first chipless tags that have been introduced (Jalaly & Robertson, 2005). The label is formed by a dielectric substrate where a ground plane covers the entire back of the substrate. Above the substrate, we can notice five conducting and disjointed metallic strips, which can be used as five uncoupled dipoles with different lengths.

The label is nothing more than a structure with resonant elements interfering with the signal emitted by the reader. Each dipole behaves as a resonant circuit, re-radiating a specific electromagnetic wave that can be captured by the reader. Thus, the geometry of each conductive strip (only different lengths in this example) determines the resonant frequencies of the label. Under operation, the reader transmits a signal whose spectrum includes all the resonant frequencies. If the label is located near to the reader, the backscattering wave spectrum contains five distinct resonances. From this spectrum, the reader detects the signal extremum at resonant frequencies, which are directly linked to each dipole length. The positions of these five resonances in the spectrum allow the reader to identify uniquely the tag.

Through this basic example, we see the general operation principle of chipless RFID, i.e. a method for encoding information from a totally

Figure 3. Diagram principle of chipless RFID system

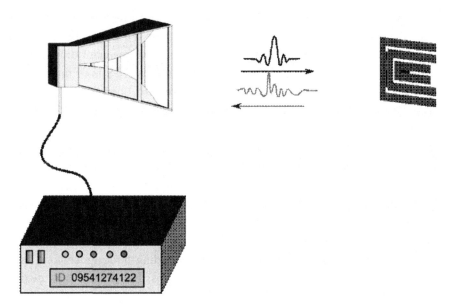

passive device and of course without any chip. We also see that the information exchange is based on the round-trip propagation of the EM wave, where the way of encoding information is radi-

Figure 4. A typical chipless tag. This dipole-type tag has been introduced by Jalaly & Robertson (2005). The dimension of the total surface of the tag is 18 x 35 mm and substrate thickness is roughly 1 mm.

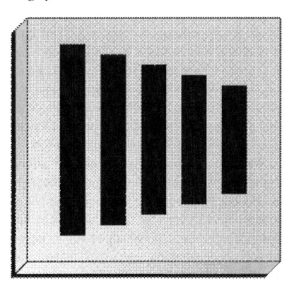

cally different from standard RFID. So, the term "RF barcode" really becomes more evident here. Similarly, we can clearly interpret it as an analogy with the principle of object radar signature. The difference here is that one must be able to link directly, as simply as possible, the EM signature of the object to its intrinsic geometry to make the identification.

To go back to the design work, this involves developing tags with a direct link between the geometry and the backscattering EM signature. It is obvious that in terms of use, it is not possible:

- To simulate all the different configurations of a tag (for instance a tag with a very low capacity of 10 bits corresponds to 1024 configurations) or
- To design each tag configuration one by one, the user should be able to get the geometry directly from the tag ID he wishes to apply to the item.

Nevertheless, it is possible to use simple considerations such as resonant circuit theory. Indeed, if we consider the case of the dipole previously introduced, by comparing the signal received from the feeding signal, it is possible to identify the resonant frequencies. It is clear through this example that there is a direct link between the geometry and the resonance frequency and thus the final tag ID. By changing the length of the dipole, it is possible to move the resonance frequency within the defined frequency band.

However, several limitations could appear. In order to increase the coding capacity of the tag, that is to say the number of configurations (different geometries and thus different resonance frequencies), it is advantageous to increase the number of resonators, i.e., in this case example, the number of dipoles as shown in Figure 4. However, to keep a direct link between the code and the geometry (here the length of each dipole), the resonators must be decoupled from each other. Otherwise the geometrical variation of the length impacts the resonant frequency associated with this element and also impacts the frequencies of the neighboring dipoles. Thus, this phenomenon makes the device useless for identification. Perfect decoupling is unfortunately not possible especially because there is a strong constraint on the size of the tags, the idea being to minimize their surface. Also, this coupling phenomenon requires associating with each chipless tag family a frequency interval. From this interval, it is possible to determine if the frequency shift observed is due to a change in the ID or not.

The second important limitation comes from the frequency band, because the available frequency band is restricted by legislation, so it is interesting to use resonators with high quality factors. Figure 5 shows different types of resonators that can be potentially used to realize chipless RFID tags. We distinguish between the short-circuited dipole antenna (Figure 5 a)), the slot antenna (Figure. 5 b)) and a coplanar line shorted at one side and opened at the other, which looks like a "C" shape (Figure 5 c)). In Figure 5 d), the radar cross section (RCS) associated with these three types of resonators are also presented.

Above all, it is clear that the presence of peaks at predetermined frequencies can be used to encode a particular ID. In addition, we also notice that the half-wave resonant dipole has a low quality factor and its RCS is of the order of -24 dBsm. The slot also operates as a half-wave antenna, but its selectivity and its RCS (-18.5 dBsm) are improved. On the contrary, the structure based on the "C" shape operates as a quarter-wave and has a good quality factor. The later structure provides an improvement in size reduction and thus in density encoding. It appears clearly in Figure 5 that the choice of the resonator determines the tag information capacity coding and its final size.

Regarding the designer work, another approach is also possible. Rather than designing specific resonators where the resonant frequency can be adjusted within a given range, it is also possible to adopt an On-Off Keying (OOK) approach, where the resonances can be present or not. In this case, we design a set of N resonators, e.g. having uniform spaced resonance frequencies and covering the entire authorized frequency band. Each resonator is optimized for a specific resonance frequency and this by tuning on many geometrical parameters as necessary. From there, to encode information, one just has to leave or prevent each resonator independently. Thus to eliminate one or more of these lines among the N lines, either the resonator under consideration should not be realized or, depending on the type of resonator used, it has to be short-circuited, so as to remove its resonant frequency (Preradovic, Roy, Karmakar, 2009). This particular approach will be discussed in the §4 of this chapter. A specific attention will be paid to the comparison of encoding densities obtained for each approach.

We focus in this paragraph on the coding method known as frequency approach, that is to say the one in connection with the example of dipoles. We will see later that other approaches exist, including those for which we will swing on the lag between multiple signals. All of these different methods will be developed in more details

Figure 5. Different types of resonators a) short-circuited dipole b) slot antenna c) C shape. These structures do not contain any ground plane. d) RCS versus frequency for the three resonators: the short-circuited dipole, the slot antenna and the C shape structure. $\lambda / 2 = 30$ cm, w = 2 mm, g = 0.5 mm.

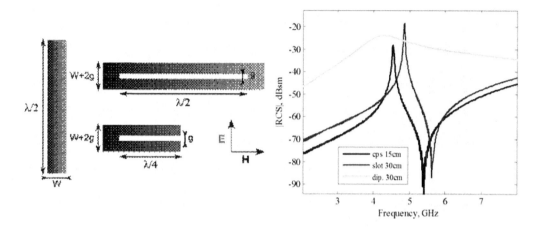

in the next paragraphs of this chapter, each time with an emphasis on the concept of information coding.

After this presentation of the basic principle of the RFID chip through an elementary example, we now focus on an optimized device that highlights the practical design work as well as the tradeoffs in terms of the coding capacity of the tag, its size and the frequency band to use.

3.1. Example of the « C » Tag

The labels consist in a common substrate without ground plane, on which conductive strips are printed in the shape of C and nested inside each other, as shown in Figure 6 a) and b). This original topology exhibits a large number of resonant frequencies, which are the basis of information coding. Indeed, the resonances are directly dependent on the geometry of the conductor patterns. This particular topology allows many slots of different lengths that resonate at specific frequencies. The slot length changes, for example by adjusting the length of short-circuit positioned in the slot, which allows to control very precisely the resonance

frequencies and thus the shape of the EM signal backscattered by the tag toward the reader (see Figure 6 d)). The label can be seen as a very specific radar target, i.e., one for which it is possible to configure precisely its EM signature from its own geometry. It plays the role of receiver, filter and reflector in the same time.

The tag shown in Figure 6 a) and b) is composed of 4 resonators numbered from 1 to 4. These resonators are totally independent from each to other apart number 3 which acts as an isolation agent between the resonators 1 and 2 on one hand and, and the resonator 4 on the other hand (Perret *et al.*, 2011 ; Perret, Tedjini, Deepu, Vena, Garet & Duvillaret, 2010). Some short circuits allow adjusting the length of the three slots numbered 1, 2 and 4 and thus their frequencies of resonance. As previously stated, the interest of such a structure is based on the fact that although the slots are very close to each other, their resonances are completely decoupled. This absence of coupling allows independent control of each resonance frequency and thus encoding some information on a small surface of 1.5 * 2 cm². This principle is illustrated in Figures 6 c) and d). Indeed, Figure

Figure 6. Structure of the "C" tag, a) schematic diagram, b) picture of the "C" tag realized on FR4 substrate,, c) Illustration of the link existing between the resonant frequencies and the geometry of the "C" tag. d) Amplitude of the tag backscattered signal versus frequency, the signal is normalized with respect to the incident signal.

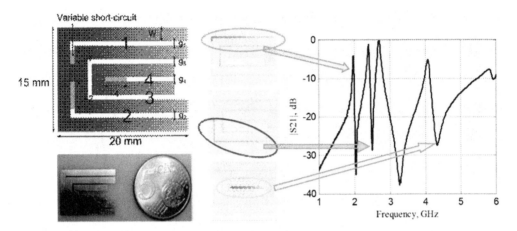

6 d) shows the three resonances at three specific frequencies corresponding to the presence of the three slots.

We are now interested in how to recover the information stored in the tag, more precisely the one embedded in its EM signature. To obtain the tag information, the backscattered signal to the reader is processed. In this example, frequencies for which signal amplitude is minimal are extracted. From there, a correlation between those physical values and a binary code, the ID tag, is defined as shown in Figure 7. In this example, a simple lookup table can give this correspondence. Indeed, this type of table is able to link each frequencies combination with a sequence of zeros and ones, corresponding to the binary code of the tag. Figure 7 illustrates the principle of encoding that can be implemented to link the EM signature to the binary code. So, as the tag accepts three slots completely decoupled, it can encode a total of 6 bits, two bits per slot. For example, if we encode the tag ID 110110, according to the tables of correspondence, the slots length have to be chosen in such way that the slot No.

1, No. 2 and No. 3 respectively have to resonate at 2.3, 2.45 and 4.8 GHz.

In summary, this chipless RFID tag is an example of an extremely compact tag, with an encoding capacity that can reach 10 bits for a small area of 1.5x2 cm^2. Moreover, as this device does not have any ground plane, it is compatible with printing technologies and conductive inks, which should substantially reduce the cost of these tags.

4. TIME SIGNATURE ENCODING

In classical telecommunication systems, time varying signal is mainly used for coding information. In the same way, classical RFID is usually based on amplitude and/or phase modulation in the time domain.

The chipless RFID tags design can be based on time domain reflectometry. Therefore, tags are interrogated by a pulse signal sent by the reader. The train of pulses generated by the tag in response is used to encode data. Generally, time domain reflections are caused by the presence of

Figure 7. Illustration of the encoding principle based on the example of the "C" tag. The correlation table shows the relationship existing between the resonance frequencies of the tag and the associated binary code. Here an elementary code containing 6 bits is described.

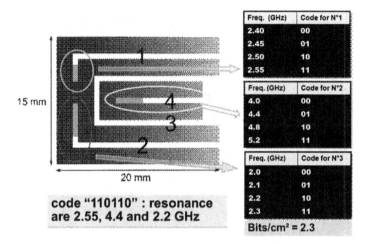

discontinuities or impedance mismatches. A mismatched termination will cause reflection of the signal. Likewise, impedance discontinuities along the transmission line cause reflections. Generally, reflections are considered as undesirable effects in microwave circuits. However, chipless RFID designers take advantage of these reflections by controlling the placement and characteristics of the impedance mismatches. For that, both lumped (usually SMT devices) and distributed capacitors were respectively used as discontinuities (Zhang, Rodriguez, Tenhunen and Zheng, 2006; Zheng, Rodriguez, Zhang, Shao ans Zheng, 2008). One of the most known examples of a time domain reflectometry based chipless RFID tag is the SAW tag (Hartmann, 2002).

There are several ways to encode data using time domain reflectometry as shown in Figure 8. A simplest way is to place the discontinuities at regular distances to represent ´1´s or ´0´s. This gives the possibility to encode data using On-Off Keying (OOK). Other option is to place a number of discontinuities at different distances in order to obtain a reflected signal that resembles pulse position modulation (PPM).

4.1. On-Off Keying (OOK)

On-Off Keying is the simplest form of amplitude-shift keying (ASK) modulation that represents digital data as the presence or absence of a specified symbol. Figure 9.a shows the schematic of the RFID tag reported by Zhang, Rodriguez, Tenhunen and Zheng (2006). It consists on a transmission line with four capacitive SMT discontinuities at regular intervals. If the capacitor is connected with transmission line, there should be pulses in the reflected pulses train, which stands "1"; if not, there is no pulse, which stands "0". Accordingly, 4 bits of information can be represented with this device. The RFID tag was realized on a Rogers 4350 substrate (ε_r = 3.48, tan (δ) = 0.004). The experimental response performed using an UWB generator is shown in Figure 9.b for the tag identified as "0101".

It is important to note that the data encoding capacity can be improved by increasing the number of capacitors. Furthermore, an 8-bit tag based on the same principle and using 8 distributed capacitors has been reported by Zheng, Rodriguez, Zhang, Shao & Zheng (2008).

Figure 8. Interrogation pulse and reflected waves. Example of different ways to encode data using: (a) the presence or absence of a specific reflector, (b) the position between reflectors. In both cases, the data encoded correspond to the same ID 1101.

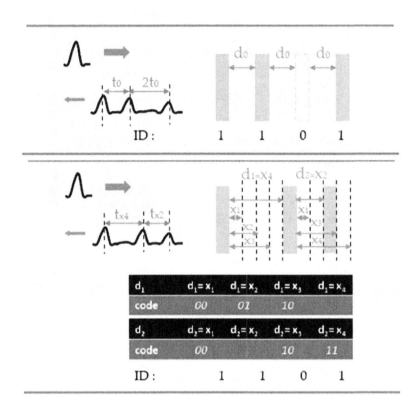

4.2. Pulse Position Modulation (PPM)

Pulse Position Modulation (PPM) is a form of signal modulation in which several message bits are encoded by transmitting a single pulse. PPM achieves higher average power efficiency than OOK at the expense of an increased bandwidth compared to OOK.

Surface Acoustic Waves (SAW) RFID tags are based on PPM principle (Hartmann, 2002). Figure 9.c shows a typical configuration of a SAW RFID tag. The functionality of SAW tags is based on the piezoelectric effect. Generally, the device consists on an Inter Digital Transducer (IDT) and a series of acoustic reflectors. The IDT is directly connected to the tag's antenna, which both receives the interrogation signal from a reader and radiates the response signal generated by the SAW tag. Each reflector generates a pulse in the SAW tag's impulse response. Information is encoded in the positions of the metallic reflectors.

Figure 9.d illustrates the pulse position encoding method used in a SAW RFID tag. In this case, a start pulse is used to provide timing synchronization for the remaining data pulses. Each data pulse can have one of 4 possible time positions and thus 2 data bits are encoded for each pulse.

4.3. Anti-Collision Principle

If multiple tags are present, anti-collision algorithms should be adopted so that each tag is read only once.

Figure 9. Time domain RFID tag (Zhang, Rodriguez, Tenhunen, & Zhengand, 2006) a) Schematic b) Measurement response of the tag "0101". c) Schematic of a SAW Tag d) Pulse Position Modulation encoding used in SAW tags.

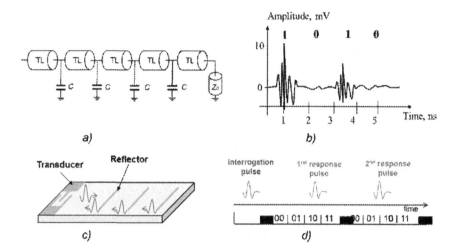

Furthermore, multiple readers may interfere each other if they are in the same field. Neighboring readers cause interference if they communicate with the same tag at the same time. Many RFID applications require the ability to read multiple tags in close proximity to one another. This is only possible if each tag has a unique ID number, which is the basis for implementing anti-collision in any RFID technology.

Temporal signature encoding allows the use of several anti-collision algorithms. Anti-collision is often considered to be an area where IC based tags have an advantage over all chipless tags. This arises because an IC tag can be deactivated on a selective basis whereas chipless tags are always "active". However, chipless tags can provide anti-collision by using several methods such as: time separation, spatial focusing and error correction code.

Time separation based anti-collision (Hartmann, Hartmann, Brown, Bellamy, Claiborne & Bonne, 2004) uses tags replying signals that occur in differing time positions (see Figure. 10.a). For example, a first tag could be designed such that its reply signal is located in the time period

"Slot1". A second tag could be designed in order to reply in another time period "Slot4" as shown in Figure 10.a. Obviously, further tags could also be designed with reply signals that occur in other non-overlapping time periods. In this system, tags of one type will not collide with tags of another type even if the same reader activates simultaneously both tags.

Besides, narrow beam antennas can be used for spatial focusing based anti-collision (Hartmann, Hartmann, Brown, Bellamy, Claiborne & Bonne, 2004). In this configuration with a group of near tags, only one will receive the incident wave and reply its time signature (Figure 10.b). In addition, in the near field, these antennas have a collimated beam, which is spatially selective. The gain inherent in narrow beam antennas has the added advantage of increasing reading range. This is particularly true when operating at 2.45 GHz since a wide variety of manufacturers produce antennas with 15 to 18 dB of gain with beam widths between 20° and 30°. Moreover, a focused beam antenna (Figure 10.b) maintaining a spot size of less than ±12 cm from 1 meter to 2.2 meters in front of the antenna was reported in (Hart-

157

Figure 10. a) Time separation based anti-collision principle; b) Spatial focusing anti-collision principle

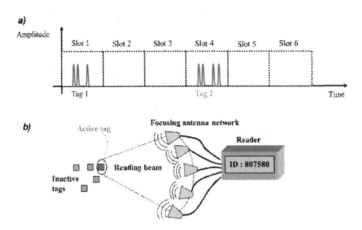

mann, Hartmann, Brown, Bellamy, Claiborne & Bonne, 2004).

In addition to previous anti-collision algorithms, linear block code was introduced by Brandl, Schuster, Scheiblhofer & Stelzer (2008). This approach allows retrieving lost data caused by destructive superposition of the response signals of multiple tags. Another advantage of this principle is that the IDs of the tags result directly without scanning all possible combinations of tags. A typical application is multi-object tagging in an industrial environment.

5. FREQUENCY SIGNATURE ENCODING

5.1. Amplitude Modulation

In frequency domain, encoding data can be achieved thanks to an amplitude modulation of the wave backscattered to the reader. Chipless tags with spectral signature encoding consist of a set of radiating elements such as antennas, resonant circuits, filters... The spectral signature is obtained by interrogating the tag with a multi-frequency signal and observing which frequencies are attenuated for example. Amplitude modulation can be obtained by using resonant elements

coupled to a transmission line (Preradovic, Balbin, Karmakar & Swiegers, 2008, April ; Preradovic, Balbin, Karmakar & Swiegers, 2008, October) or by tuning resonance frequencies of dipole arrays (Jalaly & Robertson, 2005, October; Jalaly & Robertson, 2005, June). Thus, the procedure of coding data is similar to the one used in the time domain. However, time pulse used in time domain will be replaced by a signal that exhibits the required spectral components. Therefore, OOK encoding is obtained by presence/absence of a resonance in the tag frequency response. In parallel, PPM coding is achieved thanks to a shift of the resonant frequency.

5.2. On-Off Keying (OOK)

Generally, the spectral signature is determined by the presence/absence of a set of resonance frequencies. This is an OOK modulation transposed to the frequency domain. Thus, one resonance is equivalent to one bit. The RFID reader sends out a multi-frequency interrogation signal and the chipless tag attenuates particular frequency, which corresponds to an unique spectral signature. If the signal is attenuated at a particular frequency then it represents a logic state "0" otherwise it's a logic state "1".

Jalaly reported for the first time dipole-based tags (Jalaly & Robertson, 2005, October; Jalaly & Robertson, 2005, June). The realized microstrip dipole antennas, based on the principle shown in Figure 4, were made on Taconics TLY-5 substrate ($\varepsilon_r = 2.2$, $\tan(\delta) = 0.001$). They act as selective resonators. As multibit read-only tags, the ID is determined from the presence/absence of a set of resonance frequencies. Each dipole corresponds to a single data bit. The encoding capacity can be increased using multiple frequency bands. As the dipole lengths determine the resonant frequencies, arrays of identical microstrip dipoles capacitively tuned to be resonant at different frequencies within the desired licensed-free ISM bands were realized (Jalaly & Robertson, 2005, October). The advantage is that the conductor pattern is identical for all elements, to ease fabrication.

In 2008, a multiresonator-based chipless RFID tag was introduced (Preradovic, Balbin, Karmakar & Swiegers, 2008, April; Preradovic, Balbin, Karmakar & Swiegers, 2008, October), It consists of a vertically polarized UWB disc-loaded monopole Rx tag antenna, a multiresonating circuit, and a horizontally polarized UWB disc-loaded mono-

pole Tx tag antenna (Figure 11). The Rx and Tx tag antennas are cross-polarized in order to minimize interference between the interrogation signal and the retransmitted encoded signal containing the spectral signature. The operating frequency range of this 6-bit chipless system fabricated on Taconic TLX-0 ($\varepsilon_r = 2.45$, $\tan(\delta) = 0.0019$) is 500 MHz centered on 2.25 GHz. The multiresonator tag operates like a multi-stop band filter. It is a set of cascaded spiral designed to resonate at particular frequencies and generate stop bands. These stop bands introduce attenuation and phase ripple on the interrogation signal at their resonant frequencies, which can be detected as an amplitude attenuation and phase variation by the reader as shown in Figures 11.b and 11.c).

Moreover, a chipless tag composed of 35 spiral resonators, each having a one-to-one correspondence to a particular bit of data was reported by Preradovic, Roy & Karmakar (2009). In order to minimize the size of the transponder, the microstrip line was bent and spiral resonators have been placed on both sides of the microstrip line with 3-mm separation between them.

Figure 11. 6-bit multiresonator tag published in (Preradovic, Balbin, Karmakar & Swiegers, 2008, April); a) Structure of the tag; b) Amplitude frequency response; c) Phase frequency response

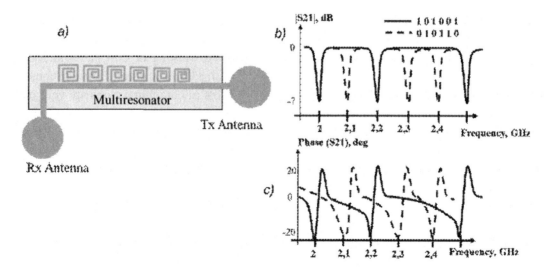

5.3. Pulse Position Modulation (PPM)

Compared to OOK, PPM encoding offers a more compact solution for the same encoding capacity. In the case of OOK encoding, one resonator is needed to encode a single data bit. Thus, the encoded bits increase linearly with the resonators numbers. However, several bits can be associated to a single resonator utilizing PPM encoding. PPM data encoding was used to design miniaturized chipless tags (Perret, Tedjini, Deepu, Vena, Garet & Duvillaret, 2010). Figure 6.b shows the 10-bit double-C resonator fabricated on FR4 substrate for the 2-5.5 GHz frequency band. Moreover, in the reading process, the backscattered signal undergoes a phase modulation determined by the phase-frequency tag characteristics, while maintaining a constant amplitude. Of course, using phase modulation compared to amplitude modulation such as identifying resonance peaks has the advantage of better overall Signal to Noise Ratio (SNR) or Bit Error Rate (BER).

6. CODING OF INFORMATION THANKS TO THE PHASE

6.1. Noise Considerations

Until recently the main coding methods for chipless tags was based on presence/absence technique, frequency shift encoding or pulse position modulation in time domain. Nevertheless, it seems that the coding in phase can bring a better reliability for tag detection (Preradovic, Roy & Karmakar, 2009), and even a better density of coding per surface than for classical presence/absence technique. When combining amplitude with phase encoding methods, a high degree of miniaturization can be reached which allows high storage capacity.

Reliability is a critical aspect of chipless tag, because it will define the performance of the used label in terms of read range and robustness to harsh environment. Before introducing the phase coding techniques, which are usable to design a chipless tag, it is necessary to identify the main sources of noise that can be found and their effects on the electromagnetic response of a tag. It is known that a transmission channel in free space can be disturbed by some effects like:

- **Fading (depending on environment):** Fading effect due to surrounding environment can change the amplitude and the phase profile as a function of the obstacles and the properties of the propagation medium.
- **Multi-path effect (sometimes):** Multi path effect is present when the tag can be read by direct path, in line of sight and also by multiple reflections. This phenomenon is present because the tag is not read in outdoor environment and some surrounding objects can reflect waves.
- **White noise (already present):** The white noise is omnipresent and is due to thermal excitation of charge carriers that compose any electrical conductor. This effect adds both amplitude and phase uncertainty to the electromagnetic signal.
- **Interference with other communication channels (nearly always presents):** We live in a world where wireless communications are everywhere. Most of the applications are using ISM bands. If designing a tag working in ISM band, interferences with adjacent channels must be considered. Most of communications are very narrowband band compared to chipless tags (few MHz). But an unwanted signal due to a Continuous Wave modulated signal can be added to the tag signature.

Moreover, the channel effect is very important for free space communications. It affects both amplitude and phase. A perfect free space transmission channel, without noise, generates attenuation effect and delay on signal. The following expressions (1) and (2) show the way that a traveling wave is affected by free space channel in terms of phase and amplitude:

$$\underline{E} = |E(\omega)| \cdot e^{j\phi(\omega)} \underbrace{\times e^{(-\alpha + j\beta)z}}_{channel\ effect} = \underbrace{|E(\omega)| \cdot e^{-\alpha \cdot z}}_{Amplitude} \times e^{\overset{j(\phi(\omega)+\beta \cdot z)}{Phase}}$$

(1)

$$\tau_d = -\frac{d(\phi(\omega) + \beta \cdot z)}{d\omega} = -\frac{d(\phi(\omega) + \frac{\omega}{c} \cdot z)}{d\omega} = \underbrace{-\frac{d\phi(\omega)}{d\omega}}_{Group\ delay} - \frac{z}{c}$$

(2)

So we notice that amplitude is attenuated as a function of distance "*z*" and phase is modified as a function of distance "*z*" too. But variation of phase is not affected. If the phase is differentiated as a function of frequency, we obtain the group delay plus a constant due to round trip time. So the shape of group delay is not affected. The constant can simply be detected and removed in order to get a group delay with zero offset. We can imagine that the group delay is not modified by distance, and so, information embedded keeps its integrity until the Signal to Noise Ratio turns to be bad enough.

To conclude, as demonstrated by Preradovic, Roy & Karmakar (2009), phase encoding allows a higher robustness to distance effect due to the transmission channel. White noise and multipath effects affect equally amplitude and phase. But some techniques exist to eliminate these unwanted effects. To remove multipath effect we can, for example, use a time windowing filter and consider only the first incoming signal toward the receiver. It avoids any destructive interference between direct path and other ones, and white noise can be attenuated by increasing the acquisition time, using averaging on several measurements.

6.2. Phase Pattern/Evolution as a Function of Frequency

The encoding of tag can be realized by controlling the phase evolution as a function of frequency. A development from Mujherjee (2007) shows this coding principle based on using a complex load connected to the tag antenna. The reading system allows getting this phase evolution depending on the complex impedance plugged to antenna. In this case the complex load is used as the unique identifier.

In the above example (see Figure 12 a) and b)), it is shown that four different loads generate four phase profiles. A database can be defined by recording all the possible phase profiles and serves as authentication system. With phase profile, using a de-embedding procedure, the reactance can be found for a large frequency span. But it is hard to predict how many combinations will give this design. So it is necessary to discretize the phase profile. To deduce combination number, the following expression can be used (3).

$$N = \left\lceil \frac{\phi_{max} - \phi_{min}}{d\phi} \right\rceil^k$$

(3)

The phase resolution $d\varphi$ should be defined and margins φ_{max} and φ_{min} for phase shift should also be set. Some of margins are given by the accuracy of practical realization and performance of the reading system. The number k, is function of the frequency of interest at which the phase is observed.

6.3. Phase Shift for Specific Frequency

Balbin and Karmakar (2009), proposed a design based on multiple patch antennas tuned at several frequencies. For each patch antenna, the phase response can be controlled between two values. The idea of this encoding method is fundamen-

Figure 12. a) Chipless tag loaded with a complex impedance; b) Phase pattern for several complex loads; c) Chipless tag using 3 stub loaded patch antennas; d) Phase Evolution for three frequencies

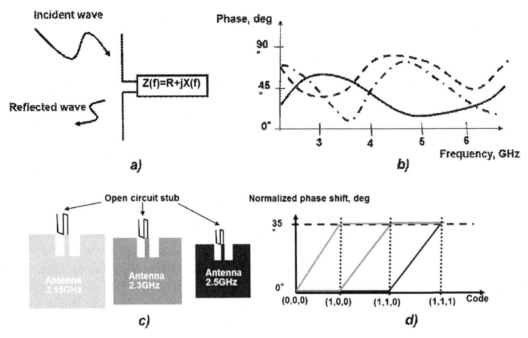

tally different from the previous one. In this case, only discrete frequency values are used to encode data, and for each frequency the phase can be independently controlled. With this method, several encoding techniques can be used as for amplitude encoding. Indeed if only two values of phase are possible for a given frequency, the coding capacity reachable is equal to that of OOK. If several values of phase shift are possible, the coding capacity is similar to the PPM case. The design shown in Figure 12 c) uses 3 patch antenna resonators loaded by a stub of variable length. In this example, the coding capacity is 1 bit per resonator because the phase takes two values, 0° and 35° as shown in Figure 12 d). So having 3 antennas, it can be encoded 3 bits. But if each stub length can take several values, assuming a phase resolution of 35°, from 0 to 180° it can be encoded 5 states, so that capacity will be equal to $\log_2(5^3) \sim 7$ bits.

6.4. Group Delay Encoding

It has been demonstrated, that measurement of group delay, can bring better results than amplitude measurement as a function of frequency (Mukherjee, Chakraborty, 2009) for an all pass device based on stacked multiple patch resonators. It is due to its better robustness to multi-path effect and clutters.

Usually it is common to use non dispersive devices in RF systems for signal integrity purposes. But dispersion characteristic can help to encode data as a function of frequency. The group delay is a parameter that informs us about the round trip of each component of signal as a function of frequency. It can be directly measured in time domain, or calculated from phase derivation with respect to frequency. In terms of coding, we could imagine a tag, having independent values of group delay for discrete frequencies. The C-section transmission line of the Figure 13 a) make possible the engineering of the dispersion curve as desired. In Figure 13 b), the group delay as a function of frequency is represented.

In this example, 3 frequencies f1 to f3 are used. For each frequency, 6 values of group delay are possible. Assuming these parameters, the number of total combinations can be found in the same way as for PPM encoding. Thus we have 6^3=216 combinations, i.e. 7.7 bits.

QPSK MODULATION SCHEME FOR CHIPLESS TAG BASED CRLH LINE

The aforementioned encoding techniques operate in frequency domain. If considering chipless tag operating in time domain, a design using phase encoding, instead of pulse position modulation can be found in (Mandel, Schüßler, Maasch & Jakoby, 2009). In this publication, Composite Right Left Handed (CRLH) concept is proposed for miniaturization purposes. As for SAW tag, the way of coding is based on setting some reflectors on a substrate to create reflected waves at specific positions. But in this case, a transmission line is used. The used reflectors exhibit several types of impedance to create different shapes of reflections. The phase is also affected and presents differences from one reflector to another one. The design idea

consists in setting lumped components on the transmission line at specific locations. The total coding capacity is N= 4^k, where "k" is the number of reflectors. With five reflectors, the coding capacity will be 10 bits. To increase this value, it is to be noticed that insertion loss introduced by each reflector is a limiting factor.

6.5. Relative Phase Shift Bandwidth Encoding

The publication (Vena, Perret & Tedjini, 2011) has demonstrated that it is possible to use the relative phase shift to encode data for a given frequency, and independently that, the frequency of resonant peak can be used to increase the coding efficiency too. This is possible because resonator behaves as a wideband phase shifter. It produces two resonant modes, a zero and a pole, which can be controlled independently. As a result, a peak followed by a dip as shown in Figure 14.a) can be seen for the amplitude. When changing the gap value G of the "C shaped" metallic strip resonator shown in Figure 14 b), the position of the dip is changed if considering the amplitude as a function of frequency. In the same time, the phase exhibits a shift width, which depends on frequency separation between the peak and the dip.

Figure 13. a) Dispersive transmission line; b) Group delay encoding principle

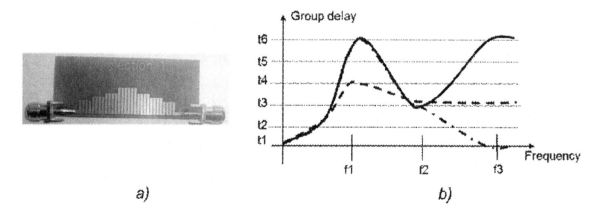

This relative phase shift width is used to encode information. In the simplest case, a narrow bandwidth gives the '0' state, while a wide bandwidth gives the '1' state. In practice, it has been demonstrated that four values of phase shift widths can be used to encode data leading to a coding efficiency of 2 bits per resonator.

7. INCREASING CODING EFFICIENCY WITH HYBRID TECHNIQUES

7.1. Coding Efficiency and Constellation Diagram

The present part will introduce some hybrid coding techniques that can be considered to increase and improve the coding efficiency of chipless tags.

As for telecommunication signals based on classical modulation schemes, a constellation diagram can be used in order to define the efficiency of coding for a given frequency instead of a given time (see Figure 15 a)). In this case, the two parameters used are amplitude and phase for a given frequency. This approach is quite similar to IQ modulation scheme in time domain.

In some special cases, more than 2 parameters can be modified for a given symbol (in time domain or in frequency domain) so a N dimensions constellation diagram can be adopted to represents N independent parameters for a given frequency as shown in Figure 15 b).

The previous constellations give information on efficiency of coding for one symbol in time domain or as introduced here, in frequency domain. In addition, a new variant of constellation integrating the frequency axis or time axis can also be introduced. In this way, constellation doesn't give the total number of combinations for each frequency or time slot, but the entire code spread on several frequencies. A graphical representation of a specific code can be plotted. Such a representation is interesting to compare several tag configurations and possible symbol (i.e. code) interference. A kind of reliability criterion can be deduced from each code graphical representation. Identification of a code can be done by graphical method if comparing the modified constellation with expected responses. In order to define a system with a high level of robustness, a subset of code can be selected to mitigate possible identification error between codes that are very similar. To do so, a kind of graphical minimal Hamming distance can be adopted.

Figure 14. a) Chipless tag based on multiple phase shifter; b) Coding principle using bandwidth of phase shift

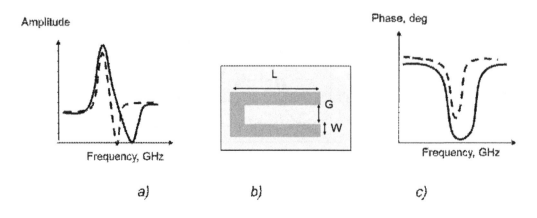

Figure 15. a) Frequency domain 2D constellation diagram; b) N dimensions constellation diagram; c) 2D identifier graphical representation; d) 3D identifier graphical representation

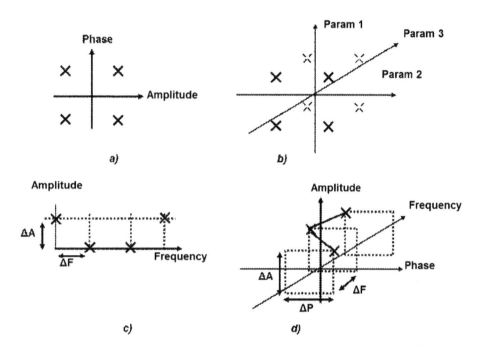

As an example, if considering the most used coding technique in frequency domain, where only the amplitude of resonance can be modified in a presence/absence way, it can be modeled by the following modified constellation in Figure 15 c). Now, if using the phase also, a 3D diagram has to be used as shown in Figure 15 d). The curve can describe a specific identifier among all the combinations.

In all previous constellation diagrams, resolution for each parameter is a limiting criterion depending on reading system accuracy and robustness to noise effect. A constellation having points very close to each other, hence needs a very accurate reading system and a low noise environment system. Conversely, a constellation having distant points is more robust, but coding efficiency is fair.

7.2. Using More than One Parameter for a Given Symbol

Most of coding techniques use only one axis of constellation and only two states per symbol as shown in Figure 16 a). This gives the lowest efficiency of coding per symbol, i.e. 1 bit per symbol. This is the case for many chipless designs that use presence/absence coding technique (Preradovic, Roy & Karmakar, 2009; Jalaly & Robertson, 2005). But this coding is robust because of strong contrast between the two combinations of each symbol.

A publication (Balbin & Karmakar, 2009), that uses CRLH transmission line and time domain reflections, goes in this way to increase the number of encoding states per axis. The presented design uses reflectors of four types instead of only one. The parameter used to encode data is the phase, and values from 0° to 270° are possible. So that 2 bits of coding per symbol is possible instead of only one for classical OOK case (see Figure 16 b).

Figure 16. a) Frequency domain 1D constellation for absence/presence coding; b) Frequency domain 1D constellation for QPSK coding; c)Frequency domain 1D constellation for frequency shift coding; d) Frequency domain 2D constellation for frequency shift + absence/presence coding

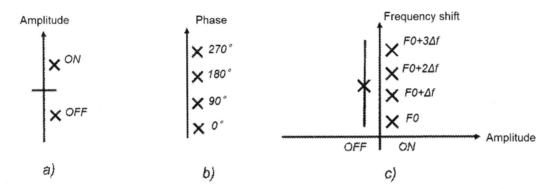

USING MORE THAN ONE AXE FOR A GIVEN SYMBOL

All previous coding technique just uses one axis of constellation. This means that only the phase shift, or amplitude keying is used in frequency domain. In time domain, only pulse position is used. Now, let's consider, more than one parameter axes. In frequency domain, it can be controlled, both the presence/absence of resonance and its frequency position in a given frequency span interval. This increases dramatically the capacity of coding for each resonance. Instead of having 1 bit per resonance as for presence/absence technique, several bits are possible. With this simple example, from a coding capacity of 4 combinations, we get an additional one, and as it is shown in Figure 16 c), this is a huge improvement in term of coding capacity. Considering a general case, with the same number of resonators, coding efficiency is very different depending on the used technique. The following diagrams show the capacity of coding in bits, as a function of the number of resonators, for OOK, frequency PPM and frequency PPM + OOK which is a hybrid coding. The following formulas have been used to calculate the number of combinations depending on the type of coding. In the case of OOK coding, the formula (4)

has to be used, while formula (5) is only valid for PPM encoding. To conclude, formula given in (6) will be the right one, when hybrid coding PPM+OOK is used.

$$N = 2^k \tag{4}$$

$$N = \prod \left[\frac{BW_i}{\Delta f} \right]^k \tag{5}$$

$$N = \prod \left[\frac{BW_i}{\Delta f} + 1 \right]^k \tag{6}$$

The curve in Figure 17 has been plotted using these formulas and shows the efficiency of each coding type as a function of resonator number for a given frequency resolution of 50 MHz and a bandwidth of 7.5 GHz compliant with UWB system (from 3.1 to 10.6 GHz) (Härmä, Plessky, Li & Hartogh, 2009).

For both PPM and PPM + OOK encoding, a sawtooth shape can be seen. This is due to ratio $BW_i/\Delta f$ in formula. To optimize the coding capacity this ratio has to be an integer, *i.e.* the local bandwidth has to be a multiple of the resolution frequency Δf.

Figure 17. Capacity in bits, in frequency domain using a resolution of 50MHz and a bandwidth of 7500 MHz

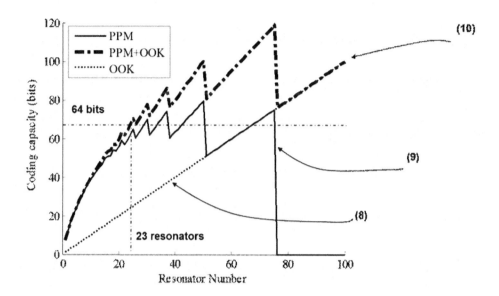

In the OOK case, i.e. absence/presence technique, one bit is equal to one resonator, so that 64 resonators are needed to get 64 bits. In the simple frequency PPM case (see Figure 17), the coding efficiency is enhanced because 28 resonators are needed if having a bandwidth of 7.5 GHz. In the last case, shown in Figure 16 c), that combines frequency PPM and OOK modulation, only 23 resonators are needed. We clearly notice here the interest of increasing, in a first step the number of combination per axis, and in a second step, using more than one axis.

Potentially other parameters such as phase and damping factor (Blischak & Manteghi, 2009) can be considered to increase the order of constellation. We can imagine for example to use a shorted

Figure 18. Selectivity based encoding in frequency domain a) Amplitude for different strip widths; b) view of dipole having two variable parameters L and G; c) Phase shape for different strip widths

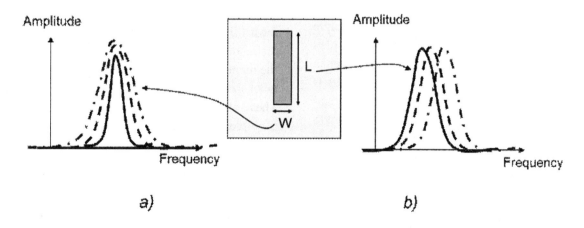

dipole (without ground plane) as shown in Figure 18, in order to combine frequency PPM encoding with a selectivity encoding if changing the width of dipole at a specific frequency.

At the present time, there is no chipless tag design that uses 3 parameters for a given frequency. But, intrinsically, as it shown by Blischak and Manteghi (2009), the analytical expression of an impulse response of any target is given by (7) and represents a sum of p damped sinusoids. The term p is the number of poles of the considered target.

$$h(t) = \sum_{k=1}^{p} \underbrace{A_k e^{j\varphi_k}}_{residue} \underbrace{e^{-m_k t} e^{j\omega_k t}}_{pole} \tag{7}$$

In this expression it can be seen that for each pole, i.e. resonant frequency, the amplitude A_k, the phase Φ_k and the damping m_k can vary. A design able to vary each of these parameters independently, will give a high coding efficiency. All presented design use for the moment only the amplitude information or phase information. The damping factor should be considered in future designs.

8. IMPACTIMPACT OF PARAMETERS ON TAG PERFORMANCE

Chipless RFID is essentially based on the radar principle, which is extremely simple to understand. However, although the operating principle is straightforward, its implementation is complex and requires a very good expertise in antenna design. Low coding capacity of chipless tags is one of the main characteristics of this identification technology. Therefore, it seems important to define some figures of merit in order to compare the performance of several coding techniques and chipless configurations. In practice, in order to store a large amount of data, a quite tedious optimization phase of the chipless tag is required. Therefore the knowledge of the key parameters of the tags is very important. So, for design purposes, it is essential to begin by inventorying the relevant variables to consider. However, many of these parameters strongly depend upon each other. Consequently, it is difficult to improve the impact of a given parameter without affecting another. This leads us to define some indicators of performance. From there, compromises must be made during the design phase in order to favor one criterion over another.

As previously introduced, a chipless tag must perform several functions such as capturing the EM wave from the reader, modifying it in order to implement the ID and finally backscattering the wave toward the reader. Obviously, the challenge here is to perform all these functions without any IC. Let's consider the most relevant parameters for any chipless RFID tag: data capacity, frequency band, label size, read range and environment.

8.1. Data Capacity

Due to the absence of IC, the data capacity storage is generally low, some tens of bits. Compared to other identification techniques, this amount of data is almost comparable to the barcode capacity, but much lower than conventional RFID tags. For instance, RFID tags can have an internal memory and so a very large capacity storage (several Kbits). In this case, from user point of view, the data capacity storage is almost transparent. At most this can slightly impact the price or even the reading time to retrieve all the data stored in the memory. However, it does not alter the final label size or the communication frequency band. These observations are very different compared to chipless RFID applications. Indeed, in chipless tags, the data capacity will directly impact many parameters such as the size of the label and the frequency band.

8.2. The Frequency Band

The frequency band choice is very important, at least for two reasons. First: large coding capacity requires large frequency bands. This is even true if we consider chipless tags based on frequency approach. Indeed, in this case, the tag capacity depends on the frequency range that is associated to each resonator. For example, if we consider a reader frequency resolution of about 50 MHz, it is possible to encode a dozen of bits per GHz. Second: legislation and regulation aspects. Indeed, the Industrial, Scientific and Medical (ISM) radio bands are far from allowing free use of several GHz frequency bands. The maximum frequency range available is rather tens to hundred of MHz. So like all communication systems, the frequency is a very important issue, and especially in our case, the coding capacity of the tag will be directly impacted. However, a solution based on the use of broadband pulse, and therefore the Ultra Wide Band (UWB) regulation for communication purposes is possible. This approach should be able to allow the market deployment of some chipless approaches. For all these reasons, it seems interesting to consider the data capacity storage of the tag versus the exploited frequency band. This criterion allows characterizing the performance of the tag regarding the frequency. So, the concept of frequency encoding density of chipless RFID tags should be introduced.

8.3. Tag Size

Many chipless configurations rely on the use of resonators (mostly frequency domain approach) or discontinuities (mostly temporal approach). Each of these elements is at the origin of the data encoding. Considering the case of resonators, it is possible to increase the coding capacity by increasing their number. For a given frequency range, this tendency is true until a certain limit. The more resonators there are, the larger the surface of the tag is. Moreover, resonators must

be sufficiently separated in order to minimize the mutual coupling and avoid spurious effects between resonators. This shows clearly why it is essential to link the concept of data capacity storage with the surface of the tags. Thus, we may introduce the concept of surface encoding density of chipless RFID tags.

8.4. Read Range

The read range of chipless tags can't be defined in the same way as for passive UHF RFID tags. Indeed, in traditional RFID UHF applications, the read range of the system is limited by the downlink (reader to tag). Once the tag is powered enough (which depend on the mismatch impedance between the antenna and the RFID chip as well as the chip minimum operating power), the reader sensitivity is generally sufficient to retrieve the information from the tag. In the case of the chipless RFID, there is no problem of mismatch impedance or chip minimum operating power. Only the tag's capacity to backscatter enough power toward the reader direction and the sensitivity of the reader are to be considered. It is comparable to the radar equation, and we will use the tag Radar Cross Section (RCS) to give an indication of the tag's ability to operate at short or long distance. For all these reasons, chipless read ranges, in free space, are greater than those of passive UHF RFID.

8.5. Tags Immunity to Operating Environment

The tag immunity to environment, which is rarely taken into account even for the traditional RFID, is very important here. Indeed, this technology has to be used in real environments, and not only in anechoic chambers. This problem of tag sensitivity to the environment is all the more important when we consider chipless frequency approaches. Indeed, to gain encoding capacity for a given frequency band, it is interesting to have resonators with a high quality factor. We can optimize the frequency band by using a large number of these resonators. Contrary to this, the presence of any object near the tag will interact with these very sensitive resonators by changing their resonant frequencies. The more important this phenomenon is, the more disturbed the reading will be. We see through this example that the tag data storage capacity is directly linked to its surrounding environment. Thus, in a very constrained environment, the tag performance in terms of number of bits should be limited in order to keep a reading rate as constant as possible. In contrast, a very sensitive tag could be used as a sensing device.

9. INDICATOR OF MERIT

As we discussed, there are several parameters and effects to consider when designing chipless tags. Most of these parameters are interrelated, and usually only compromises between different criteria should be made for specific application. From there, it is interesting to build some global factors that can give an idea of the various compromises that we may have to do. If we consider the first three parameters described above, that is to say the data storage capacity, the frequency band, and the label dimension, it is clear that these three parameters are intimately interlinked, even if they describe different aspects. Therefore, it is important to introduce an indicator of merit that will take into account these three criteria. To this end, it is useful to introduce the concept of coding Density Per Surface (DPS) and coding Density Per Frequency band (DPF). Thus, it is possible to classify each approach in terms of DPS and DPF. From here, to gather these concepts on a same graph, we can represent the coding density per surface versus the coding density per frequency for each chipless approach. Figure 19 shows such a representation where the main published chipless approaches have been carried out. We can notice from Figure 19 that several trends have

emerged. One approach focuses on the coding DPS (Vena, Perret & Tedjini, 2011). In this case, the coding density per surface is very high. Other approaches target the coding DPF (Deepu, Vena, Perret & Tedjini, 2010). Finally, others opt for a compromise (Preradovic, Roy & Karmakar, 2009; Jalaly & Robertson, 2005) between the surface and the frequency band. We have also to keep in mind that the coding density per surface and frequency is strongly linked to the resolution of reading system.

10. EXPERIMENTAL MEASUREMENT SET-UP

As far as the experimental characterization of chipless device is considered, two main measurement set-up can be used. The first set-up is mainly suitable for tags encoded in frequency domain. In this case, to recover data it can be used a Frequency Stepped Continuous Wave system to get the amplitude and the phase response as a function of the frequency. The second set-up, more general, is a reading system able to sample the transient response and process it

In most cases, to detect the Frequency response of a tag, a VNA (Vectorial Network Analyzer) connected to one or two broadband antennas can be used. So, the scattering parameters such as S11 and S21 (see Figure 20 a)) are measured. The use of a VNA allows detecting very weak signals backscattered from a chipless tag since a VNA has a typical dynamic range in the order of 110dB and more. Typically some tags having an RCS (Radar Cross Section) of -20dBsm can be detected at 50 cm from the antennas with an emitting power of 0dBm. Usually it is used some directive antennas having a gain close to 10dBi in the UWB (Ultra Wide Band), from 3.1 to 10.6GHz, since most of the designs based on frequency encoding operate in this band. To demonstrate that the chipless RFID technology can operate outside a labora-

Figure 19. Density of coding per surface as a function of density of coding per frequency for several designs

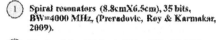

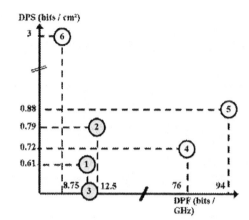

1. Spiral resonators (8.8cmX6.5cm), 35 bits, BW=4000 MHz, (Preradovic, Roy & Karmakar, 2009).

2. Dipole, (1.8cmX3.5cm), 5 bits, BW=400 MHz, (Jalaly & Robertson, 2005).

3. Phase encoded tag, (12cmX5cm), 3 bits, BW=« 350MHz », (Balbin & Karmakar, 2009).

4. Cavity tag (Δf=2MHz) (3x6cm), 13 bits, BW=170 MHz, (Deepu, Vena, Perret & Tedjini, 2010).

5. Cavity tag (Δf=1MHz) (3x6cm), 16 bits, BW=170 MHz, (Deepu, Vena, Perret & Tedjini, 2010).

6. "C" tag (Δf=50MHz) (1.5x2cm), 9 bits, BW=2200 MHz, (Vena, Perret, Tedjini, 2011).

tory, Preradovic & Karmakar (May, 2010) have designed a reading system based on this principle.

However, in order to detect all chipless tags (whatever is the encoding approach) it can be used a measurement system able to sample the transient response as depicted in Figure 20 b). It is generally composed of a pulse generator and a DSO (Digital Sampling Oscilloscope). In this case, the bi-static configuration using two antennas is preferred to isolate the received signal from the transmitted one. Since a DSO has a much lower sensitivity than for a VNA (typically -50dBm), a low noise amplifier has to be used before the DSO. To this end, RF compact amplifiers with a gain of 40dB operating in the UWB band are commercially available. This type of measurement system has a key role, since it is the way to develop some chipless tag systems able

Figure 20. a) Measurement set-up operating in the frequency domain; b) Reading system, based on the analysis of the transient response of a tag

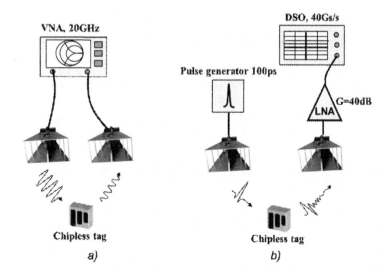

to fit the requirements settled by the FCC and ECC regulations for UWB systems. In order to detect a tag response with harmonics until 10GHz, the DSO sampling rate has to be as high as 20Gs/s. Regarding, the source pulse, to cover the band from 3.1 to 10.6GHz, a pulse width lower than 100ps is needed.

Finally, some works have also shown the possibility to use a waveguide set-up (Jang, Lim, Oh, Moon and Yu, 2010) or a cavity system (Deepu, Vena, Perret & Tedjini, 2010) to detect the resonance frequencies of a chipless tag encoded in the frequency domain. This way can overcome some practical issues such as the emitting power mask. But one of the main advantages of the chipless RFID technology that is the possibility to detect an object remotely and Non Line Of Sight is lost.

CONCLUSION

Even if the first chipless tag appeared in 1945, still now, this technology is in its infancy age. Chipless technology exhibits many practical advantages when compared to the traditional RFID based on IC chip and normalized communication protocols. Cost and robustness are among the most relevant advantages of chipless solutions. However, the coding capacity of chipless tags is quite low when compared to competing solutions such as barcode and traditional RFID. Many groups worldwide are conducting research projects in order to improve the overall performance of chipless solution and more specifically the amount of data that can be stored into the tag structure.

In this chapter we discussed different techniques for the implementation of a given ID. Some techniques are based on time response meanwhile some others exploit frequency response. In both cases the information can be coded under schemes similar to the well-known modulation technique PPM & OOK. Most of the published works exploit only one parameter for coding. Using more than one parameter and combining modulation PPM and OOK in the same device allow a significant improvement in coding capacity. Potentially, chipless tags storing tens to hundred bits are possible when exploiting UWB bands. The coding capacity of any chipless tag depends on its surface, the used frequency band and the ID implementation technique. To be able to address a quantitative comparison between several solutions it is helpful to introduce some global parameters such as the coding Density Per Surface (DPS) and the coding Density Per Frequency (DPF).

In this chapter we compared several chipless tags discussed in numerous publications. In most cases, the investigation was focused DPF or DPS parameters. New designs should consider both DPS and DPF parameters. In general this implementation requires a tedious phase of 3D-electromagnetic simulation and design. This is one of the challenges facing the development of chipless technology. However the availability of powerful 3D simulators as well as high performance computation platforms will contribute to accelerate the emerging of chipless solutions. Indeed, chipless is seen as an RF barcode solution, which exhibits very interesting advantages with respect to optical barcode such as read range and non line-of-sight capabilities.

Finally, in this chapter we did not consider the implementation of the ID (or a part of it) into the volume of the tag. Such an approach could be considered but at very high frequency i.e. in terahertz bands. Chipless tags operating in the THz range could be developed thanks to multilayer substrates and thus are compatible with microelectronic process. Some pioneering development can be found (Tedjini, Perret, Deepu & Bernier, 2009).

REFERENCES

Balbin, I., & Karmakar, N. C. (2009, August). Phase-encoded chipless RFID transponder for large-scale low-cost applications. *IEEE Microwave and Wireless Components Letters, 19*(8).

Blischak, A., & Manteghi, M. (2009, June). *Pole residue techniques for chipless RFID detection.* IEEE Antennas and Propagation Society International Symposium, Charleston.

Brandl, M., Schuster, S., Scheiblhofer, S., & Stelzer, A. (2008). *A new anti-collision method for SAW tags using linear block codes.* IEEE International Frequency Control Symposium.

Cumming, D. R. S., & Drysdale, T. D. (2003, March). *Security tag.* (Patent, GB 0305606.6).

Deepu, V., Vena, A., Perret, E., & Tedjini, S. (2010, June). *New RF identification technology for secure applications.* IEEE International Conference on RFID-Technology and Applications 2010.

Finkenzeller, K. (2004). *RFID handbook: Fundamentals and applications.* Wiley.

Giusto, D., Iera, A., Morabito, G., & Atzori, L. (2010). *The internet of things.* 20th Tyrrhenian Workshop on Digital Communications. Springer.

Glinsky, A. (2005, February). Theremin: Ether music and espionage. Urbana, IL: University of Illinois Press.

Härmä, S., Plessky, V. P., Li, X., & Hartogh, P. (2009, April). Feasibility of ultra-wideband SAW RFID tags meeting FCC rules. *IEEE Transactions on Ultrasonics, Ferroelectrics, and Frequency Control, 56*(4). doi:10.1109/TUFFC.2009.1104

Hartmann, C., Hartmann, P., Brown, P., Bellamy, J., Claiborne, L., & Bonne, W. (2004). *Anti-collision methods for global SAW RFID tag systems.* IEEE Ultrasonics Symposium.

Hartmann, C. S. (2002, October). A global SAW ID tag with large data capacity. *IEEE Ultrasonics Symposium,* (Vol. 1, pp. 65–69), Munich, Germany.

Jalaly, I., & Robertson, D. (2005, June). RF barcodes using multiple frequency bands. *IEEE MTT-S Digest.*

Jalaly, I., & Robertson, I. D. (2005, October). Capacitively-tuned split microstrip resonators for RFID barcodes. *European Microwave Conference* (Vol. 2).

Jang, H.-S., Lim, W.-G. Oh, K.-S., Moon, S-.M., & Yu, J.-W. (November, 2010). Design of low-cost chipless system using printable chipless tag with electromagnetic code. *IEEE Microwave and Wireless Components Letters, 20.*

Landt, J. (2001, October). *Shrouds of time: The history of RFID.* Retrieved from www.aimglobal.org/technologies/rfid/resources/shrouds_of_time.pdf

Mandel, C., Schüßler, M., Maasch, M., & Jakoby, R. (2009). *A novel passive phase modulator based on LH delay lines for chipless microwave RFID applications.* International Microwave Workshop Series, Croatia.

Mukherjee, S. (2007). Chipless radio frequency identification by remote measurement of complex impedance. *Proceedings of the 37th European Microwave Conference,* Munich, Germany.

Mukherjee, S., & Chakraborty, G. (2009, December). *Chipless RFID using stacked multiple patches.* Applied Electromagnetics Conference, Kolkata.

Perret, E., Hamdi, M., Vena, A., Garet, F., Bernier, M., Duvillaret, L., & Tedjini, S. (2011). *RF and THz identification using a new generation of chipless RFID tags. The Radioengineering Journal - Towards EuCAP 2012: Emerging Materials.* Methods, and Technologies in Antenna & Propagation.

Perret, E., Tedjini, S., Deepu, V., Vena, A., Garet, & F., Duvillaret, L. (2010, February). *Etiquette RFID passive sans puce.* Patent, FR, 10/50971, N. REF: B100087.

Preradovic, S., Balbin, I., Karmakar, N. C., & Swiegers, G. (2008, April). A novel chipless RFID system based on planar multiresonators for barcode replacement. *IEEE RFID Conference*, (pp. 289 – 296). Las Vegas, Nevada, USA.

Preradovic, S., Balbin, I., Karmakar, N. C., & Swiegers, G. (2008, October). Chipless frequency signature based RFID transponders. *38th European Microwave Conference*, (pp. 1723-1726). Amsterdam, Netherlands.

Preradovic, S., Karmakar, N., & Balbin, I. (2008, October). RFID transponders. *IEEE Microwave Magazine*, *9*(5), 90–103. doi:10.1109/MMM.2008.927637

Preradovic, S., Karmakar, N., & Balbin, I. (2010, December). Chipless RFID, Bar code of the future. *IEEE Microwave Magazine*, *11*(7), 87–97. doi:10.1109/MMM.2010.938571

Preradovic, S., & Karmakar, N. C. (2010, May). *Multiresonator based chipless RFID tag and dedicated RFID reader*. IEEE MTT-S IMS, Anaheim, USA.

Preradovic, S., Roy, S., & Karmakar, N. C. (2009). *Fully printable multi-bit chipless RFID transponder on flexible laminate*. Asia Pacific Microwave Conference.

Stockman, H. (1948). *Communication by means of reflected power* (pp. 1196–1204). IRE Proceedings.

Taylor, D. (2009, April). Introducing SAR code – A unique chipless RFID technology. *RFID Journal, 7th Annual Conference,* Orlando.

Tedjini, S., Perret, E., Deepu, V., & Bernier, M. (2009, September). *Chipless tags, the RFID next frontier.* Invited paper, 20th Tyrrhenian Workshop on Digital Communications, Sardina, Italy.

Vena, A., Perret, E., & Tedjini, S. (2011, March). Novel compact RFID chipless tag. *Progress in Electromagnetics Research Symposium 2011*, Marrakesh, Morocco.

Vena, A., Perret, E., & Tedjini, S. (2011, June). *RFID chipless tag based on multiple phase shifters.* IEEE MTT-S IMS, Baltimore, USA.

Zhang, L., Rodriguez, S., Tenhunen H., & Zheng, L. R. (2006, June). An innovative fully printable RFID technology based on high speed time-domain reflections. *High Density Microsystem Design and Packaging and Component Failure Analysis,* (pp. 166 – 170).

Zheng, L., Rodriguez, S., Zhang, L., Shao, B., & Zheng, L. R. (2008, May). Design and implementation of a fully reconfigurable chipless RFID tag using inkjet printing technology. *IEEE International Symposium on Circuits and Systems*, (pp. 1524 – 1527).

KEY TERMS AND DEFINITIONS

Coding Capacity: The capacity of the bits that can be held on the chipless RFID tag.

DPF: Density per Frequency

DPS: Density per Surface

Electromagnetic Signature: The unique identity using electromagnetic wave on every single tag to differentiate between multiple tags.

Frequency Domain: A term used to describe the domain for analysis of mathematical functions or signals with respect to frequency.

Modulation: The process of varying one or more properties of a high-frequency periodic waveform, called the carrier signal, with a modulating signal which typically contains information to be transmitted.

Time Domain: A term used to describe the domain for analysis of mathematical functions or signals with respect to time.

Chapter 9
Printing Techniques and Performance of Chipless Tag Design on Flexible Low-Cost Thin-Film Substrates

Rubayet-E-Azim Anee
Monash University, Australia

Nemai Chandra Karmakar
Monash University, Australia

Sushim Mukul Roy
Monash University, Australia

Ramprakash Yerramilli
Securency International Pty Ltd, Australia

Gerhard F. Swiegers
University of Wollongong, Australia

ABSTRACT

Radio Frequency Identification (RFID) is an emerging technology playing a vital role in modern automatic identification system. Chipless RFID is a new dimension in the field of radio-frequency application systems with immense potential to manufacture low-cost, multi-bit RFID tags for potential barcode replacement on polymer, paper, and other flexible substrates. In this chapter, the authors present a detailed overview of the printing methods, substrates, and materials used for printing chipless RFID tags. Based on the available literature, an attempt is made to review the printing and performance related issues of printed RF devices that are currently published. The basic aspects of printing of chipless tags with conductive inks are discussed in brief.

1. INTRODUCTION

RFID stands for Radio Frequency Identification which uses the radio frequency for communication between tags and reader through free space. A typical RFID system consists of a Reader, transponder (RFID tags) and processing unit for processing the data from the transponder as shown in Figure 1.

Automatic identification procedure has become popular in this age of modern technology and is playing a vital role in recent business activities of global economy. In coping up with the fast

DOI: 10.4018/978-1-4666-1616-5.ch009

Figure 1. Schematic of a generalized radio frequency identification system

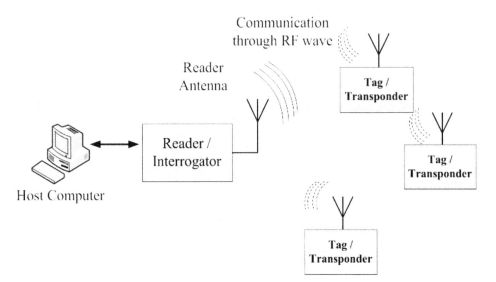

paced lifestyle of this Century, the demand for automation is on the increase day by day. In this respect, Radio Frequency Identification (RFID) finds an ever increasing application and has only been emerging recently as a powerful alternative to conventional security technology for product authentication. RFID systems have been gaining more and more popularity in areas such as supply chain management, inventory control, automated identification systems, and any place requiring identifications of products or people. RFID technology is better than barcode in many ways (Non Line of Sight Communication, longer operating range, quick reading of tags etc.)(Preradovic & Karmakar, 2010)), and may totally replace the older 'barcode identification system' all together in the future, if the printing of tags on flexible low-cost substrates hits commercial success. A typical RFID (Radio Frequency Identification) system (Figure1) includes RFID tag (Transponders), reader (Interrogation Unit) and Data Processing Unit (Host Computer for necessary processing on received data to decode the Tag ID). Tags are attached to objects/persons, and readers communicate with tags in their transmission ranges via radio signals. Yang Xiao et al (Xiao et

al., 2007) discussed in depth about the technologies, applications and research issues of RFID. Here it can be concluded that issues relating to the cost efficiency of RFID equipment, other issues such as energy efficiency, interference from multiple readers and security issues need further investigation.

The main hindrance to commercialization of RFID technology comes from the price of the tag which is much higher compared to the traditional barcode feature. The bulk of the cost of RFID tag comes from the silicon chip that is used to store data. Significant research and development investments are in place worldwide to replace the silicon chip with printed circuitry (printed transistors / resonating circuits) on the same substrate. The antenna is either printed together with the circuitry or separately which is then attached to the circuit with some low-cost technique (Subramanian et al., 2010). Chip-less tag does not require any chip/IC hence there is no need of a battery to turn on the chip. Such tags can be printed on low cost substrates using conventional printing technology with conductive ink for use in item level tagging and other applications. As chipless tags do not require any power supply to turn on its response, they can work with low power and even from long distances.

2. OVERVIEW OF RFID TAGS: THE CHALLENGE OF BARCODE REPLACEMENT

2.1. RFID Tags

RFID tags at first can be divided into two broad categories as Chip-based RFID tags and Chipless RFID tags. The traditional chip-based tags can be classified as active, passive and semi-passive tags and they are consisting of Integrated Circuit (IC) and antenna. Active and semi-passive tags have on-board battery for power supply for the tag operation. Active tag uses its battery to broadcast the radio wave to reader whereas the semi-passive tag relies on reader to supply power to turn on the IC of the tag thus broadcasting the tag ID. Passive RFID tag relies completely on the transmitted power form the reader as it does not have any on-board power supply. These traditional chip-based tags are costly mainly because of the IC. Chipless tags do not use any IC or battery. Hence it is less costly than their chip-based counterpart. Chipless tags that have been reported in literature can be classified by their operating principle into various groups. They can be broadly categorized by time-domain and frequency-domain based tags. Time domain based tags includes SAW tags (Along et al., 2010), delay-line based tags (Chamarti & Varahramyan, 2006; Shrestha et al., 2009) and group-delay based tags (Gupta et al., 2010). SAW tags use piezoelectric crystal to create surface acoustic wave. Hence they are not fully printable. The delay line based tags and group delay based tags have lower number of bit capacity though they are printable. Frequency domain based tags encode data in frequency domain using resonant structures. They include magnetically coupled L-C resonators (Jalaly & Robertson, 2005) where resonances are created by an N-element dipole array. Space-filling curve (McVay et al., Aug. 2006)

used as spectral signature has also been reported where the tag is a frequency selective surface and is manipulated by the Peano and Hilbert curve to give resonances.

2.2. Barcoding Technologies

While RFID systems have made remarkable strides in the identification market, they have yet to achieve a significant displacement of barcode technology. Barcode identification is a line-of-sight process that involves scanning a printed code with a small laser apparatus. In its most common embodiment, the code uses a sequence of printed vertical bars and spaces to represent numbers and symbols. The laser beam is reflected or absorbed by the bars and spaces, with the resulting patterns detected by the reader and converted into digital data according to a conversion protocol. This data is then transferred to a computer for immediate action or stored for later use. Several barcode reading protocols (called "symbologies") exist. Since 1973, the Uniform Product Code (UPC) has converted 1D (or "one-dimensional") barcodes used in most retail stores. The European Article Numbering (EAN) system employed in Australia was developed by Joe Woodland, the inventor of the first barcode system. EAN is a superset of the UPC that allows extra digits for country identification. Barcodes have traditionally contained relatively limited information – typically a single code for each type of item to which they are affixed. For example, Code 30 1-D barcodes, which are widely used to mark supermarket goods, convert to 10 data characters, each of which may be numeric or alphabetic. The alphabetic characters are only single case (usually the upper case). Since each of the 10 positions in the code may be occupied by an alphanumeric character (26 possibilities) or a numeral (10 possibilities), such barcodes offer a maximum of $(26 + 10)^{10} = 3.66 \times 10^{15}$ possible permutations. This is equivalent to a 51-bit binary code. So-called 2D barcodes may encode substantially more information. For example, a PDF417 2D barcode provides 62 data characters, each of which may be alphabetic (upper or lower case) or numeric.

2.3. Chipless RFID for Barcode Replacement

While exceedingly effective and market proven, barcodes have the significant limitation that they are a line-of-sight technology. This makes automatic reading problematic; a human operator is typically required. RFID systems potentially overcome this key disadvantage. Because they can be read remotely, over an intervening distance, a human operator is typically not required. This makes it practical to read RFID tags at many more points in a logistic- or other distribution chain. They can also be read more often, thereby locating chain failures more rapidly. So, why have RFID tags not replaced barcodes? The main impediment as alluded above is the cost of the tags. For most logistics purposes, RFID tags must offer at least 1 billion possible unique addresses in order to compete with barcode technology. However, the only to produce such a code on a RFID transponder has traditionally been by incorporation of a silicon chip. The cost of fabricating and affixing such a chip is substantial (>10 cents per tag), even when it is manufactured in multi-billion volumes. By contrast, barcodes cost fractions of a cent to print. The solution to this dilemma is to develop a fully-printable, multi-bit RFID tag. The multiresonator-based chipless RFID tag (Preradovic et al., Apr. 2008) designed and patented by S. Preradovic *et al.* is a fully planar chipless tag of the type required. It is easy to print on polymer or paper substrate with conductive ink and has been shown to be capable of encoding with at least 36-bits of information. Another fully planar and printable tag has been reported (Balbin & Karmakar, 2009) which is based on backscattered signal from the tag. Both of these planar tags have high data capacity making them suitable for printing on polymer, paper and other flexible substrates.

3. PRINTING REQUIREMENT

The ability to print a complete wireless RFID tag on flexible substrate opens the door for its use in many low cost applications. Printing is a fully additive process enabling direct deposition of ink on the substrate without the need to go through steps such as etching, stripping and cleaning which are most common in photolithography process. The direct nature of printing with low ink volumes at high speed reduces the manufacturing cost for commercial markets (Subramanian, et al., 2010). The requirements for mass production of printed RFID tags (Blayo & Pineaux, 2005) are outlined in Table 1.

Careful considerations of the above requirements lead to the choice of a narrow window of printing techniques and conductive inks. The printing technique and the conductive ink that is used have its own merits and demerits. In principle, the conductivity of the printed metal track should be as close as possible to the bulk conductivity of the metal, but in practice this is difficult to achieve, considering various constraints of the printing processes with conductive inks.

4. PRINTING TECHNIQUES

The availability and the method of application of suitable inks therefore occupy a very prominent place in tag printing (Subramanian, et al., 2010). Literature (Subramanian, et al., 2010) shows that different kinds of conductive pastes have been used to print conductive tracks, antennas etc. on flexible polymer and paper substrates using ink-jet, roto-gravure, screen printing and other techniques. The subject of gravure printing is dealt in detail by Alejandro (Vornbrock, 2009) in a technical report. Anne Blayo and Bernard Pineaux (Blayo & Pineaux, 2005; Huebler et al., 2002; Sangoi et al., 2005) covered a brief account of commonly used printing methods and their role in RFID printing. An overview of these techniques is given here.

Table 1. Essential requirement for printed RFID tags

Accuracy
• Aspect ratio of printed slot and lines should closely correspond to the design.
• The accuracy of the printed form of the design depends on the accuracy of photographic reproduction of the original design and other process parameters of related cylinder engraving (etching) in the case of gravure printing.
Resolution
• The line and slot widths of RF tags in the GHz frequency range will be small in size, usually in the micrometer range. This poses a printing challenge due to difficulties in obtaining engraved cylinders with high accuracy and conductive inks that are suitable for high speed printing. This is especially true for gravure printing.
Conductivity
• The electrical conductivity of the printed conductor directly affects the RF performance of the tag. The conductivity that can be obtained in practice depends upon the sintering temperature. The sintering conditions are chosen based on the type and material of the substrate. For example Polypropylene has a low melting point and therefore sintering temperatures cannot go beyond 120°C for this material. Silver coatings on PET (polythylene terephthalate) can be sintered at 150°C.
• The sintering temperature of various inks depend on the size of the conductive particles in them. The sintering temperature is low and is approximately between 70-100°C for nanoparticle inks. For micron sized particles, the required temperature is 100~200°C.
Thickness
• Signal strength and propagation depends on the conductor thickness that often needs to exceed the skin depth for obtaining acceptable performance of the tag.
• With certain types of conductive paste or inks, the electrical onductivity shows a dependence on the print thickness. In this case, thicker tags are to be printed to obtain high electrical conductivity and improved RF performance. In comparison, nanoparticle sized inks can be printed in lower thickness to obtain similar levels of RF performance.
Longevity
• It depends on the robustness of the printed tag. This brings in several requirements such as adhesion, flexibility and stability of the printed inks with respect to the substrate and operating environment.
• Tags must have long shelf life to be used for long time on the target substances without deterioration. In the case of banknotes, use of overlayers on top of the tags may be considered to increase their life in circulation.

4.1. Flexography

Flexography is essentially a rubber stamp on a roll of the selected substrate. In this technique, the pattern to be printed is raised out of the surface of the printing form. This raised surface then gets inked, and the ink is then transferred to the substrate. Solvents are removed with heat applied via IR radiating lamps or UV lamps arranged under the moving web. Flexography can be used as a laboratory printing technique because of the ease of manufacture and low cost of the rubber printing plates. The main concern for printed RFID tags lies is the drying of the ink around the raised edges leading to narrowing of slots and widening of the lines. Water based inks are difficult to work as they dry quickly around the raised edges of the tracks than solvent based inks. Solvent based inks might cause swelling or deterioration of the plate due to reaction between the plate material and the solvent. The collection and drying of the ink around raised edges during a long print run ultimately leads to shorting between two adjacent tracks separated by a slot. This can be avoided by careful selection of anilox roller size and print speed suitable for the printing ink.

4.2. Offset Lithography

Printed pattern with ink and water surrounding it is impressed on the paper which absorbs water and prevents ink from spreading. Two transferring processes are involved: one from master to offset roll and the other from the offset roll to the substrate.

Lithography, specifically offset lithography, is one of the most widely used printing techniques in newspaper industry. It is a chemical process.

Lithography depends on the principle that oil and water do not mix. Images (words and art) are put on plates which are dampened first by water, then ink. The ink sticks to the image area, the water to the non-image area. Then the image is transferred to a rubber blanket and from the rubber blanket to paper. Before the job can be printed, the document must be converted to film and "plates." Images from the negatives are transferred to printing plates in much the same way as photographs are developed. A measured amount of light is allowed to pass through the film negatives to expose the printing plate. When the plates are exposed to light, a chemical reaction occurs that allows an ink-receptive coating to be activated. For electronics, issues generally arise due to cross-contamination that is difficult to control and use of substrates that do not absorb water. The use of two patterns transfer will also lead to additional loss of pattern fidelity which is clearly undesirable in RFID device printing. Figure 2 presents the process of Offset Lithography Printing in brief.

4.3. Ink-Jet

Ink-jet is the most developed and evolved printing process in last few years and is very much suitable for research and prototyping. Ink may be water-based, solvent-based, hot melt or UV curable. Curing of the ink is usually accomplished during printing by heating the substrate to the desired temperature for non-UV curable inks. The ease of printing directly from the design file onto substrate of choice and size without much wastage of ink makes this technique unique and widely used in printed electronics. It goes without saying that very complex ink formulations are required to accomplish printing of RFID patterns. Typical ink formulations have their viscosities around 10 mPa.s. The ink jet process may be of two types: continuous ink-jet and "drop on demand" (DOD) ink jet. In the DOD process one single ink droplet can be jetted from the reservoir through the nozzle

Figure 2. Offset lithography printing process (Kipphan, 2001)

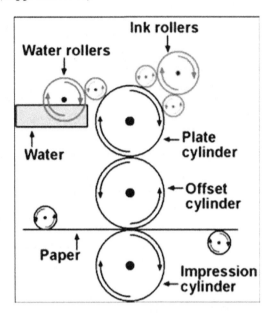

often by the vibration of a piezo element (piezo system). The sizes of DOD droplets can be very small, and tracks of 25-50µ wide can be printed with ease. Problems due to clogged cartridges are very common in ink-jet printing. This is often due to incorrect ink formulation which does not match with the type of cartridge that is specified for the particular printer model. Cleaning of clogged cartridges is a laborious job. For this reason, it is recommended to use new cartridges whenever this happens.

4.4. Screen Print

The ink is dragged along the stencil with a squeegee, forcing the ink to transfer on to the substrate. The wet prints are usually dried in an oven through solvent evaporation. UV, IR or electron beam curing are also used. Basic screen printing consists of a frame onto which a mesh is stretched on one face, hinged to a baseboard, and a flexible squeegee which forces the ink through the clear areas of the mesh and onto the substrate positioned on the baseboard. Choosing the right mesh to suit a particular job can be just as critical as choosing the right ink. The primary purpose of the mesh is to support the stencil, but beyond that the choice of mesh has considerable influence on print definition, ink usage, and stencil durability. This is more so in the case of printing RFID tags. The screen mesh count depends on the conductive ink used. It is different for flake and nanoparticle inks and changes with the target printing resolution that needs to be obtained for a particular RF design. Screens made of well stretched fabric are generally suitable for flake particle inks. To prints with nanoparticle inks, stainless steel screens are specially used by many suppliers. It is always cost effective to get the screens made by professional suppliers. The information regarding ink formulations such as viscosity, type of solvent and particle size may be required to supply for accurately determining the process of stencil making.

Some of the problems associated with screen printing of RFID patterns are the suitability of the viscosity of the ink to the printing process and the amount of squeeze force applied while dragging the ink to fill up the pores in the stencil followed by another squeeze movement to push the ink off the pores for transfer to the substrate. The squeeze force applied depends on the operator in manual screen printing trials. With automatic printing machines, this problem will not exist. However, enough proofs have to be printed for checking the quality of the printing to ensure correct settings before proceeding with long print runs.

Screen printing although is used by many to print RFID patterns, is not economical for expensive Nano-inks. Often it is necessary to print thicker tracks to keep the conductivity high enough for signals to be received and retransmitted back to the reader. Typical conductor thicknesses are 5 to 10μm for most of the conductive inks. Printing of Nano-inks in this thickness is not economical due to their high cost.

4.5. Gravure

A good coverage of the gravure printing process as applied to printing electronics is given by Donovan Sung et al (Sung et al., 2010). Gravure is a roll-to-roll printing technique which has high throughput, long print runs, uniformity, and versatility. In gravure printing, an engraved or chemically etched cylinder is rolled over a moving substrate, typically paper or plastic. A doctor blade, wipes off excess ink from the non-image areas of the cylinder surface before contact with the substrate. The important aspects of Roto-gravure printing is shown in Figure 3. Gravure process is attractive for printing electronics including RFID patterns in high volumes compared to any other printing process.

According to Donovan Sung et al (Sung, et al., 2010), the two very important actions steps in a gravure printing process is 1) cell emptying and 2) drop spreading. The printed dot size does not linearly increase with increase in the cell size. This puts a limitation of the cell size in terms of the width and cell to cell spacing for obtaining accuracy in printing. The authors (Sung, et al., 2010) conducted a thorough in-depth research on gravure printing and observed that the cell emptying is affected by a number of variables such as cell width, cell aspect ratio, print speed, ink viscosity, and ink/substrate surface energies. They observed that wider cells are preferred as they empty out more ink; high surface energies of the substrate pull the ink more effectively out of the cell, low viscosity inks offer less resistance to shear than high viscosity inks. The print speed also determines the amount of ink transfer from the cell to the substrate. At low prints speeds (10 meters per minute), with high viscosity inks, more ink would transfer than at higher speeds (50-150 meters per minute). Low viscosity inks tend to flood or 'bleed out' the substrate and reduce image sharpness. This is more pronounced at low speeds than at higher print speeds. The gravure

Figure 3. Roto-gravure printing process

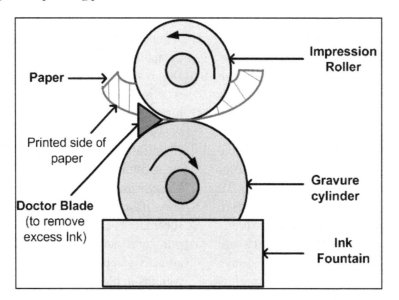

industry generally has its own viscosity standards for printing inks. Sung (Sung, et al., 2010) concludes that it is possible to print uniform lines whenever the ratio of cell spacing to cell width is between 1.06 and 1.40. If this ratio is much larger than 1.40, then scalloped lines will be produced. The cell spacing is defined as the distance measured from the center of one cell to the center of the adjacent cell.

A thorough understanding of the parameters that control gravure printing of conductive inks is required for researchers working with this process. It is also necessary to optimize the process parameters to obtain high print quality. Some of the process parameters are: ink preparation, substrate preparation, drier setup, printer setup, doctor blade angles, impression pressure on the cylinder, print speed etc.

Figure 4. Inks for printed electronic devices

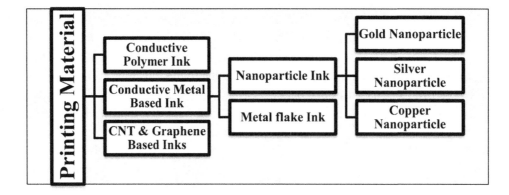

5. SUBSTRATE MATERIAL AND CONDUCTIVE INKS

The classification shown in Figure 4 gives a broad picture of the variety of materials that can be used in conductive inks for printed electronics. The subject of conductive inks is discussed in detail in the following paragraphs with a brief relevance to the RF performance.

5.1. Conducting Polymer Substrate

Polymers are usually non-conductive in nature and suitable for application as insulating materials. But because of their ease of fabrication and flexibility, they stand as excellent candidates in various applications where both flexibility and conductivity is required. Polymers are modified to make them electrically conductive (Bechevet et al., 2005; Verma et al., 2010) for a variety of applications. One of these applications is for printing RFID tag antennas on a choice of flexible substrates as presented in the literature (Verma et al., 2010). A recent article (Borowiec, 2010) described an RF antenna made from a conducting polymer with 900MHz, 2 and 6 GHz resonant frequency. There are several conducting polymers such as polyaniline, polypyrole, polythiophenes, and poly (3, 4-ethylenedioxythiophene) (PEDOT) etc. (Borowiec, 2010). Among these, PEDOT is found to exhibit relatively high conductivity and stability (Sankir, 2008).

Sankir (Sankir, 2008) reported the fabrication and performance of a 6 GHz microstrip patch antenna with PEDOT and Polypyrrole. Though the performance is low corresponding to Copper based antenna, nevertheless, the performance of all the patterns tested, showed almost similar kind of radiation pattern with nearly identical directivity (Verma, et al., 2010). Conducting polymers (CP) are also known for their optical transparency. An optically transparent CP RFID meandering dipole antenna has been fabricated (Kirsch et al., 2009) using Clevios PH500 PEDOT-PSS and has a

conductivity of 300 S/m. The conductivity was improved to 5×10^5 S/m with surface modification to lower the surface tension. Another CP antenna was reported (Kirsch, et al., 2009) on PET (Polyethylene Terephthalate) and was designed for 900MHz center frequency but failed to reveal distinct resonant frequency. In spite of the distinct advantages of flexibility, optical transparency and low cost, conducting polymers to date have not been able to demonstrate optimum performance as antennas. This has largely been due to their lower gain and conductivity compared to any antennas printed in conductive metal inks.

5.2. Conductive Inks

Conductive ink in either metal or polymer form is required for printing electronic devices on paper and flexible plastic substrates with a goal to minimize the production cost of end products. Conductive metal inks are available in three forms: solvent with binder form, water based stabilized for a specific pH range and UV-curable (Magdassi, 2009; Oldenzijl et al., 2010) for printing electronics. The main constituents of conductive inks are: pigment (metal) which determines the conductivity of the printed track, binder that keep the pigment particles together, solvent for pigment dispersion and additives to improve the adhesion, rheology, free flow etc. (Oldenzijl, et al., 2010). Conductive inks can be formulated from flake shaped metal particles or nanometer sized particles.

5.2.1. Metal-Based Nanoparticle Inks

Conductive inks available today mostly use Silver (Ag) pigment for conductivity. Silver is used in micron, sub-micron or nanoparticle size. Occasionally gold nanoparticle inks are also employed. Copper and aluminum are inexpensive compared to gold and silver but due to their quick oxidation nature to form insulating oxides, found very limited use in printed electronics (Vornbrock,

2009). Carbon based nanoparticle inks are low cost, but show higher resistivity than metal inks (Magdassi, 2009). Silver based inks are known for their stability and higher conductivity than either copper or aluminum (http://www.coolmagnetman. com/). It does not show reduction in conductivity due to oxidation (silver oxide is conductive) (http://www.qsl.net/). The price of silver metal varies from time to time increasing the cost of silver inks.

Metal nanoparticles sinter at lower temperatures than their bulk form (HUANG et al., 2003; Magdassi, 2009) and therefore are more suited to certain class of plastic substrates, well known example being polypropylene with a distortion temperature of $120^{\circ}C$. Silver nanoparticle inks with a range of mass loadings from < 10% to ~80% are available for different printing processes, also in a range of viscosities.

Printing on flexible plastics with nanoparticle inks was first reported in 2003 where organic encapsulated gold nanoparticles were printed on a substrate using ink-jet printing technique. By optimizing the size of the nanoparticle and also the length of encapsulant, continuous gold films with low resistance, annealed at less than $150^{\circ}C$ were obtained (HUANG, et al., 2003). There are reports (Volkman et al., 2004) on Ink-jet printing with silver and copper nanoparticle inks. Volkman (Volkman, et al., 2004) found that with gold, the "conduction temperature" which is the minimum temperature at which the nanoparticles sinter to form a continuous line could be reduced using shorter alkane thiols. He did not observe (Volkman, et al., 2004) any such phenomenon with silver nanoparticles and found no relationship between the conduction temperature and the length of encapsulant. A CPW-fed rectangular monopole antenna was printed (Rida, et al., 2009) using an inkjet printer with silver nanoparticle ink on LCP (Liquid Crystal Polymer). The conductivity of the paste varied from $0.4 \sim 2.5 \times 10^7$ S/m depending on the curing temperature (Rida, et al., 2009) used. According to the literature the measured result showed good agreement with simulation except for minor shift in resonant frequency.

Shaker (Shaker et al., 2011) reported ink-jet printing of a planar monopole ultra wideband (UWB) antenna on paper. Cabot conductive ink CCI-300 was used for the printing. According to Magdassi et al., (MagdassiGrouchko et al., 2010), this particular ink was curable at $100^{\circ}C$ due to surface modification effect of the ultra-fine silver nanoparticles (http://www.cabot-corp.com). The dc conductivity reported was $\sim 9 \times 10^6$ to 1.1×10^7 S/m. The printed antenna showed a good correlation between simulation and measurements up to 16GHz. The antenna exhibited Omni-directional radiation pattern up to 10.5 GHz and efficiency better than 80% throughout the whole frequency band as mentioned in the reported literature.

Narkcharoen and Pranonsatit reported (Narkcharoen & Pranonsatit, 2011) a printed chipless RFID multi-resonator tag. It was printed on a Teflon substrate with water-based silver ink through an additive technique Fill-until-Full (FuF). The printed tag showed performance as expected from simulation results.

The copper conductive inks will offer a low cost solution to printed electronics compared to silver inks. Copper oxidizes quickly in ambient conditions to form a layer of non-conducting oxide reducing the printed track conductivity (Magdassi, et al., 2010). The oxidation of copper can be prevented if it can be coated with another material for protection against oxidation. Among the materials investigated for oxidation protection are carbon, graphite, polymers, silica and metals in general (Magdassi, et al., 2010). Graphene coated copper nanoparticles are reported (Luechinger et al., 2008) to be stable up to $165^{\circ}C$ and due to the exceptionally high conductivity of graphene, the copper ink coatings are expected to retain conductivity due to graphene over layer. Polymer over coating of copper nanoparticles caused increase in resistance (Magdassi, et al., 2010), by 3 to 5 orders of magnitude than bulk copper. Kobayashi and Sakuraba reported results of silica coating of copper nanoparticles (Kobayashi & Sakuraba, 2008). Being an insulating material, silica coating is not attractive for printing electronics. Another approach (Grouchko et al., 2009) is to modify the copper core-shell structure with a coating of another metal. The modification can be accomplished by various methods (Magdassi, et al., 2010). Conductive tracks and RFID antennas printed using such bimetallic inks (Magdassi, et al., 2010) showed poor performance due to the presence of organic stabilizers in the ink. The measured resistivity was high due to the incompatibility of materials during the sintering process (Magdassi, et al., 2010). A new sintering process for silver nanoparticles that enabled sintering at room temperature has been reported (Magdassi, Grouchko et al., 2010). A drastic decrease in resistance was observed. It is therefore concluded that among bimetallic inks, silver modified copper nano-ink offers high electrical conductivity (Grouchko, et al., 2009). This can open up new possibilities in the field of printed electronics by reducing the manufacturing cost for printed electronic products.

5.2.2. Inks Containing Metal Flakes

Metal flake based inks can also be used in printing RFID tags and other printed electronic products. The essential pigment is in the shape of flakes, which are particles typically $> 1\mu$. Flakes of various sizes are used (HUANG, et al., 2003; Pudas et al., 2005) to maximize conduction. The higher the number of contact points among the flakes, higher will be the conductivity of the track. Pudas (Pudas, et al., 2005) reported an experiment where ink containing silver flakes in a polymer binder was used to print various patterns on plastic and paper using R2R printing. The result showed higher resistance for patterns printed on paper than plastic due to the difference in their surface roughness. Binder based inks using flake shaped metal particles are known to exhibit lower conductivity of the printed traces largely due to

poor contact between metal particles. In order to increase the conductivity, printing of thicker ink layers is required (Siden et al., 2007). Siden (Siden, et al., 2007) reported results on flexography printed antennas with flake particle inks. The thickness was varied to achieve increase in the conductivity. The low cost of flake inks compared to nanoparticle based inks is always attractive to keep the cost of RFID tags low for commercial markets. It must be remembered that RFID tags require high electrical conductivity for best performance. Often this requirement is to be met for low melting ($< 120°C$) plastics. In this case, flake based inks are not much useful as their curing temperatures generally exceed $120°C$ to enable efficient sintering of the particles. It has to be mentioned here that polyethylene terephthlate (PET) is an exceptionally stable material and has a higher temperature resistance ($\sim 160°C$). Obviously the only choice is nanoparticle based inks for printing on plastics that have low temperature resistance ($<120°C$).

5.3. Graphene

Graphene is a new class of material that has recently been developed and offers a potentially low-cost conducting substrate. The element carbon (symbol: C) exists in Nature in several different forms, of which the best known is coal. Graphite is a form of coal in which the carbon atoms exist in a multi-layered arrangement. Because the layers readily slide over each other, graphite is mainly used as a solid-state lubricant in, for example, pencil leads. Graphite is also highly conducting; it is used as an electrode in several electrochemical processes. Recent research has developed a method to separate the layers of graphite and suspend them in liquid solution. When the liquid is printed, the individual layers of carbon may be deposited in an extremely thin arrangement that is, nevertheless, highly conductive. This individual layer, some of which may be overlapping, is called Graphene. Graphene is characterized by superior electrical and mechanical performance (Dragoman et al., 2010). Documented research (Deligeorgis et al., 2009; Neculoiu et al., 2010) proves that graphene in its nanoparticle form can be used in microwave applications and is tunable to 50Ω impedance. Graphene based inks have opened up a new dimension in printable conductive inks. According to a report (Sanjay Monie), Vorbeck has developed a conductive ink based on crumpled graphene which is very cheaper alternative to expensive silver inks. The main advantage of graphene-based ink is that it does not require any heat-treatment after printing unlike silver inks which need to be sintered at suitable temperatures. Graphene is undoubtedly, a potential candidate for low cost printed electronic products. It has also been reported (Loo et al., 2008) that graphene-based inks are flexible and allow multiple folding of printed tracks without any reduction of the conductivity.

Table 2 presents a summary of the conductive Inks outlining their advantages and disadvantages.

6. PRINTING ISSUES AND TAG PERFORMANCE

The functionality of printed RFID tags depends on various factors that control the printing quality. Among them, the achievable printing tolerance plays a key role and is often dependent upon design complexity. For best performance, the printing tolerances should be within ± 1μ of the designs. This is difficult to obtain in practice, particularly with flexo or gravure methods of printing. The printing defects that occur therefore depend on the type of printing process. Each printing process has its own limitations due to factors that are often controlled by ink related properties. As an example, screen printing of Ag based inks require a specific viscosity range than gravure and flexo based inks. Ink transfer in gravure printing is a function of groove depth, ink viscosity, speed of printing, solvent evaporation rate etc. In flexo,

ink transfer depends on the anilox roller (bcm count), the print speed and printer settings. In screen printing, printing accuracy depends on the screen resolution which varies with ink particle size. With Nano inks, high resolution screens are generally preferred. With flake inks, low resolution screens can be used. The printing quality also depends on the adoption of manual or automatic process of printing. Manual printing depends on the operator's expertise to obtain good quality prints. The effects of some of these factors are discussed below with pictures wherever available.

6.1. Printing Issues

The printing defects can be broadly put into three different categories. They are described below.

6.1.1. Ink Overflow

Track widening occurs due to ink overflow. This occurs at low speed printing in flexo and gravure processes. After the completion of ink transfer process, the ink remains wet and flowing until fully dried hence causes widening of the lines and narrowing of the slots. Ink overflow results in a dimension change of the RF antenna and resonating structures causing a shift in the performance relative to simulation. There are various other reasons of ink overflow leading to spreading but the main factor is the viscosity of the ink and the evaporation rate of the solvent from the ink (Sung, et al., 2010). Optimisation of the viscosity, printer settings and speed of printing will improve printing accuracy. Figure 5(b) shows a picture of printed slots and lines (original design and dimension shown in Figure 5(a)) on a polymer substrate with conductive silver paste. In the original design, the width of lines and slots is kept equal. It is clearly evident that the printed width of the lines and slots (Figure 6(b)) is not equal and do not correspond to the original design. The printed lines showed constrictions at several places and also breaks in printing and shortening in some points. Figure

Table 2. Summary of the conductive inks

Conductive Polymer
Advantages
• The flexibility of printing on plastics and flexible substrates.
• Possibility to combine transparency and conductivity for increasing its suitability for security against counterfeiting.
Limitations
• Printed antennas show lower gain and conductivity.
• Thicker layers are required for increased performance.
Nanoparticle Ink (Gold,Copper,Silver)
Advantages
• The required sintering temperature is low which makes it suitable for printing on paper and polymer substrates.
• Silver nanoparticle inks are widely used due to its high conductivity and less oxidation tendency than copper.
• Copper nanoinks can be used with a suitable oxidation protection coatings.
Limitations
• The inks are available in low viscosity making it difficult to use in R2R manufacturing employing high speed gravure printng.
Ink with metal flakes
Advantages
• Silver flakes of various sizes are used as pigment with suitable binders.
• The inks generally require two or three coats with accurate registration control, for example, in gravure printing. This approach is not recommended for RFID devices which require high printing tolerances.
Limitations
• A single layer of printing is generally not sufficient to obtain required conductivity. Adding another printed overlayer may suffer from inaccurate registration control.
Graphene based Ink
Advantages
• Graphene is monolayer carbon which is characterized by superior electric performance.
• Graphene based inks are being manufactured and availability is low.
Limitations
• Graphene based coatings with high conductivities required for RFID antennas are not available.

5(c) shows the lines and slots for screen printing where ink overflow causes broadening of lines and as a consequence contraction of slots. Figure 5(d) shows a better quality printing in terms of overflow as here the overflowing is less and hence the dimensions are undergoing litle change. Screen printing also causes shortening of adjacent lines and in some cases broaken line as can be seen from Figure 5(e). Figure 5(f) shows a flexo printed line which shows multiple scratches on the surface. These scratches may create electrically open circuits which may restrict the transmission of RF signal from one point to another. But Flexo printing offers better resolution which can be viewed in Figure 5(g). In Gravure with lower dimension the printing quality deteriorates as seen in Figure 5(h). Also contraction occurs in the lines.

6.1.2. Surface Roughness

Surface Roughness is the measure of surface irregularities. Increased surface roughness will increase effective signal path length and decrease conductivity. The particle shape, size and loading will to some extent determine the surface roughness of the ink after transfer. The substrate surface roughness also contributes to coating surface roughness. The effect of surface roughness can be significant when working in GHz frequency range. It affects the ac resistance of the printed structures and consequently the RF performance. It has been reported that a mean surface roughness of 0.54μ may cause 9.1% increase per unit length for printed silver antennas using nanoinks (Shin et al., 2009). The effect is more of a concern with

Figure 5. Pictures of printed lines and slots using various printing process and with various ink types (the specification is indicated under each image)

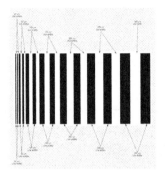

(a) Picture of lines and slots as per original design

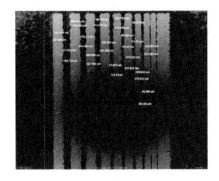

(b) Gravure printed lines showing jagged edges

(c) Screen Printed lines (Low Resolution)

(d) Screen printed lines with conducting Nano-paste

(e) Picture of a screen printed design showing broken lines and necking

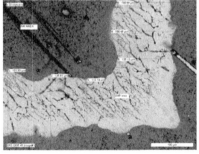

(f) A flexo printed line in Nano-particle ink 1

(g) Flexo printed lines in Nano-particle Ink 2

(h) gravure printed lines in Nano-ink with printing defects

Figure 6. (a) Spiral in CPW configuration (b) Zoomed view of spiral (L=4.5mm, W=4.4mm, w$_{cond}$=0.5mm, w$_{gap}$=0.2mm) (c) Simulated insertion loss for Taconic TLX-0 ($_r$=2.45, tanδ=0.019, h (thickness of substrate)=0.5mm and copper cladding, t=0.017mm) and Polymer substrate ($_r$=2.15, tanδ=0.004, h (thickness of substarte)=0.070mm and thickness of ink,t=0.003mm)

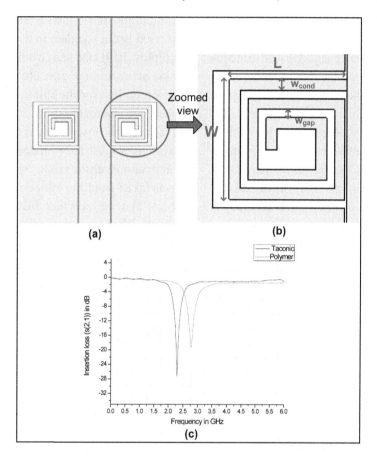

flake inks (Vornbrock, 2009) than nanoparticle inks. Nanoinks are known to show low surface roughness, in the order of nanometers (Shin, et al., 2009; Vornbrock, 2009).

6.1.3. Jagged Edges and Discontinuity

Imperfect or jagged edges can cause shorting between adjacent tracks, particularly between the CPW line and Ground plane on both sides of the CPW line. Cylinder engraving, etching resolution and quality have a direct influence over appearance of jagged edges. Printing discontinuities are breaks in the conductive tracks and an example

is shown in Figure 5(h). Tags that contain printing discontinuities will not be able to efficiently transmit RF power from one end to other end. The desired output is therefore not achievable in such tags. Smoothness of edges and electrical continuity of tracks are important to obtain an optimum RF performance. In conventional printing, it is not uncommon to spot print defects such as jagged edges, breaks in tracks and craks in the coatings. The reasons for the observed effects lie in non-uniform sintering, rapid quenching or quick drying of the ink (Vornbrock, 2009). As discussed above, the presence of jagged edges has their origin in printing resolution for gravure process.

Ink overflow also contributes to the appearance of jagged edges (Sung, et al., 2010).

6.2. Tag Performance on Polymer Substrate

Imperfect printing of the RF tags will lead to inconsistency and difficulties in the measurement of RF signals. The measurement difficulties often arise due to printing inaccuracies, low conductivities and other directionality issues related to printing. An example is the well-known gravure printing process. In gravure process, the lines oriented parallel to the print direction always print differently than those in the cross direction. The accuracy is better for lines oriented in the print direction. So the performance of the printed RFID tags varies with the orientation of printing. Environmental noise also contributes to RF performance degradation and is identified during actual measurements on the printed tags.

Printing of a chipless RFID tags in the GHz frequency range, with high printing accuracies still is a subject of intense research. The main concern has always been the conductor losses of the printed tracks (Mäntysalo & Mansikkamäki, 2009). As reported by Mantesalo (Mäntysalo & Mansikkamäki, 2009), conductor loss occurs due to various imperfections such as surface roughness, air pockets, line edge definition, lack of electrical contacts among the conductive particles of the ink, presence of binders (Mäntysalo & Mansikkamäki, 2009; Merilampi et al., 2009) leading to reduced dc conductivity etc. The thickness of the printed track also contributes to the conductor losses at frequencies where the skin depth is of the order of print thickness (Mäntysalo & Mansikkamäki, 2009).

The degradation may be described from three different directions as discussed below.

6.2.1. Shift in Resonance Frequency

The resonant frequency of a printed tag may shift from the simulated frequency to a different position. Figure 6(a) shows a resonating spiral which is used to encode data in a multiresonator-based chipless RFID tag and (b) shows the zoomed version of the spiral. Figure 6(c) shows the simulated insertion loss for the spiral. The black line is the insertion loss for the spiral on Taconic TLX-0. The red line is the insertion loss for the same spiral on a polymer substrate with a conductive ink. For the simulation with the printed spiral, the required data was taken from the ink conductivity (σ=666667 S/m) and the polymer substrate technical data sheets. The resonance frequency spiral was found to be f_{r1}=2.46GHz and f_{r2}=2.77GHz respectively on Taconic and polymer.. The simulations showed a variation of 300MHz between the two substrates. The actual masuremnet of printed tag shows variation from the simulation result because of printing imperfections, background nosie and low conductivity of the ink used for printing.

6.2.2. Degradation in Q-factor (Pulse Widening and Decrease in Steepness)

Q-factor determines the sharpness of response. In the case of printed resonators, the Q-factor changes due to printing inaccuracies, and thereby causes a widening of the response. Again in Figure 6(c), the 3-dB bandwidth of the curve (red) is higher than the curve in black. The 3-dB bandwidth of the entire structure on Taconic is 530MHz whereas on polymer it is 700MHZ. The authors observed bandwidth widening for the same RF structures (Figure 6(a)) printed with conductive nanoinks on thin polymer sheets. A broad explanation for the increase in bandwidth is printing inaccuracies and low electrical conductivity of the inks used. High Q factors are obtained when the printed coatings show high conductivities nearly approaching that of pure metal. In the case of silver nanoinks, the conductivity of printed coating should be nearly equal to pure silver conductivity.

6.2.3. Degradation in Rejection Level

This may change the the insertion loss at the resonance frequency. As spectral-signature based chipless RFID tags are decoded by setting a pre-defined threshold to identify a resonance in the received signal, the change in rejection level will affect the decoding performance. In the above simulation, the rejection level changed from -27 to -19 dB.

7. CONCLUSION

The chapter has summarized the printing issues and techniques regarding printed RFID tags based on available literature. It has also presented some experimental results with printed antenna and chipless tags. The identified areas of concern in printing are low track conductivities, printing process limitations to obtain high printing tolerances, unavailability of fast drying inks for high speed printing, environmental and economy considerations. There will be cost savings associated with ink consumption, if RF structures can be printed in grid shapes. Typical applications of RFID tags are in item level tagging, document identification, anti-counterfeiting for banknote industry, inventory management and many other areas not covered here. Undoubtedly the best RFID tags are silver ink-jet printed and are available on low cost substrates. RF design engineers need to consider limitations of the printing processes and wherever possible try to compensate for the printing inaccuracies. The RFID reader also require optimisation and tuning to show capability of reading and identifying the data that is often received with a background of noise from a tag that has low printing tolerances. The ability to manufacture high volume low cost RFID tags by roto-gravure and flexography is always attrac-

tive. Currently the market is dominated by silver conductive inks exclusive to screen printing and ink-jet printing. Inks for flexo and gravure are on short supply.

ACKNOWLEDGMENT

The work is supported by the Australian Research Council's Linkage Project Grant (LP0989652: printable, Multi-Bit RFID for Banknotes.) and Securency International Pty Ltd.

REFERENCES

Along, K., Chenrui, Z., Luo, Z., Xiaozheng, L., & Tao, H. (2010, 17-19 June 2010). *SAW RFID enabled multi-functional sensors for food safety applications.* In The IEEE International Conference on RFID-Technology and Applications (RFID-TA), 2010.

Balbin, I., & Karmakar, N. C. (2009, September). *Novel chipless RFID tag for conveyer belt tracking using multi-resonant dipole antenna.* In The 39th European Microwave Conference, Rome, Italy.

Bechevet, D., Tan-Phu, V., & Tedjini, S. (2005, 3-8 July). *Design and measurements of antennas for RFID, made by conductive ink on plastics.* In The IEEE Antennas and Propagation Society International Symposium, 2005.

Blayo, A., & Pineaux, B. (2005). Printing processes and their potential for RFID printing. In The *Proceedings of the 2005 Joint Conference on Smart Objects and Ambient Intelligence: Innovative Context-Aware Services: Usages and Technologies.*

Borowiec, R. (2010, 14-16 June 2010). *A printed RFID Tag antenna for metallic objects operating in UHF band*. In The 18th International Conference on Microwave Radar and Wireless Communications (MIKON), 2010. *Cabot Corp*. Retrieved 24 August, 2011, from http://www.cabot-corp.com/New-Product-Development/Printed-Electronics/Products

Chamarti, A., & Varahramyan, K. (2006). Transmission delay line based ID generation circuit for RFID applications. *IEEE Microwave and Wireless Components Letters*, *16*(11), 588–590. doi:10.1109/LMWC.2006.884897

Cool Magnet Man. (n.d.). Retrieved 26 August, 2011, from http://www.coolmagnetman.com/magcondb.htm

Deligeorgis, G., Dragoman, M., Neculoiu, D., Dragoman, D., Konstantinidis, G., & Cismaru, A. (2009). Microwave propagation in graphene. *Applied Physics Letters*, *95*(7). doi:10.1063/1.3202413

Dragoman, M., Neculoiu, D., Dragoman, D., Deligeorgis, G., Konstantinidis, G., & Cismaru, A. (2010). Graphene for microwaves. *IEEE Microwave Magazine*, *11*(7), 81–86. doi:10.1109/MMM.2010.938568

GBPPR. (n.d.). Retrieved 26 August, 2011, from http://www.qsl.net/n9zia/why_silver_plate.html

Grouchko, M., Kamyshny, A., & Magdassi, S. (2009). Formation of air-stable copper-silver core-shell nanoparticles for inkjet printing. *Journal of Materials Chemistry*, *19*(19), 3057–3062. doi:10.1039/b821327e

Gupta, S., Nikfal, B., & Caloz, C. (2010, 7-10 Dec. 2010). *RFID system based on pulse-position modulation using group delay engineered microwave C-sections*. In The Asia-Pacific Microwave Conference Proceedings (APMC), 2010.

Huang, D., Liao, F., et al. (2003). *Plastic-compatible low resistance printable gold nanoparticle conductors for flexible electronics* (Vol. 150). Pennington, NJ: ETATS-UNIS: Electrochemical Society.

Huebler, A., Hahn, U., Beier, W., Lasch, N., & Fischer, T. (2002, 2002). *High volume printing technologies for the production of polymer electronic structures*. In The 2nd International IEEE Conference on Polymers and Adhesives in Microelectronics and Photonics, 2002. POLYTRONIC 2002.

Jalaly, I., & Robertson, I. D. (2005, 12-17 June). *RF barcodes using multiple frequency bands*. In The IEEE MTT-S International Microwave Symposium Digest, 2005.

Kipphan, H. (2001). *Handbook of print media: Technologies and production methods*. Springer.

Kirsch, N. J., Vacirca, N. A., Plowman, E. E., Kurzweg, T. P., Fontecchio, A. K., & Dandekar, K. R. (2009). *Optically transparent conductive polymer RFID meandering dipole antenna*. In The IEEE International Conference on RFID.

Kobayashi, Y., & Sakuraba, T. (2008). Silica-coating of metallic copper nanoparticles in aqueous solution. *Colloids and Surfaces A: Physicochemical and Engineering Aspects*, *317*(1-3), 756–759. doi:10.1016/j.colsurfa.2007.11.009

Loo, C. H., Elmahgoub, K., Yang, F., Elsherbeni, A., Kajfez, D., & Kishk, A. (2008). Chip impedance matching for UHF RFID tag antenna design. *Progress in Electromagnetics Research*, *81*, 359–370. doi:10.2528/PIER08011804

Luechinger, N. A., Athanassiou, E. K., & Stark, W. J. (2008). Graphene-stabilized copper nanoparticles as an air-stable substitute for silver and gold in low-cost ink-jet printable electronics. *Nanotechnology*, *19*(44). doi:10.1088/0957-4484/19/44/445201

Magdassi, S. (Ed.). (2009). *The chemistry of inkjet inks*. Hackensack, NJ: World Scientific. doi:10.1142/9789812818225

Magdassi, S., Grouchko, M., Berezin, O., & Kamyshny, A. (2010). Triggering the sintering of silver nanoparticles at room temperature. *ACS Nano, 4*(4), 1943–1948. doi:10.1021/nn901868t

Magdassi, S., Grouchko, M., & Kamyshny, A. (2010). Copper nanoparticles for printed electronics: routes towards achieving oxidation stability. *Materials, 3*(9), 4626–4638. doi:10.3390/ma3094626

Mäntysalo, M., & Mansikkamäki, P. (2009). An inkjet-deposited antenna for 2.4 GHz applications. *AEÜ. International Journal of Electronics and Communications, 63*(1), 31–35. doi:10.1016/j.aeue.2007.10.004

McVay, J., Hoorfar, A., & Engheta, N. (Aug. 2006). Theory and experiment on Peano and Hilbert curve RFID tags. In The *Proceedings of the Wireless Sensing and Processing*, San Diego, USA.

Merilampi, S., Laine-Ma, T., & Ruuskanen, P. (2009). The characterization of electrically conductive silver ink patterns on flexible substrates. *Microelectronics and Reliability, 49*(7), 782–790. doi:10.1016/j.microrel.2009.04.004

NanoPrintTech. (n.d.). Retrieved 24 August, 2011, from http://www.nanoprinttech.com/old/printing-technology.html

Narkcharoen, P., & Pranonsatit, S. (2011, 17-19 May). *The applications of fill until full (FuF) for multiresonator-based chipless RFID system*. In The 8th International Conference on Electrical Engineering/Electronics, Computer, Telecommunications and Information Technology (ECTI-CON), 2011.

Neculoiu, D., Deligeorgis, G., Dragoman, M., Dragoman, D., Konstantinidis, G., Cismaru, A., et al. (2010). *Electromagnetic propagation in graphene in the mm-wave frequency range*. In The 40th European Microwave Conference, EuMC 2010, Paris.

Oldenzijl, R., Gaitens, G., & Dixon, D. (2010). *Conduct radio frequencies with Inkpp*. Retrieved from http://www.intechopen.com/articles/show/title/conduct-radio-frequencies-with-inks

Preradovic, S., Balbin, I., Karmakar, N. C., & Swiegers, G. (2008, 16-17 April). *A novel chipless RFID system based on planar multiresonators for barcode replacement*. In The IEEE International Conference on RFID, 2008.

Preradovic, S., Balbin, I., Karmakar, N. C., & Swiegers, G. (Apr. 2008). *A. P. P. Application*.

Preradovic, S., Balbin, I., Roy, S. M., Karmakar, N. C., & Swiegers, G. (2008). *Radio frequency transponder*. (Australian Provisional Patent Application P30228AUPI, Apr. 2008).

Preradovic, S., & Karmakar, N. C. (2009, 20-22 Aug. 2009). Design of fully printable chipless RFID tag on flexible substrate for secure banknote applications. In *The 3rd International Conference on Anti-counterfeiting, Security, and Identification in Communication, ASID 2009*.

Preradovic, S., & Karmakar, N. C. (2010). Chipless RFID: Bar code of the future. *Microwave Magazine, 11*(7), 87–97. doi:10.1109/MMM.2010.938571

Pudas, M., Halonen, N., Granat, P., & Vähäkangas, J. (2005). Gravure printing of conductive particulate polymer inks on flexible substrates. *Progress in Organic Coatings, 54*(4), 310–316. doi:10.1016/j.porgcoat.2005.07.008

Rida, A., Li, Y., Reynolds, T., Tan, E., Nikolaou, S., & Tentzeris, M. M. (2009, 1-5 June 2009). *Inkjet-printing UHF antenna for RFID and sensing applications on liquid crystal polymer.* In The Antennas and Propagation Society International Symposium, APSURSI '09.

Sangoi, R., Smith, C. G., Seymour, M. D., Venkataraman, J. N., Clark, D. M., & Kleper, M. L. (2005). Printing radio frequency identification (RFID) tag antennas using inks containing silver dispersions. *Journal of Dispersion Science and Technology, 25*(4), 513–521. doi:10.1081/DIS-200025721

Sanjay Monie, P. (n.d.). *Developments in conductive inks.* Retrieved 29 August, 2011, from http://www.vorbeck.com/news.html

Sankir, N. D. (2008). Selective deposition of PEDOT/PSS on to flexible substrates and tailoring the electrical resistivity by post treatment. *Circuit World, 34*(4), 32–37. doi:10.1108/03056120810918105

Shaker, G., Safavi-Naeini, S., Sangary, N., & Tentzeris, M. M. (2011). Inkjet printing of ultra-wideband (UWB) antennas on paper-based substrates. *Antennas and Wireless Propagation Letters, 99,* 1–1.

Shin, D.-Y., Lee, Y., & Kim, C. H. (2009). Performance characterization of screen printed radio frequency identification antennas with silver nanopaste. *Thin Solid Films, 517*(21), 6112–6118. doi:10.1016/j.tsf.2009.05.019

Shrestha, S., Balachandran, M., Agarwal, M., Phoha, V. V., & Varahramyan, K. (2009). A chipless RFID sensor system for cyber centric monitoring applications. *IEEE Transactions on Microwave Theory and Techniques, 57*(5), 1303–1309. doi:10.1109/TMTT.2009.2017298

Siden, J., Fein, M. K., Koptyug, A., & Nilsson, H. E. (2007). Printed antennas with variable conductive ink layer thickness. *Microwaves. Antennas & Propagation, 1*(2), 401–407.

Subramanian, V., Liao, F., & Huai-Yuan, T. (2010, 27-28 September). *Printed RF tags and sensors: The confluence of printing and semiconductors.* In The European Microwave Integrated Circuits Conference (EuMIC), 2010.

Sung, D., de la Fuente Vornbrock, A., & Subramanian, V. (2010). Scaling and optimization of gravure-printed silver nanoparticle lines for printed electronics. *IEEE Transactions on Components and Packaging Technologies, 33*(1), 105–114. doi:10.1109/TCAPT.2009.2021464

Verma, A., Bo, W., Shepherd, R., Fumeaux, C., Van-Tan, T., Wallace, G. G., et al. (2010, 20-24 September). *6 GHz microstrip patch antennas with PEDOT and polypyrrole conducting polymers.* In The International Conference on Electromagnetics in Advanced Applications (ICEAA), 2010.

Volkman, S. K., Pei, Y., Redinger, D., Yin, S., & Subramanian, V. (2004). Ink-jetted silver/copper conductors for printed RFID applications. In *The Materials Research Society Symposium Proceedings.*

Vornbrock, A. F. (2009). *Roll printed electronics: Development and scaling of gravure printing techniques.* Berkeley, CA: University of California.

Xiao, Y., Yu, S., Wu, K., Ni, Q., Janecek, C., & Nordstad, J. (2007). Radio frequency identification: Technologies, applications, and research issues: Research articles. *Wireless Communication and Mobile Comptuing, 7*(4), 457–472. doi:10.1002/wcm.365

XPEDX. (n.d.). Retrieved 24 August, 2011, from http://xpedx.edviser.com/default.asp?req=knowledge/article/151

ADDITIONAL READING

Ah, C. S., Hong, S. D., & Jang, D.-J. (2001). Preparation of Aucore Agshell nanorods and characterization of their surface plasmon resonances. *The Journal of Physical Chemistry B, 105*(33), 7871–7873. doi:10.1021/jp0113578

Bo, G., & Yuen, M. M. F. (2009, 10-13 August). *Optimization of silver paste printed passive UHF RFID tags*. In International Conference on the Electronic Packaging Technology & High Density Packaging, ICEPT-HDP '09.

Chandrasekhar, P. (1999). *Conducting polymers, fundamentals and applications: A practical approach*. Kluwer Academic. doi:10.1007/978-1-4615-5245-1

Deligeorgis, G., Dragoman, M., Neculoiu, D., Dragoman, D., Konstantinidis, G., & Cismaru, A. (2009). Microwave propagation in graphene. *Applied Physics Letters, 95*(7). doi:10.1063/1.3202413

Hines, D. R., Southard, A. E., Tunnell, A., Sangwan, V., Moore, T., Chen, J. H., et al. (2007). *Transfer printing as a method for fabricating hybrid devices on flexible substrates*.

Hrehorova, E., Pekarovicova, A., & Fleming, P. D. (2006). *Gravure printability of conducting polymer inks*.

Karmakar, N. C. (2010). *Handbook of smart antennas for RFID systems*. John Wiley & Sons. doi:10.1002/9780470872178

Kim, W. K., Jung, Y. M., Cho, J. H., Kang, J. Y., Oh, J. Y., & Kang, H. (2010). Radio-frequency characteristics of graphene oxide. *Applied Physics Letters, 97*(19), 193103–193103-193103. doi:10.1063/1.3506468

Koh, K. S., Hu, S., Law, C. L., & Dou, W. (2009). *A planar monopole antenna integrated with delay line for passive UWB-RFID tag applications*.

Leung, S. Y. Y., & Lam, D. C. C. (2008). Geometric and compaction dependence of printed polymer-based RFID tag antenna performance. *IEEE Transactions on Electronics Packaging Manufacturing, 31*(2), 120–125. doi:10.1109/TEPM.2008.919334

Merilampi, S., Björninen, T., Ukkonen, L., Ruuskanen, P., & Sydänheimo, L. (2011). Embedded wireless strain sensors based on printed RFID tag. *Sensor Review, 31*(1), 32–40. doi:10.1108/02602281111099062

Molesa, S. E. (2006). *Ultra-low-cost printed electronics*. Berkeley: University of California.

Mukherjee, S. (2007, 5-6 Sept. 2007). *Chipless radio frequency identification (RFID) device*. In The RFID Eurasia, 2007 1st Annual.

KEY TERMS AND DEFINITIONS

Chipless Tag: An RFID transponder without any integrated circuit (IC).

Flake: Metal particle with less than 5 μm of diameter, it can be of sphrical or any other shape. Different types of shapes are used to increase contacts among the particles.

Graphene: It is obtained from Graphite by separating the multiple layers of graphite and suspended it in liquid solution. It is highly conductive.

Nanoparticle: Metal particle of nano-meter sized.

Q-Factor: Q-factor (Quality factor) defines the sharpness of response of resonator or filter. It is a dimensionless quantity which is the ratio of resonant frequency with the 3-dB Bandwidth. RFID Tag: Transponder, conatins a unique identification number which is the ID. RFID: Radio Frequency Identification, wireless data communication tehnique through radio frequency signals.

Section 4
RFID System and Detection of RFID Tags

Chapter 10
The Multitag Microwave RFID System with Extended Operation Range

Igor B. Shirokov
Sevastopol National Technical University, Ukraine

ABSTRACT

The problems of radio frequency identification are discussed. It was shown that the use of passive transponders is preferable, but weak energy of system in this case reduces the operation distance and decreases the noise-immunity of the system. The problems of traditional radio-frequency identification systems are discussed. In this chapter the use of homodyne method of useful signal selecting was proposed. The augmentation signal of transponder was obtained by means of frequency shift with the help of controlled phase shifter. This solution allows to increase the energy and the noise-immunity of the system (the operation distance is increased). Furthermore, the interrogator can treat several transponders simultaneously in this case. Additionally the use of one-port transistor amplifier for increasing of operation range was proposed. The energy consumption of such amplifier and its cost are very low, but the gain of amplifier can reach 20 dB and more.

INTRODUCTION

Radio-Frequency Identification (RFID) is of interest for specialists in transport, logistics and other related areas where the identification of object is very important and must be carried out as soon as possible.

The significant advantage of all types of RFID systems is the noncontact matter of their operation and non-line-of-sight nature of the technology.

Tags can be read through a variety of substances, where barcodes or other technologies of optical reading would be useless. Though it is a cheaper technology, RFID has become indispensable for a wide range of automated data collection and identification applications that would not be possible otherwise.

Any RFID system consists of basic unit and mobile unit (tag or transponder), which position and unique code must be identified (Sharpe,

DOI: 10.4018/978-1-4666-1616-5.ch010

1995). The terms "basic" and "mobile" were assumed conditionally (the tag may be immobile but the seeking device (basic unit) mobile). The basic unit, in turn, consists of transceiver with the decoder and antenna.

The antenna emits signals to activate the tag and read and write the data to it. Antennas are available in a variety of shapes and sizes; they can be placed in a door frame to receive the tag data from people or things passing through the door or it can be mounted on an interstate toll plaza to monitor the traffic on the highway. The electromagnetic field produced by the antenna must constantly take place while multiple tags are expected continuously. If constant interrogation is not required, the field can be activated by a sensor device. The antenna is packaged often with the transceiver and decoder to become a reader (so-called interrogator).

RFID tags are categorized as either active or passive. Active RFID tags are powered by an internal battery and are typically read/written, i.e., tag data can be rewritten and/or modified. The battery-supplied power of an active tag generally gives it a longer reading range. The shortcomings of such solution are greater size, greater cost and the limited operational life.

Passive RFID tags operate without a separate external power source and obtain operating power, generated from the reader. Passive tags are consequently much lighter than active tags, less expensive, and offer a virtually unlimited operational lifetime. The shortcoming is that they have shorter reading ranges than active tags and require a higher output power of reader.

In the chapter we will discuss active tags only, as the operation range of RFID system is of main interest. Furthermore, several transponders can be located within the system service area and the interrogator must recognize each.

UNDERSTANDING OF PROBLEMS OF CONVENTIONAL RFID SYSTEMS

Harmonic RFID System

Functioning of the RFID system assumes a sending of a probing pulse with the help of transceiver of basic unit which activates a transponder, established on the object. Transponder replies by the reciprocal radio-frequency pulse modulated by a unique identification code. The use of passive transponder represents the greater interest for consumers, as its power supplying is carried out due to the energy of a signal of basic unit within the communication time. Under clear reasons the link of such system should be duplex. Due to the increased power of such communication link, there is an opportunity in this case to send several commands to transponder and to organize, thus, any necessary exchange protocol.

The organization of duplex channel involves the engineers in complicated design of transponder and increases its cost. In practice there are widely used so-called harmonic RFID systems (Colpitts et al, 2004) and (Vanjari et al, 2007) in which the transceiver of the interrogator sends initial signal on one radio frequency, and transponder responds on frequency, which is multiple to the initial one. As a rule, frequency rate is equal to two. Such approach satisfies the main principle of duplex communication. On the other hand the doubling of frequency allows us to use the same known antenna for reception and radiation. Multiplication of frequency of the transceiver can be achieved by the elementary technical solution, for example by connection to the terminal of the transponder's antenna of Shottki diodes, as it is shown on Figure1 a.

Besides the multiplication of interrogator signal frequency, they make its rectifying for transponder modulator feeding. Obviously, the energy of signal, which is received from interrogator and rectified, is weak, and this signal must be presented during entire communication session.

Figure 1. Transponder circuit scheme a), Spectrogram of combinatorial RFID system b)

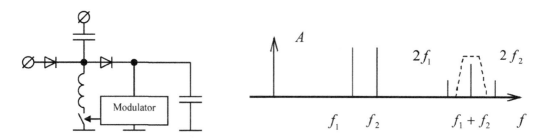

The Shottki diodes let us minimize the losses and the real value of rectified voltage can reach few volts. That is enough for the feeding of modulator.

Shortcomings of Harmonic RFID System

However, the use of harmonic RFID systems is inevitably accompanied with a number of difficulties of the technical nature. Let's discuss some of them.

The signal power, which is received by transponder, can be written down with well known equation:

$$P_{RX} = P_{TX} + G_{TX} + G_{RX} + L + 10 \log \left(\frac{\lambda}{4 \pi R} \right)^2, \quad (1)$$

where P_{TX} is the transmitter output power, G_{TX}, G_{RX} are the antennas gains, L is all feeders losses, λ is the wavelength, R is the operation distance.

For the operating wavelength about 0.3 m the transmitter antenna gain usually does not exceed 12 dB. We must assume the antenna gain of transponder to be equal to 0 dB or even less, taking into account the antenna size, which is really smaller than wavelengths for upward and downward links.

So far as we use passive transponder with the ordinary Shottky diode as frequency multiplier, the power attenuation in transponder itself can reach

–20 dB and more. Thus, taking into consideration the interrogator transmitter output power in 30 dBm, all feeder losses are about –2 dB, and the interrogator receiver sensitivity is about –110 dBm, the maximum operating distance will be near 30 m, where the free space attenuation of upward link will be –62 dB and the downward link attenuation will be –68 dB due to the frequency doubling.

Thus, for the ensuring of desired system operation distance in 30 m at least, we must ensure the radiated power of interrogator transmitter in 30 dBm. It's not a good solution!

Further, the difference between power levels of radiated and received signals on a link of such length will be 140 dB. Taking into account that the operation frequency of interrogator receiver is the double frequency of the transmitter, the second harmonic of the transmitter signal reacts on the receiver directly. Taking into consideration the mentioned above attenuations of transmitter power on the upward and downward links in 140 dB, the level of second harmonic of transmitter signal must be depressed on the value of 160 dB at least. The use of combined two-band antenna in this case is impossible, because none the filter of harmonic can suppress the second harmonic of transmitter signal to such level. The only use of two narrowband antennas can solve the problem with the lot of restrictions.

All of mentioned above is true if there are no any other semiconductors in environment except for Shottky diode of transponder. For example, the car body encloses a lot of semiconductors of natural origin. It may be the places of metal welding, various oxides etc. The scattering cross-section of the car body is much more effective than the aperture of transponder antenna. So, the level of second harmonic of signal, which is radiated by the car body itself, can be comparable with transponder signal. All of those facts deteriorate the work of harmonic RFID system.

And at last, the discussed system can serve the single transponder at certain time moment, that transponder which signal has the larger level. This large-level signal depresses other signals from other transponders. The interrogator "loses" these transponders. If signals from several transponders are comparable upon the level (it can take place if we'll take into account the system operation distance), the interrogator can lose all of transponders.

Summarizing all of mentioned above we can affirm following:

1. The radiated power of interrogator transmitter must be essential.
2. The spectrum cleanliness of interrogator transmitter must be high; especially it concerns the level of the second (multiple) harmonic of radiation signal.
3. As the signal power of second harmonic, which is retransmitted by the transponder, is not high the operating distance of system is short.
4. The semiconductors of natural origin (various oxides etc), which exist in natural environment and on the objects, decrease the system noise-immunity.
5. The interrogator can serve the single transponder in certain time interval on a certain range of distances.
6. The energy of RFID link is weak and operation distance is small.

Thus, discussed harmonic RFID system has no ubiquitous use. The electromagnetic compatibility (EMC) of system is weak and system does not satisfy to the demands of state-of-art tendencies of so called "green communication."

Combinatorial RFID System

The combinatorial RFID system has some advantages in comparison with previous systems (Shirokov et al, 2003). In this system the transmitter simultaneously sends two signals with different but close frequencies. By a transponder design and functioning principle such decision is similar to harmonic RFID system. The electric scheme of transponder of combinatorial RFID system is completely identical to one which we considered above (see figure 1 a). However in system operation and obtained quality parameters the essential differences take place in discussed systems.

For the understanding of advantages of combinatorial RFID system, let's analyze the spectrogram shown in figure 1 b. Here two transmitters of interrogator by means of the pair (for each transmitter its own) or single antenna radiate two electromagnetic waves. One of transmitter signals can be modulated by certain command sequence, which is sent to transponder. Frequencies of signals are accordingly equal to f_1 and f_2.

The presence of nonlinear elements causes the occurrence on transponder antenna clips of ensemble of combinational components. Three of them, which are in operating frequency band of transponder antenna and interrogator receiver, will be of our interest only. They are the component of doubled frequency of first transmitter $2f_1$, the component of doubled frequency of second transmitter $2f_2$, and the component of frequencies sum f_1+f_2.

The input chains of interrogator receiver pass only combinatorial component of frequency sum, which is processed afterward. The gain-frequency characteristic of mentioned chains is shown in figure 1 b) with the dashed line. The main filtration of useful signal, basically, can be implemented in the chains of intermediate frequency of receiver. It is defined by the receiver circuit technology only.

Thus, second harmonic of both transmitter signals do not affect the interrogator receiver essentially. We can speak about the magnitude of suppression of parasitic signal. Not expensive technical solutions let us do it.

The energy of combinatorial signal of transponder is quadrupled (twice voltage amplitude) in comparison with the energy of signal of second harmonic because the signals of both transmitters form this component.

Let's summarize the advantages in comparison with the previous system:

1. The demands to the transmitter spectrum cleanness are not rigid, because none harmonic component of transmitter does block the interrogator receiver.
2. The presence of signals pair with different frequencies doesn't result in deep signal fading because the interference behaviors for two different oscillations are different in principle.
3. The energy of combinatorial system is higher than the energy of harmonic one, so the system operation distance is large.

Shortcomings of Combinatorial RFID System

The combinatorial system has the improved characteristics but remains not perfect still:

1. The radiated power of both interrogator transmitters must be essential as well.
2. The RFID system operation distance remains low still (little bit increasing).
3. The interrogator can serve the single transponder, which is placed in a zone of system operation.
4. The radiation characteristics of transponder antenna (in case of use of single antenna) will not be equal in different frequency bands.

In case of use of two separate antennas for transmitters and receiver respectively the system cost increases.

5. The problem of natural-origin semiconductor remains still.
6. The EMC of combinatorial system is even worse than mentioned above harmonic one.

Thus, the combinatorial RFID system is not suitable for "green communication" as well. The presence of two transmitters increases the system cost and decreases the system EMC.

The main disadvantage of both systems is the absence of possibility to treat several signals arriving from several transponders simultaneously. Taking into account the stated system operation range in 30 m the occurrence of mentioned situation is the normal practice.

NEW APPROACH TO A PROBLEM

Homodyne Method

We propose to use the homodyne method of useful signal selecting (Koshurinov, 2003), where the initial microwave signal from transmitter is the heterodyne one for a signal, which is received from tag. However, such approach assumes the obtaining of response signal at the same frequency. This circumstance involves into signal processing on direct current. Dynamic range and accuracy of such system are not high.

In present work we propose to use the frequency shift of response signal (Shirokov, 2007, 2009, and 2010). Signal processing is carried out at the frequency of combinatorial component in this case, ensuring the obtaining of good system characteristics. This frequency shift is chosen relatively small with respect to microwave link frequency and it can be about 100 ppm. With such assumption the interrogator and transponder antennas, both operate in the same frequency band and with very narrow bandwidth. For both antennas we can realize the arbitrary radiation characteristics we need in. The system is duplex in principle, but duplex filter is not used in this case (see below). Furthermore, the interrogator operates with single antenna.

The frequency shift of microwave signal is obtained by means of monotonous shift of its phase (Jaffe et al, 1965). The change of microwave-oscillations phase over the period T of the low-frequency oscillations by 2π is tantamount to the frequency shift of the microwave signal by the frequency $\Omega = 2\pi / T$. This approach is equivalent to Doppler's frequency shift, which takes place in a case of irradiation of moving object. The technique of Doppler's signal selecting is very simple and it is widely used in numerous radar applications.

Thus, the use of conventional controlled phase shifter in transponder let us to obtain the frequency shift of initial microwave signal on small value. The amplitude of frequency-transformed microwave signal remains the same if the losses in phase shifter are not high. In this case the energy of microwave link is 20 dB higher than for link of previous systems, letting us to reduce the power, which is radiated by the interrogator transmitter on the same value at the same system operation range. The use of controlled phase shifter of reflection type let us minimize the design and the cost of transponder (Shirokov, 2009).

Signal Amplification by Transponder

However, the energy of microwave link of described system remains weak, so the operation distance will not be high. Classical approach to elimination of this problem is the use of microwave amplifier in conjunction with the microwave circulator (Shirokov, 2006). But the gain of transponder will not be essential in this case. The circulator decoupling will restrict the loop gain of transponder in a narrow frequency band. The loop gain can reach 20 dB only. That will be true

for perfect matching of transponder antenna with the circulator. The real value of antenna voltage standing-wave ratio (VSWR) will decrease the loop gain of transponder additionally. Further, good decoupling of circulator can be implemented in a narrow frequency band only. So, the use of microwave band-pass-filter in a transponder loop will be obligatory in this case.

The perfect technical solution in this case is the use of reflection microwave amplifier (Shirokov, 2011). Such approach to the problem decision allows excluding the microwave circulator from the transponder design with the series of positive factors. First of all the design of transponder will be very simple and its cost will be low respectively. Besides the circulator the microwave band-pass-filter will be excluded as well. The transponder gain will not be determined by the circulator decoupling in a certain frequency band.

Reflection amplifiers for X band operation have been developed in the late seventies of last century by utilizing a circulator and a Gunn diode oscillator, providing 10 dB gain with a noise figure of 15 dB. Authors (Venguer et al, 2002 and 2004) proposed reflection amplifier for 1.5 GHz frequency band. The proposed amplifier also consists of a circulator but connected to a one-port circuit that uses a gate connected HJ FET biased in a nonlinear region of the I-V curves to operate in the negative resistance mode. This new type of reflection amplifiers has very low noise and high gain properties as it is demonstrated by theoretical and experimental results. The use of circulator is not obligatory in this case and its decoupling will not restrict the gain of transponder.

A low noise one-port transistor amplifier (OPTA) has been developed in the 1.4-1.6 GHz frequency band to study the capabilities of this kind of amplifiers resulting in a lower cost and simplified research (Venguer et al, 2002). The electrical scheme of amplifier and its gain characteristics are shown in figure2.

The active element chosen is a GaAs Hetero Junction FET with a gate length $L_g = 0.3 \ \mu$m and a gate width $W_g = 280 \ \mu\text{m}$. This device has a minimum noise factor $N_f = 0.3$ dB and an associated gain $G = 19$ dB at 1.5 GHz when it is biased at $V_{ds} = 2$ V and $I_{ds} = 10$ mA ($P = 20$ mW).

This microwave amplifier possesses the highest simplicity of design implementation, has very low power consumption, and has excellent noise characteristics. Described amplifier operates in narrow frequency band, but this feature is not dramatic one in our case. Furthermore, the perfect antenna matching can be implemented in a narrow frequency band as well. Thus, instead of the attenuation in -20 dB, which takes place in case of harmonic or combinatorial RFID system, we obtain the signal amplification in 20-30 dB.

And at last, the gain of mentioned amplifier can be changed in a simplest manner by the changing of the bias voltage on the gate of FET. In other words there is a good opportunity to modulate the transponder signal by the unique code in an easy way. Such approach (Shirokov, 2011) was put on a basis of described RFID system.

System Design

The block diagram of multitag microwave RFID system with extended operation range is shown in Figure 3.

The system consists of the interrogator equipment and the transponders (Tn) themselves. The interrogator equipment consists of the microwave oscillator (MW OSC), the microwave circulator, the microwave antenna (A), the microwave mixer (MIX), the low-frequency band-pass filters (BPFn), the low-frequency amplifiers (AMPn), and the demodulators (DMDn). Each transponder consists of the receiving microwave antenna, the controlled microwave transmission phase shifter (CPS), the reflection microwave amplifier (RMAn), the low-frequency oscillator (LFOn), and the source of unique code (UCSnj).

Figure 2. The scheme of OPTA and its gain characteristics

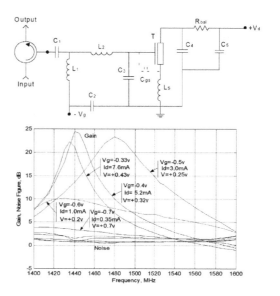

Microwave oscillator generates continuous microwave oscillations in a suitable frequency band, for example on a single frequency in 1.5 GHz with the output power, approximately in 10 dBm.

Microwave circulator ensures the decoupling of incident (upward link) and reflected (downward link) waves. The standard value of losses is 0.1-0.5 dB and decoupling is about –20 dB. It is ironic that the better value of decoupling is undesired in our case. The "parasitic" decoupled power will be the heterodyne signal for useful one of downward link.

Microwave interrogator antenna is the conventional one with desired characteristics. The antenna bandwidth may be very narrow, about 0.1%. The arbitrary antenna pattern and perfect VSWR can be implemented in this case. The antenna gain 10-15 dB and suitable main beam form for our specific application can be realized without any problems.

The mixer may be the conventional microwave one-port diode mixer. In case of use of two-port IC microwave mixer the additional microwave directional coupler will be of need. The dynamic

Figure 3. The block diagram of multitag microwave RFID system

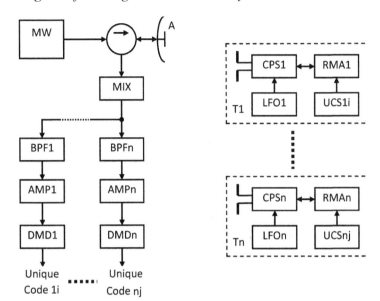

range of mixer must be high because the transponders may be located on different distances and they can form different signal levels.

The band-pass filter jointly with low-frequency amplifier in each channel selects and amplifies the component of difference of initial microwave signal and frequency-transformed one. The central frequency of each channel must correspond to the frequency shift of each transponder. The bandwidth of filter must correspond to the spectrum width of received signal, which is modulated by the unique code sequence. There are no problems to implement arbitrary gain of each low-frequency channel. The system of automatic gain control can be realized as well. It will be especially actual for various distances of transponder location.

On the demodulator outputs we obtain the unique code sequence from each of transponders.

In the transponder the controlled microwave transmission phase shifter implements the phase change law mentioned above. We understand the microwave signal passes the phase shifter twice. So, on each pass the phase of microwave signal must be changed from 0 to π on the period T of controlling signal. The summary change of the

phase of microwave signal in 2π ensures the mentioned above microwave signal frequency shift in $\Omega = 2\pi / T$.

The low-frequency oscillator of each transponder forms one of the several low-frequency signals, which controls the phase shifter. On the other hand, this unique low-frequency signal may be the transponder ID signal. But the number of these frequencies cannot be arbitrary high. Restrictions on this number will be discussed later. This "preliminary" ID doesn't identify the transponders themselves, but does allow the treating the signals from several transponders simultaneously.

The reflection microwave amplifier ensures the increasing of operation range of RFID system. The design of this OPTA was discussed above. The gain of amplifier can be changed by simple manner by the changing of its bias voltage (Venguer, 2002). So, the amplitude modulation of microwave signal by the unique code sequence can be implemented with such approach.

The source of unique code sequence forms real transponder ID. The number of real IDs can be arbitrary. The only thing we will be talking about will be the time of sequence transmitting or the time of transponder identification. The speed of sequence transmitting is limited and these restrictions will be discussed later as well.

Base Equations

The microwave signal, which is generated by the microwave oscillator, can be described, as

$$u_1(t) = U_0 \sin\left(\omega_0 t + \varphi_0\right),$$

where U_0 is the amplitude, ω_0 is the frequency, φ_0 is the initial phase of microwave signal.

The signal received by transponder antenna is characterized by following equation:

$$u_2(t) = U_0 A \sin\left(\omega_0 t + \varphi_0 + \Delta\varphi\right),$$

where A is the generalized attenuation, that takes into account all of feeder transfer coefficients, antenna gain, microwave free-space attenuation. The term $\Delta\varphi$ is a generalized phase shift, which characterizes the phase progression during microwave propagation and phase shift in all feeders and chains.

In transponder the microwave signal passes through the transmission phase shifter twice, which is controlled by the low-frequency oscillator. Thus, the microwave signal obtains the frequency shift according to the following equation:

$$u_3(t) = U_0 A \sin\left[\left(\omega_0 + \Omega_n\right)t + \varphi_0 + \Delta\varphi + \varphi_n\right],$$

where Ω_n is the frequency of one of several low-frequency signals and φ_n is the initial phase of this low-frequency signal.

The microwave signal, which is transformed on the frequency, is amplified in the reflection microwave amplifier, is modulated by the unique code, and then it is radiated back in a direction of the interrogator. There this signal is received

with interrogator antenna. The secondary received microwave signal will be:

$$u_4(t) = U_0 A^2 G F_M(t) \sin\left[\left(\omega_0 + \Omega_n\right)t + \varphi_0 + 2\Delta\varphi + \varphi_n\right],$$

where $F_M(t)$ is the modulation function (unique code i), G is the OPTA gain.

The secondary received microwave signal is multiplied with the origin microwave one in the mixer. The combinatorial component of these microwave signals is described as:

$$u_5(t) = U_0 A^2 G F_M(t) \sin\left(\Omega_n t + \Delta\varphi_\Sigma\right), \qquad (2)$$

where $\Delta\varphi_\Sigma$ is the sum of phase shifts that takes into account all of constant phase shifts in all of parts of equipment and both links.

From (2) it is clear that the frequency ω_0 of initial microwave signal is absent, the factor $U_0 A^2 G$ defines the amplitude, the factor $F_M(t)$ defines the modulation function with unique code, and the frequency Ω_n defines one of low-frequency signals or "preliminary" identification of transponder.

This component of differential frequency Ω_n is selected by one of band-pass filters and is amplified by one of low-frequency amplifiers. If there are several transponders in a range of system operation distance with different frequencies shifts Ω_n, all of them are received, selected and demodulated in corresponding channel. The number of low-frequency oscillations is chosen taking into account tag traffic and probability of coincidence of low-frequencies of several tags (transponders).

Features

As the change of the phase of microwave signal in transponder does not change its amplitude, the power of the signal, which is radiated with transponder, is the same as the power of the signal, which is received by it. In this case there is no power attenuation in –20 dB, as it was in harmonic RFID system. Furthermore, the additional amplifying in 20-30 dB takes place. The operating distance of presented system according the Equation (1) is much more. On the other hand we can reduce the transmitted microwave power for yielding the same operation distance, and solve the problems of electromagnetic compatibility (EMC).

The special rigid restrictions to the level of second harmonic as well as all other spectral components of transmitter signal are absent. All of harmonic components are subtracted in the microwave mixer, where the initial microwave signal and frequency-transformed microwave signal are multiplied. The restrictions to the level of these spectral components are standard from the EMC point of view.

The transmitter and the receiver of interrogator both operate with the same antenna. No duplex filter in system is used in principle. The interrogator and transponder antennas will be narrowband, so, they can implement good radiation characteristics and it is possible to improve the EMC additionally.

The noise-immunity of presented RFID system is high because relatively narrowband filters are used jointly with the low-frequency amplifiers. Semiconductors of environment do not exert influence upon the system. The only near zone noise of microwave oscillator must be taken into consideration. But, there are no problems to realize perfect characteristics of microwave oscillator.

The presented RFID system can treat the signals from several tags simultaneously on an extended operation distance. All of tags can be placed within operation zone and the level of signals received from them can be quite different. Any of the received signals does not depress other signals in principle because the frequency shift of the microwave signal is different in different tags (transponders). All of them are presented at the mixer output as the particular spectral component of difference of initial microwave signal and

frequency-transformed one. Each low-frequency signal will be selected by corresponding band pass filter, will be amplified and processed separately from each other. The only thing we must do is to ensure the high enough dynamic range of the mixer. There are no obstacles to fulfill this demand.

Peculiarities of Realization of the Phase Shifter

General Assumptions

The main advantage of the equipment, which is designed on the basis of homodyne methods, is the simplicity of hardware realization with high metrological characteristics. The essence of homodyne methods consists in the use of the same generator for formation both: a probing signal, and a heterodyne signal (a reference signal).

In the previous works (Gimpilevich et al, 2007) and (Jandiery et al, 2007) the case of periodic change of probing signal phase on linear law (within the period) was considered. The analytical formulae for calculations of amplitudes and initial phases of harmonic of the combinatorial current of homodyne signal converter were received. These equations allow us to estimate the results of homodyne frequency transformations for various practically important cases. In works (Shirokov et al, 2006) and (Gimpilevich et al, 2006) it was shown, that for the obtaining of comprehensible characteristics it is necessary to provide the high accuracy of phase shift installation of probing signal and high linearity of dependence of this shift in time domain. At violations of these requirements the errors of homodyne frequency transformations appear which reduce the accuracy of radio engineering system at whole. In practice the creation of linear law of phase change with comprehensible characteristics represents the certain difficulties and not always it solves the problem, especially in a band of centimeter and millimeter waves.

For elimination of mentioned shortcomings we propose to change the continuous by discrete phase change of probe signal. This statement is based on a possibility of approximation of linear function by step one. Such approach allows us to establish the discrete values of phase shift with high accuracy on each step of phase change at an adjustment of controlled phase shifter. As a result these arrangements will provide the high metrological characteristics of radio engineering system.

In presented work we are tending to obtain the generalized equations for calculation of amplitudes and initial phases of all harmonics of difference current of homodyne frequency converter at discrete changes of probe signal phase. The analysis will be carried out for any number of discrete values of phase shift within the period of controlled signal.

Analysis of Homodyne Converter at Arbitrary Number of Phase Steps

Let's consider the case of periodic discrete phase change of probe signal. This case is realized by implementation of the discrete phase shifter with periodic law of phase change $\theta(t)$ (Figure 1). We will consider the general case, at which the number of discrete values of phase shift within the period T is equal to m (m is the integer), and the phase $\theta(t)$ is changed with step in $\Delta\theta = 2\pi/m$. Step approximation of linear function is thus reached.

Let's write down the analytical expression, which describes the periodic step function $\theta(t)$ on an time interval T, which is equal to the period of control signal:

$$\theta(t) = \begin{cases} 0 & if & 0 < t < \dfrac{T}{m}; \\ \Delta\theta & if & \dfrac{T}{m} < t < \dfrac{2T}{m}; \\ 2\Delta\theta & if & \dfrac{2T}{m} < t < \dfrac{3T}{m}; \\ \cdots\cdots\cdots\cdots\cdots\cdots\cdots\cdots\cdots\cdots \\ (m-1)\Delta\theta & if & \dfrac{(m-1)T}{m} < t < T. \end{cases} \tag{3}$$

The combinatorial component of a current through a nonlinear element of homodyne frequency converter we will represent in a form of (Shirokov, 2010):

$$i_d(t) = k_x \cos[\theta(t) + \varphi_x], \tag{4}$$

where $k_x = kU_xU_r$; k is the constant factor; U_x, φ_x are the amplitude and the phase, which are appeared at the passing or reflection of measuring signal thru (from) object under the test accordingly; U_r is the amplitude of a reference signal.

Let's substitute (4) in (3). In a result we will obtain:

$$\theta(t) = \begin{cases} I_1 = k_x \cos\varphi_x & if\ 0 < t < \dfrac{T}{m}; \\ I_2 = k_x \cos(\Delta\theta + \varphi_x) & if\ \dfrac{T}{m} < t < \dfrac{2T}{m}; \\ I_3 = k_x \cos(2\Delta\theta + \varphi_x) & if\ \dfrac{2T}{m} < t < \dfrac{3T}{m}; \\ \cdots & \cdots \\ I_m = k_x \cos[(m-1)\Delta\theta + \varphi_x] & if\ \dfrac{(m-1)T}{m} < t < T. \end{cases} \tag{5}$$

Taking into account the spectral density of a single rectangular impulse and the theorem of shift, and considering the fragment of the signal (5) on time interval ($0 \ldots T$), the complex amplitudes of periodic process spectrum harmonicas we will write down in a kind of

Figure 4. The phase change law of homodyne frequency converter

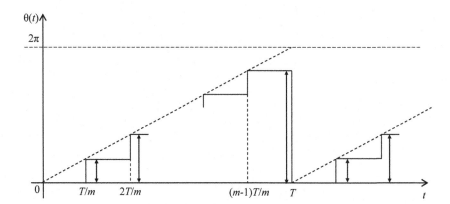

$$\dot{I}_n = \frac{2k_x}{m} \frac{\sin\left(\frac{\pi n}{m}\right)}{\frac{\pi n}{m}} [I_1 + I_2 e^{-j\frac{2\pi n}{m}}$$

$$+ I_3 e^{-j\frac{4\pi n}{m}} + ... + I_m e^{-j\frac{2\pi(m-1)n}{m}}] e^{-j\frac{\pi n}{m}}, \quad (6)$$

where n is the harmonic number; $\Omega = 2\pi/T$ is the angular frequency.

Let's rewrite (6) in a form:

$$\dot{i}_n = \frac{2k_x}{m} \frac{\sin\left(\frac{\pi n}{m}\right)}{\frac{\pi n}{m}} \left\{ \left[I_1 + I_2 \cos\left(\frac{2\pi n}{m}\right) + I_3 \cos\left(\frac{4\pi n}{m}\right) + ... + I_m \cos\left(\frac{2\pi(m-1)n}{m}\right) \right] + \right.$$

$$\left. + j \left[I_2 \sin\left(\frac{2\pi n}{m}\right) + I_3 \sin\left(\frac{4\pi n}{m}\right) + ... + I_m \sin\left(\frac{2\pi(m-1)n}{m}\right) \right] \right\} e^{-j\frac{\pi n}{m}}.$$

Using values for $I_1, I_2, ... I_m$ (5), we will mark out the real and the imaginary parts of expression in curly brackets:

$$\text{Re} = \cos\varphi_x + \cos\left(\frac{2\pi n}{m}\right)\cos\left(\frac{2\pi}{m} + \varphi_x\right) + \cos\left(\frac{4\pi n}{m}\right)\cos\left(\frac{4\pi}{m} + \varphi_x\right) + ...$$

$$... + \cos\left(\frac{2\pi(m-1)n}{m}\right)\cos\left(\frac{2\pi(m-1)}{m} + \varphi_x\right);$$

$$\text{Im} = \sin\left(\frac{2\pi n}{m}\right)\cos\left(\frac{2\pi}{m} + \varphi_x\right) + \sin\left(\frac{4\pi n}{m}\right)\cos\left(\frac{4\pi}{m} + \varphi_x\right) + ...$$

$$... + \sin\left(\frac{2\pi(m-1)n}{m}\right)\cos\left(\frac{2\pi(m-1)}{m} + \varphi_x\right).$$

Let's transform these equations using the multiplying of sinuses and cosines formulae and entering following designations:

$$\frac{2\pi(n-1)}{m} = \alpha \; ; \; \frac{2\pi(n+1)}{m} = \beta \, .$$

As a result we will receive: (see Exhibit 1)

Taking into account well known trigonometric equations:

$$\cos\alpha + \cos 2\alpha + ... + \cos k\alpha = \frac{\cos\dfrac{\alpha}{2} - \cos\dfrac{(2k+1)\alpha}{2}}{2\sin\dfrac{\alpha}{2}} \, ;$$

$$\sin\alpha + \sin 2\alpha + ... + \sin k\alpha = \frac{\sin\dfrac{(2k+1)\alpha}{2} - \sin\dfrac{\alpha}{2}}{2\sin\dfrac{\alpha}{2}} \, ,$$

and supposing $k = (m-1)$, previous expressions will be: (see Exhibit 2)

Exhibit 1.

$$\mathrm{Re} = \cos\varphi_{\mathrm{x}} + \frac{1}{2}\left\{\cos\varphi_{\mathrm{x}}\left[\cos\alpha + \cos 2\alpha + ... + \cos(m-1)\alpha\right] + \cos\varphi_{\mathrm{x}}\left[\cos\beta + \cos 2\beta + ... + \cos(m-1)\beta\right] + \right.$$

$$\left. + \sin\varphi_{\mathrm{x}}\left[\sin\alpha + \sin 2\alpha + ... + \sin(m-1)\alpha\right] - \sin\varphi_{\mathrm{x}}\left[\sin\beta + \sin 2\beta + ... + \sin(m-1)\beta\right]\right\};$$

$$\mathrm{Im} = \frac{1}{2}\left\{\cos\varphi_{\mathrm{x}}\left[\sin\alpha + \sin 2\alpha + ... + \sin(m-1)\alpha\right] + \cos\varphi_{\mathrm{x}}\left[\sin\beta + \sin 2\beta + ... + \sin(m-1)\beta\right] - \right.$$

$$\left. - \sin\varphi_{\mathrm{x}}\left[\cos\alpha + \cos 2\alpha + ... + \cos(m-1)\alpha\right] + \sin\varphi_{\mathrm{x}}\left[\cos\beta + \cos 2\beta + ... + \cos(m-1)\beta\right]\right\}.$$

Exhibit 2.

$$\mathrm{Re} = \cos\varphi_{\mathrm{x}}\left[1 + \frac{\cos(\pi(n-1))\sin\left(\frac{\pi}{m}(m-1)(n-1)\right)}{2\sin\left(\frac{\pi}{m}(n-1)\right)} + \frac{\cos(\pi(n+1))\sin\left(\frac{\pi}{m}(m-1)(n+1)\right)}{2\sin\left(\frac{\pi}{m}(n+1)\right)}\right] +$$

$$+ \sin\varphi_{x}\left[\frac{\sin(\pi(n-1))\sin\left(\frac{\pi}{m}(m-1)(n-1)\right)}{2\sin\left(\frac{\pi}{m}(n-1)\right)} - \frac{\sin(\pi(n+1))\sin\left(\frac{\pi}{m}(m-1)(n+1)\right)}{2\sin\left(\frac{\pi}{m}(n+1)\right)}\right];$$

$$\mathrm{Im} = \cos\varphi_{x}\left[\frac{\sin(\pi(n-1))\sin\left(\frac{\pi}{m}(m-1)(n-1)\right)}{2\sin\left(\frac{\pi}{m}(n-1)\right)} + \frac{\sin(\pi(n+1))\sin\left(\frac{\pi}{m}(m-1)(n+1)\right)}{2\sin\left(\frac{\pi}{m}(n+1)\right)}\right] -$$

$$- \sin\varphi_{x}\left[\frac{\cos(\pi(n-1))\sin\left(\frac{\pi}{m}(m-1)(n-1)\right)}{2\sin\left(\frac{\pi}{m}(n-1)\right)} - \frac{\cos(\pi(n+1))\sin\left(\frac{\pi}{m}(m-1)(n+1)\right)}{2\sin\left(\frac{\pi}{m}(n+1)\right)}\right].$$

Let's consider, that:

$$\cos\left[\pi(n-1)\right] = \cos\left[\pi(n+1)\right] = (-1)^{n+1}$$

and

$$\sin\left[\pi(n-1)\right] = \sin\left[\pi(n+1)\right] = 0.$$

Thus, the real and the imaginary parts of expression will become: (see Exhibit 3)

Let's transform these equations taking into account well known trigonometric formulae. Furthermore, we will notice, that in expressions at $n = qm + 1$ and $n = qm - 1$ (q is any positive integer) the uncertainty of type $0/0$ arises, which we will solve by means of L'Hôpital's rule:

Exhibit 3.

$$\mathrm{Re} = \cos\varphi_x \left\{ 1 + (-1)^{n+1}\frac{1}{2}\left[\frac{\sin\left(\frac{\pi}{m}(m-1)(n-1)\right)}{\sin\left(\frac{\pi}{m}(n-1)\right)} + \frac{\sin\left(\frac{\pi}{m}(m-1)(n+1)\right)}{\sin\left(\frac{\pi}{m}(n+1)\right)} \right] \right\};$$

$$\mathrm{Im} = -\sin\varphi_x(-1)^{n+1}\frac{1}{2}\left[\frac{\sin\left(\frac{\pi}{m}(m-1)(n-1)\right)}{\sin\left(\frac{\pi}{m}(n-1)\right)} - \frac{\sin\left(\frac{\pi}{m}(m-1)(n+1)\right)}{\sin\left(\frac{\pi}{m}(n+1)\right)} \right]$$

$$\frac{\sin\left(\frac{\pi}{m}(m-1)(n-1)\right)}{\sin\left(\frac{\pi}{m}(n-1)\right)} = \begin{cases} (-1)^n & if\ n \neq qm+1 \\ (-1)^{n+1}(m-1) & if\ n = qm+1 \end{cases}$$

$$\frac{\sin\left(\frac{\pi}{m}(m-1)(n+1)\right)}{\sin\left(\frac{\pi}{m}(n+1)\right)} = \begin{cases} (-1)^n & if\ n \neq qm-1 \\ (-1)^{n+1}(m-1) & if\ n = qm-1 \end{cases}$$

As a result we will receive:

$$\mathrm{Re} = \begin{cases} 0 & if\ n \neq qm+1\ ;\ n \neq qm-1 \\ \frac{m}{2}\cos\varphi_8 & if\ n = qm+1\ ,\ n = qm-1 \end{cases};$$

(7)

$$\mathrm{Im} = \begin{cases} 0 & if\ n \neq qm+1\ ;\ n \neq qm-1 \\ \frac{m}{2}\sin\varphi_8 & if\ n = qm+1\ ,\ n = qm-1 \end{cases}.$$

(8)

Substituting (7) and (8) in the expression for amplitude of harmonic, we will receive:

$$i_n = \begin{cases} 0 & if\ n \neq qm+1\ ;\ n \neq qm-1; \\ k_x\dfrac{\sin\left(\frac{\pi n}{m}\right)}{\frac{\pi n}{m}}(\cos\varphi_x + j\sin\varphi_x)e^{-j\frac{\pi n}{m}} & if\ n = qm+1\ ;\ n = qm-1. \end{cases}$$

(9)

From (9) we can see there is no harmonic except for harmonic with numbers $n = qm \pm 1$ in a signal spectrum. Using (9) and taking into account the expression for k_x, we will write down expressions for modules and for initial phases of spectrum components for difference current of homodyne frequency converter:

$$I_n = \begin{cases} 0 & if\ n \neq qm+1\ ;\ n \neq qm-1; \\ kU_xU_r\left|\dfrac{\sin\left(\frac{\pi n}{m}\right)}{\frac{\pi n}{m}}\right| & if\ n = qm+1\ ,\ n = qm-1. \end{cases}$$

(10)

$$\psi_n = \begin{cases} not\ defined & if\ n \neq qm+1\ ;\ n \neq qm-1; \\ \varphi_x - \dfrac{\pi n}{m} + \arg\left[\sin\left(\frac{\pi n}{m}\right)\right] & if\ n = qm+1\ ,\ n = qm-1. \end{cases}$$

(11)

Let's analyze some properties of the obtained spectrum.

The first harmonica ($q = 0$) and also harmonicas with numbers $m \pm 1$ ($q = 1$), $2m \pm 1$ ($q = 2$), $3m \pm 1$ ($q = 3$) etc are presented in the spectrum.

Amplitudes of harmonicas are directly proportional to the amplitude of a measuring signal, and the initial phases are equal to the initial phase of a measuring signal with a constant coincidence.

Figure 5. Spectrograms of amplitudes and initial phases for a case of m=5

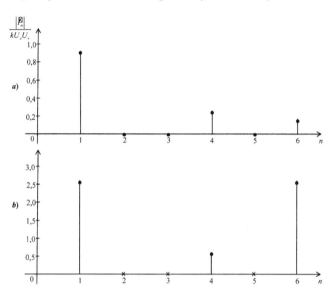

Amplitudes of spectrum harmonicas decrease with harmonica number increasing.

When we increase the discrete phase steps m, the frequency distance between the first harmonic and the nearest to it, is increased as well. This fact improves the selecting conditions for the first-harmonica component with the help of band-pass filter.

If the number of discrete values of phase shift m increases beyond all bounds ($m \rightarrow \infty$) it follows, that the first harmonica with maximum amplitude $I_{1\max} = kU_xU_r$ and initial phase $\psi_1 = \varphi_x$ is presented in a spectrum only, what corresponds to the linear law of phase shift change.

Example of the Spectral Analysis at Five-Step Change of Signal Phase

As an example, the spectrograms of amplitudes and initial phases for a case of $m = 5$ are shown in figure 5a and 5b.

Spectrograms are calculated by means of formulae (10) and (11). The amplitude spectrogram is plotted in normalized kind. The normalizing was carried out concerning the maximum value of the first harmonica amplitude at the linear change of phase shift ($I_{1\max} = kU_xU_r$). The spectrogram of initial phases is plotted for $\varphi_x = \pi$.

From figure 5 it follows, that in a spectrum at $m = 5$ there are harmonicas with the numbers 1, 4, 6, 9, 11 etc. Thus, the amplitude of the first harmonica is $0,935 I_{1\max}$. The frequency distance between the first harmonica and the nearest fourth one is 3Ω .

Overview of Other Cases

As we have pointed out above, the practical implementation of phase shifter, which realizes the linear law of phase changing, is a complex problem. It is impossible to ensure the stable characteristics of phase shifter in a wide temperature range and in presence of other disturbing factors. From (10) and (11) and figure 5 it is clear the discrete phase shifter with number of steps higher than 2 can be used for homodyne frequency converter. The basic question, which is to be solved, is the number of steps there must be in phase shifter.

Figure 6. Step approximation of sinusoidal current of homodyne frequency converter

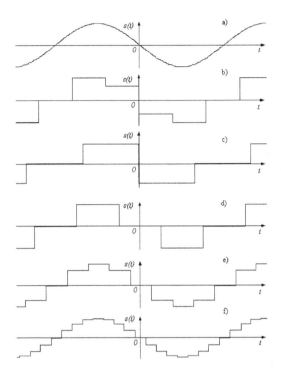

The spectrum analysis for different number of steps of phase shifter was carried out. The level of the first harmonica of signal, which approximates the sinusoid oscillation by 3, 4, 5, 8 and 16 steps, was obtained (Shirokov, 2004). Results of calculations are summarized in table. The step approximation of sinusoidal current of homodyne frequency converter at the number of phase shift step in 5, 3, 4, 8, and 16 is shown in figure 6.

Choosing of the Number of Steps

The number of steps, obviously, must satisfy the binary law. Consequently, the use of 4-steps, 8-steps, 16-steps, and so on of phase shifter is much optimal for use in homodyne detection systems.

From the other hand, we can see that if number of steps is 4 or higher, the level of the first harmonic is more than 90 percent from theoretical possibility. If we increase the number of steps from 3 to 4, the level of first harmonic increases as well and

reaches 8 percent. The further increasing of the number of steps up to 8 results in the increasing of level about 5 percent. When the number of steps increases from 8 to 16, the increasing of the level of first harmonic achieves 1 percent only.

Thus, if critical condition is the simplicity of transponder at normal quality, it is recommended to use 4-step shifter. If critical condition is the quality of signal, it is recommended to use 8-step shifter. The application of 16 and more steps of phase shifter complicates the transponder, but it does not give considerable advantages and it is unjustified.

From Table 1 one more law is traced. Besides the basic harmonicas, the nearest harmonica component with an essential level, has a serial number $m - 1$, what absolutely corresponds to mentioned above conclusions. This fact allows us to determine the unequivocally requirements to the filtering units of interrogator. And with the growth of number of steps the cutoff frequency of the filter can be increased adequately in relation to the frequency of the basic harmonica.

Some Aspects of Phase Shifter Realization

When we choose the number of steps in 8, the phase discrete will be 45° respectively. It involves in implementation of 3 binary digits. The most significant bit controls the phase shift in 180°, another in 90° and the last in 45°. The microwave signal phase sequence must be 0°, 45°, 90°, 135° etc. or 0°, 315°, 270°, 225° etc. We must take into consideration that the controlled microwave transmission phase shifter is used and the microwave passes the phase-shifting chains twice. So, the phase-shifting chains must be distributed as 90°, 45° and 22.5° respectively, taking into account the two ways of microwave signal passing. The specific technical realizations of microwave transmission phase shifter are well known and they are not discussed here.

Table 1. Dependence of harmonic level from number of steps

Number of steps	Level of harmonic							
	1	2	3	4	5	6	7	8
3	0,82	0,41	0	0,21	0,16	0	0,12	0,1
4	0,90	0	0,30	0	0,18	0	0,13	0
5	0,93	0	0	0,23	0	0,16	0	0
8	0,98	0	0	0	0	0	0,14	0
16	0,99	0	0	0	0	0	0	0

The change of phase of microwave signal over the period T of the controlling signal with the lowest frequency (for the cell of 90° phase shift, most significant bit or 180° for double passing) by 2π is tantamount to the frequency shift in $\Omega = 2\pi/T$ of the initial signal. The increasing law of phase change results in forming of the transformed signal with the frequency in $\omega_0 - \Omega$, the decreasing law in $\omega_0 + \Omega$.

SIMULATION

The simulation of frequency transformations in discussed RFID system was carried out. The scheme of simulation is shown in figure 7.

Here (MO) is the microwave oscillator, (MIX) is the mixer. These two units represent the part of interrogator. Microwave oscillator has two out-

puts. The signal from the first output is the heterodyne one. This signal feeds the reference input of mixer. The signal from the second output is the initial signal which is radiated in the direction of transponder.

For simulation the initial frequency f_0 in 1 GHz was chosen. Possible attenuation of signal on direct and return paths and the gain of antennas of interrogator and transponder are not taken into account. These factors do not affect upon the frequencies transformation in a system.

The controlled phase shifter (CPHS) and the amplitude modulator (AM) both simulate the work of transponder. For the simulation one path controlled microwave transmission phase shifter was used. It is easily to describe and to simulate the work of such phase shifter without the changing of approach in general. The controlling signal results in changing of phase of initial microwave oscillation by 2π over the controlling signal period T. For simulation the period T in 1 μs was chosen. The resulting frequency shift F_n will be 1 MHz. The number of steps of controlled phase shifter was chosen equal to 8 for simulation. The simulation was carried out in environment MathCAD. For simulation the initial phase of microwave oscillations was equal to 0 and the amplitude factor was equal to 1.

The law of phase changing of microwave oscillations is described by the following equation:

Figure 7. Block diagram of simulation process

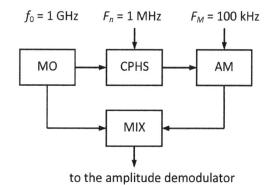

to the amplitude demodulator

Figure 8. The law of phase changing of microwave oscillations in time domain and signal spectrum at the phase shifter output

Figure 9. The spectrums of signal from single a) and pair of transponders b) on the mixer output

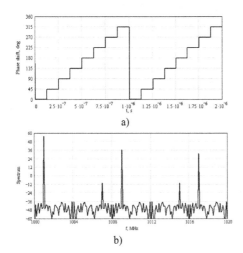

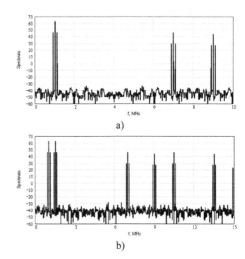

$$\Phi(t) = 180 \left\{ 0.5 \operatorname{sign} \left[-\sin \left(2\pi F_n t \right) \right] + 0.5 \right\}$$
$$+ 90 \left\{ 0.5 \operatorname{sign} \left[-\sin \left(4\pi F_n t \right) \right] + 0.5 \right\}$$
$$+ 45 \left\{ 0.5 \operatorname{sign} \left[-\sin \left(8\pi F_n t \right) \right] + 0.5 \right\}.$$

This law of microwave signal phase changing is shown in Figure 8 a).

Signal on the output of controlled phase shifter will be:

$$u_{\text{CPHS}}(t) = \sin \left[2\pi f_0 t + \Phi(t) \frac{\pi}{180} \right]. \qquad (12)$$

Here the initial phase of controlling signal was accepted equal to 0.

The signal spectrum at the phase shifter output (12) is shown in Figure 8 b).

The plots show the microwave oscillations obtain the frequency shift in 1 MHz and the frequency of main harmonica of transformed microwave oscillations at the phase shifter output is equal to 1001 GHz. The order of nearest harmonica with essential level is equal to 7, as it was pointed out in

previous section. The frequency of this harmonica is equal to 1007 GHz.

Further, in transponder the frequency-transformed signal was modulated by the amplitude with unique code sequence $F_M(t)$. For simulation, the amplitude modulation with harmonious single-tone signal was chosen. The modulation depth was chosen equal to 30% and the frequency of modulating single-tone signal was chosen equal to 100 kHz.

The signal on the output of amplitude modulator will be described by the following equation:

$$u_{AM}(t) = \left[1 + 0.3 \sin \left(2\pi F_M t \right) \right] \sin \left[2\pi f_0 t + \Phi(t) \frac{\pi}{180} \right].$$

This signal is reradiated with the transponder and received with interrogator. In interrogator this received signal is mixed with heterodyne signal (initial microwave oscillations) in the mixer. For simulation the conventional multiplier was chosen as the mixer. The operation of conventional multiplication results in the appearance of two components: with the frequency of sum and the frequency of difference. The component with highest frequency (2GHz) is filtered out and the component with the frequency of difference is of interest. One is described as:

$$u_{MIX}(t) = \left[1 + 0.3\sin\left(2\pi F_M t\right)\right]\cos\left[\Phi(t)\frac{\pi}{180}\right].$$

The spectrum of this signal is shown in Figure 9 a).

Thus, on the mixer output we can watch (Figure 9 a) the amplitude-modulated signal with well known spectrum. Whereas the spectrum of this signal contains several components (each is amplitude-modulated), we can select desirable (main) component with the band-pass filter.

In a case of presence of several transponders in the system operating area, on the output of mixer there will be appeared several components with certain unique frequencies F_m. The spectrum of signals from two transponders on the output of mixer is shown in Figure 9 b). Here the first transponder forms the frequency shift of initial microwave oscillations on 1 MHz, the second transponder forms the frequency shift on 1.5 MHz. As we can see, both main spectral components are presented on the output of the mixer. Each component can be selected with corresponding band-pass filter with the certain central frequency (1 MHz and 1.5 MHz respectively). Obviously, the width of pass-band of each filter must satisfy to the width of spectrum of desirable corresponding component (with modulation).

The number of transponders, which can be treated simultaneously, depends from the spectrum width of each main spectral component and from the difference of frequencies of main spectral component and its nearest parasitic spectral component with essential level. Certainly, this difference of frequencies must be taken into account for the lowest of frequency shifts of initial microwave oscillations. For our case the lowest frequency shift is equal to 1 MHz, the difference of frequencies of its spectral components is equal to 6 MHz

(see Figure 6) and for the width of spectrum of useful component with modulation in 200 kHz, the number of transponders, which can be treated with interrogator simultaneously is equal to 30.

This number of transponders can be increased easily by the increasing of number of steps in phase shifters. The difference of frequencies of the main and the first parasitic harmonic is increased respectively. In a limit, with the use of continuous phase shifter rather than discrete one, this difference of frequencies will be increased unlimitedly. But it has no special sense as the restrictions on a frequency band will give the antennas of transponder and interrogator.

When we increase the frequency shift itself, we increase the number of treated transponders too.

After the desirable component has been chosen, the obtained signal can be demodulated and it is possible to determine the unique code sequence $F_M(t)$.

RESTRICTIONS

Certainly, the number of frequency shifts is strictly limited because of the frequency band reason, that is used. In fact this number cannot exceed few hundreds. This frequency band is limited by the frequency-related characteristics of used antennas of tags and interrogator. Further, the phase shifter band-pass features limit mentioned frequency band. And at last, we must take into account the problems of EMC in general.

The coincidence of different transponder frequency shifts results in depressing of the signal with lower level. Thus, for the estimation of system efficiency, the probability characteristics must be taken into account.

Further, the demodulator represents the modulation function $F_M(t)$. This function is the time-domain one. The maximum pulse-repetition frequency of unique sequence, described with this function, is limited with the frequency characteristics of interrogator low-frequency band-pass filter. In fact this pulse-repetition frequency will not be greater than few kbits/s. Thus, the length of this sequence can not be arbitrary large. That is why the information capacity of system will not be large too. However, in a certain applications it will be quite enough. From the other hand, we are talking about the time of tag identification. If this time is not of interest for our application (immobile tags or very low speed of its moving) the information capacity of system will be high.

All of mentioned restrictions can not eliminate the main advantage of presented system: the interrogator can treat large number of tags simultaneously.

CONCLUSION

Thereby, presented RFID system can be used in all aspects of human life. For example it can organize the enterprise admission work, the biological and medical investigations etc. This system will be especially useful when there is large traffic and EMC problems are significant.

The suggested technical solution has the following advantages:

1. Requirements to the cleanliness of interrogator transmitter spectrum are not high, all of spectral components of transmitter are mutually subtracted and eliminated;
2. The energy of communication channel of presented RFID system is essentially much higher, the operation distance is large.
3. The noise-immunity of the system is high because the relatively narrow band-pass filters are used in a low-frequency path of interrogator and no semiconductors of natural and artificial origin react on the system;
4. The EMC of system is good;
5. The interrogator can treat the signals from several transponders simultaneously; this is the main advantage of discussed RFID system.

REFERENCES

Colpitts, B. G., & Boiteau, G. (2004). Harmonic radar transceiver design: Miniature tags for insect tracking. *IEEE Transactions on Antennas and Propagation*, *52*(11), 2825–2832. doi:10.1109/TAP.2004.835166

Gimpilevich, Y. B., & Shirokov, I. B. (2006). The homodyne frequency transformation at inexact installation of phase shift change range of probing signal. *News of Higher Educational Institutions. Radio Electronics Kiev*, *49*(10), 54–63.

Gimpilevich, Y. B., & Shirokov, I. B. (2007). Generalized mathematical model of homodyne frequency conversion method under a periodic variation in sounding signal phase shift. *Telecommunications and Radio Engineering*, *66*(12), 1057–1065. doi:10.1615/TelecomRadEng.v66.i12.20

Jaffe, J. S., & Mackey, R. C. (1965). Microwave frequency translator. *IEEE Transactions on Microwave Theory and Techniques*, *13*, 371–378. doi:10.1109/TMTT.1965.1126002

Jandieri, G. V., Shirokov, I. B., & Gimpilevich, Yu. B. (2007). The analysis of metrological features of homodyne method of frequency transformation. *Georgian Engineering News*, *2*, 38–45.

Koshurinov, E. I. (2003). The homodyne radar with optimal signal processing. *IEEE Proceedings of Crimean Microwave Conference*, (pp. 737-739). Sevastopol, Ukraine, 8-12 Sep. 2003.

Sharpe, C. A. (1995). *Wireless automatic vehicle identification* (pp. 39–58). Applied Microwave & Wireless FALL.

Shirokov, I. B. (2009). The multitag microwave RFID system. *IEEE Transactions on Microwave Theory and Techniques*, *57*(5), 1362–1369. doi:10.1109/TMTT.2009.2017315

Shirokov, I. B. (2010, September 10). *The method of radio frequency identification.* (Patent of Ukraine, #91937, MPC G01R 29/08, 6 p).

Shirokov, I. B. (2011, January 5). *The method of operation range increasing of RFID system.* (Patent Application of Ukraine, #a201100198, MPC G01R 29/08, 7 p.).

Shirokov, I. B., & Durmanov, M. A. (2007). Multi-Abonent RFID system. *IEEE Proceedings of Crimean Microwave Conference* (pp. 756-757). Sevastopol, Ukraine, 10-14 Sep. 2007.

Shirokov, I. B., & Gimpilevich, Yu. B. (2006). Influence of phase shift change nonlinearity of probing signal on an error of homodyne frequency converter. *News of Higher Educational Institutions. Radio Electronics Kiev, 49*(12), 20–28.

Shirokov, I. B., Gimpilevich, Y. B., & Polivkin, S. N. (2010). The generalized mathematical model of homodyne frequency converter at discrete change of probing signal phase. *Radio Engineering: All-Ukrainian Interdependent Science-Technology Magazine, 161,* 119-125. Kharkov 2010.

Shirokov, I. B., & Isofatov, V. V. (2003). The increasing of noise immunity of automatic system of RFID of vehicles and cargos. *IEEE Proc. of Crimean Microwave Conference* (pp. 718-719). Sevastopol, Ukraine, 8-12 Sep. 2003.

Shirokov, I. B., Jandieri, G. V., & Sinitzyn, D. V. (2006). Multipath angle-of-arrival, amplitude and phase progression measurements on microwave line-of-sight links. *European Space Agency, (Special Publication). ESA SP, 626.*

Shirokov, I. B., & Polivkin, S. N. (2004). The selection of the number of phase shifter steps in tasks of investigations of channel characteristics, carried out with homodyne methods (in Russian). *Radio Engineering: All-Ukrainian Interdependent Science-Technology Magazine, 137,* 36-43. Kharkov 2004.

Vanjari, S. V., Krogmeier, J. V., & Bell, M. R. (2007). Remote data sensing using SAR and harmonic reradiators. *IEEE Transactions on Aerospace and Electronic Systems, 43*(4), 1426–1440. doi:10.1109/TAES.2007.4441749

Venguer, A. P., Medina, J. L., Chávez, R. A., & Velázquez, A. (2002). Low noise one-port microwave transistor amplifier. *Microwave and Optical Technology Letters, 33*(2), 100–104. doi:10.1002/mop.10236

Venguer, A. P., Medina, J. L., Chávez, R. A., Velázquez, A., Zamudio, A., & Il'in, G. N. (2004). The theoretical and experimental analysis of resonant microwave reflection amplifiers. *Microwave Journal, 47*(1), 80–93.

KEY TERMS AND DEFINITIONS

Homodyne Detection: A method of detecting frequency-modulated radiation by non-linear mixing with radiation of a reference frequency, the same principle as for heterodyne detection.

Microwave Amplifier: A device for increasing the power of a microwave signal.

Microwave Antenna: A transducer designed to transmit or receive electromagnetic waves at microwave frequency.

Microwave Mixer: A device that combines two or more microwave signals into one or two composite output signals.

Microwave Oscillator: An electronic circuit that produces a repetitive microwave signal.

Microwave Phase Shifter: A microwave network which provides a controllable phase shift of the RF signal.

Transponder: A receiver-transmitter that will generate a reply signal upon proper electronic interrogation.

Chapter 11
A Novel Approach in the Detection of Chipless RFID

Prasanna Kalansuriya
Monash University, Australia

Nemai Chandra Karmakar
Monash University, Australia

Emanuele Viterbo
Monash University, Australia

ABSTRACT

This chapter presents a different perspective on the chipless RFID system where the chipless RFID detection problem is viewed in terms of a digital communication point of view. A novel mathematical model is presented, and a novel approach to detection is formulated based on the model. The chipless RFID tag frequency signatures are visualized as points in a signal space. Although data bits are stored in the tags using unconventional techniques, the proposed model enables the detection of these data bits through conventional robust detection methods. Through simulations it is shown that the proposed detection method has better performance compared to contemporary detection approaches.

INTRODUCTION

Radio frequency identification (RFID) is widely used in many aspects of modern society where applications range from farming and food industry to finance and banking (Weinstein, 2005). The key feature that makes RFID such an attractive technology in many applications is the ability of quick, automatic and wireless extraction of information. This saves huge amounts of time and labor required for the monotonous procedures involving information retrieval, data entry and inventory management in many applications.

Despite the superior technological capacity, RFID technology has not yet fully penetrated into applications involving large scale item tagging such as library management systems, retail industry, logistics etc. (IDTechEx, 2006; S. Preradovic & Karmakar, 2010). This is due to the fact that the printable and integrated circuit technology enabling the advance features of the RFID tag has not proven to be cost effective when competing with existing technologies such as the one and two dimensional optical barcodes.

However, recent advances in RFID technology have promised a significant reduction in the cost

DOI: 10.4018/978-1-4666-1616-5.ch011

of RFID tags. These tags, commonly referred to as chipless RFID tags in literature (S. Preradovic & Karmakar, 2010), possess no integrated circuitry (chip) and are essentially passive reflectors or absorbers of electromagnetic radiation. Due to the absence of any electronic circuitry or any intelligent signal processing a chipless RFID is essentially the radio frequency counterpart of the ordinary optical barcode. This enables mass scale production of these tags at a very low cost comparable with optical barcodes but with some of the attractive features and benefits of the conventional RFID technology.

The conventional chipped RFID tag uses standard methods to wirelessly transfer the information it holds to an RFID reader. These methods vary in complexity from communication protocols such as WLAN, Zigbee and Bluetooth used for active RFID tags to simple back-scatter modulation techniques using amplitude shift keying (ASK), frequency shift keying (FSK) and phase shift keying (PSK) in passive and semi-passive RFID tags (Bolic, Simplot-Ryl, & Stojmenovic, 2010). The functionality of the chip and the wireless transfer of information are both powered by energy extracted from an interrogation signal sent by the RFID reader or by energy contained in an energy storage such as a battery. In the case of a chipless RFID tag, since it has no chip to facilitate active means of wireless transmission of digital data, it relies on passive approaches to convey the information it holds. As oppose to actively digitally re-modulating the interrogation signal sent by the RFID reader a chipless RFID tag passively transforms the interrogation signal characteristics to carry data back to the reader. This transformation occurs when the interrogation signal hits the tag and it is characterized by the passive microwave properties of the chipless RFID tag. Sharp and abruptly changing features in the amplitude (S. Preradovic, Balbin, Karmakar, & Swiegers, 2008; S. Preradovic & N. C. Karmakar, 2009), phase (Balbin & Karmakar, 2009) or time of arrival (Chamarti & Varahramyan, 2006; Hu, Law, &

Dou, 2008; Shao, et al., 2010) of the modified interrogation signal are used in order to represent data bits. A portion of the modified interrogation signal is reflected back towards the RFID reader where it is used to detect the information carried by the chipless tag. The detection performed by the RFID reader is essentially comparable to a miniature RADAR system attempting to distinguish different data carrying RADAR signatures produced by chipless RFID tags. Transferring data using analog and passive means has shifted the intelligence required at the tag into the reader and lowered the cost per tag. However, it has significantly increased the processing requirements at the RFID reader.

Detection of the data stored in the chipless RFID is performed by analyzing the received modified interrogation signal. The signal processing involved is quite challenging since the received signal is very weak and is also affected by unwanted interference and noise such as, mutual coupling between antennas at the receiver, clutter in the environment, mulitpath etc. Different techniques are reported in research literature for the detection of information contained in the received signals from chipless RFID tags. In (Koswatta & Karmakar, 2010; S. Preradovic & N.C. Karmakar, 2009) chipless RFID reader designs are presented which utilize threshold based detection through the use of calibration tags. Here, the effect of clutter and antenna coupling is also removed by using the calibration tags. However, since this approach is based on hard thresholds derived through calibration it does not posses the flexibility and adaptability required in the detection process to address errors due to a dynamic environment. The authors of (Blischak & Manteghi, 2009; Manteghi, 2010) characterize the backscattered radiation from a chipless tag using a set of poles and residues. The data is stored using the pole locations which are changed by altering the structure of the chipless RFID tag. Detection is performed by extracting the poles and residues from the backscattered signal using the Matrix Pencil Algorithm. This method does not rely on calibration tags for detection. But the method has only been presented as a proof of concept approach for encoding information and detection for a low number of bits. In (Dullaert, Reichardt, & H., 2011) the phase signature is used to detect the resonances encoding information bits in a multiresonator based chipless RFID tag. In order to mitigate the effect of noise on the measured phase response it is expressed using a series of prolate spheroidal wave functions (PSWF). Through this approach the direct detection range (range of detection without using a calibration tag) can be improved up to 10cm. Continuous wavelet transform (CWT) is proposed in (Lazaro, Ramos, Girbau, & R., 2011) in order to reduce the effect of noise and improve reader range in ultra wideband (UWB) RADAR based chipless RFID readers for reading time coded time domain reflectometry (TDR) based chipless RIFD tags. With this approach the authors have been able to significantly improve detection range (1.5 m). From the contemporary research reported in literature it is clear that the chipless RFID system has only been investigated from a microwave and RADAR theoretical perspective. A different view of the problem could prove advantageous in solving the challenges faced. Therefore, the main focus of this chapter is based on presenting the chipless RFID system through a wireless and digital communication perspective.

As opposed to chipless RFID, the detection involved with conventional chipped RFID use standard detection methods often used in digital and wireless communication. However, these approaches to detection cannot directly be used in the domain of chipless RFID. In a wireless communication system the main deterrents to the successful point to point transmission of information is the wireless channel lying between them. This channel is dynamic and causes signal levels received at the receiver to fluctuate rapidly (Goldsmith, 2005) causing erroneous detection. Hence, in such a situation the channel is estimated through pilot data signals and its effect on the

received signal is removed at the receiver. Figure 1 (a) shows data transmission in a conventional digital communication system. In a chipless RFID system the interrogation signal transmitted by the RFID reader is transformed by the tag to carry data. Here, as shown in Figure 1 (b), the transmitter and receiver are both collocated at the RFID reader which transmits a known signal which is equivalent to a pilot data signal and estimates how the RFID tag has modified the interrogation signal to detect the data stored in it. This process is comparable to the channel estimation done in a conventional wireless communication system and the chipless RFID tag serves as the channel between the transmitter and receiver. The difference is that the data is not carried in the signals transferred between the transmitter and receiver but in the channel which is estimated. Even though the channel shown in Figure 1 (b) consists of only the chipless tag, in reality the channel comprises of two components. One is the channel due to desired response of the chipless RFID tag. The other is the undesirable channel due to pathloss, multipath fading as experienced in any wireless transmission of data. The problem is to remove the affects of the undesirable channel and accurately estimate the desirable channel due to the chipless RFID tag.

Additive white Gaussian noise (AWGN) or thermal noise is one of the main sources of noise in digital communication. Apart from the errors caused by AWGN, data carried in chipless RFIDs are also affected by errors in fabrication of the microwave structures. Most of the chipless RFID techniques reported in research (Hu, et al., 2008; Jang, Lim, Oh, Moon, & Yu, 2010; S. Preradovic, et al., 2008; S. Preradovic & N. C. Karmakar, 2009; S. Preradovic & Karmakar, 2010) are operating beyond the S – band frequencies (above 2 GHz). Microwave structures built on the chipless tags in these frequencies are very sensitive to small errors in their dimensions. Hence, minute structural inconsistencies and imperfections resulting in mass scale production of these chipless

Figure 1. (a) Conventional wireless communication system (b) chipless RFID system

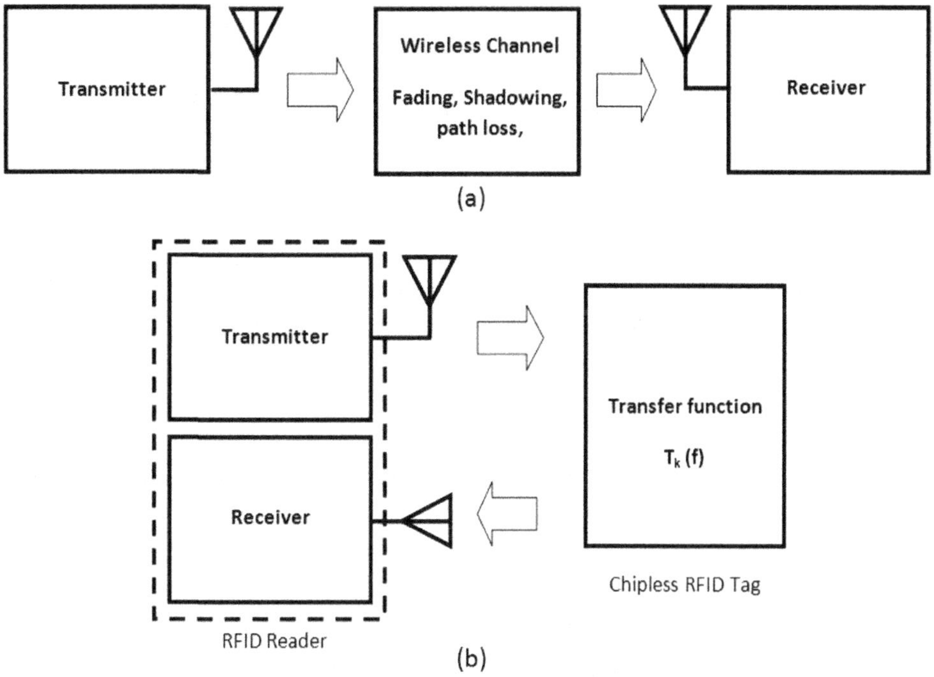

RFID tags would yield errors in the desired tag performance. Also, commercial production of chipless tags will be performed using printable conductive ink materials having lower conductivity than pure copper or silver. This would also result in further loss in their performance.

To address the aforementioned problems in detection, algorithms need to be developed which are able to detect data bits accurately amidst distortions caused by the wireless channel and noise. Although mature and robust detection approaches are widely used in digital communication they cannot be directly applied due to the non-standard forms of data representation used in the chipless RFID tags. Therefore, a mathematical model is required that represents the data of chipless RFID tags in a way which allows the use of those detection methods. This chapter presents a novel mathematical model for the detection of chipless RFID tags based on signal space representation (SSR) (Haykin, 2009) of tag frequency signatures. Here, the frequency signatures from tags are represented as signal points in a signal space and detection of data bits is based on minimum distance detection.

SYSTEM DESCRIPTION

Chipless RFID systems operate based on different principles. Two widely reported principles of operation in literature are i.) retransmission based chipless tags (S. Preradovic & N. C. Karmakar, 2009) and ii.) back-scatter based chipless tags (Balbin & Karmakar, 2009). For the purpose of demonstrating the novel detection scheme based on SSR, a retransmission based chipless RFID tag is considered. However, the theory and concepts introduced here are not limited to chipless RFID systems based on this operating principle. The proposed method can be extended to cater for chipless RFID systems which rely on other types of operating principles.

Figure 2. System model of a retransmission based chipless RFID system

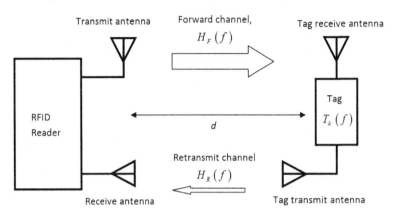

Figure 2 depicts the retransmission based chipless RFID system investigated. A retransmission based chipless RFID tag comprises of two components, i.) antenna elements for receiving and retransmitting signals ii.) a passive microwave filter capable of transforming the received interrogation signal. The operation of the tag is quite straight forward. The interrogation signal transmitted by the RFID reader travels through the forward transmission channel $H_F(f)$ and is picked up by the receiving antenna of the tag. This signal travels through the passive microwave filter of the tag and is transformed to carry some information. The transformation is characterized by the transfer function $T_k(f)$ of the microwave filter. Features present in the amplitude or phase of the transfer function are used in representing information bits. Hence, depending on the data stored in the tag the transfer function will vary. The transformed interrogation signal coming out of the microwave filter is retransmitted via the transmitting antenna of the chipless tag. At this point the signal is quite weak and it also needs to travel through the retransmission channel $H_R(f)$. Once it reaches the receive antenna of the RFID reader the signal level is very weak and it will also be adversely affected by the thermal noise. Apart from the effects of the forward and retransmit channels and thermal noise there exists another deterrent to the accurate detection of information

contained in the chipless tag. This is the interference between the antennae of the RFID reader and clutter produced by the surrounding environment. Taking all these factors into consideration the total received signal at the reader receiving antenna can be expressed as,

$$R(f) = H_F(f) T_k(f) H_R(f) S(f) + I(f) S(f) + N(f), \quad (1)$$

where $S(f)$ and $R(f)$ are the Fourier transforms of the transmitted interrogation signal and received signal respectively, $I(f)$ is the transfer function describing the behavior of the interference and $N(f)$ is the total noise affecting the received signal. From (1) it is clear that in order to obtain an estimate of the information carrying transfer function $T_k(f)$ the other factors affecting the received signal need to be removed. The channel transfer functions $H_F(f)$ and $H_R(f)$ consists of the effects of non-uniform antenna gain functions (reader antennas and tag antennas) and the frequency selective pathloss incurred due to electromagnetic waves travelling a distance "*d*" to the tag. The effect of these terms can be removed accurately to some extent. However, removal or mitigation of the effect of interference is quite a challenging task.

Exhibit 1.

$$\mathbf{T} = \begin{bmatrix} \mathbf{t}_1 & \mathbf{t}_2 & \cdots & \mathbf{t}_k & \cdots & \mathbf{t}_n \end{bmatrix} = \begin{bmatrix} \mathbf{u}_1 & \mathbf{u_2} & \cdots & \cdots & \mathbf{u}_n \end{bmatrix} \begin{bmatrix} \sigma_1 & & & 0 \\ & \sigma_2 & & \\ & & \ddots & \\ 0 & & & \sigma_n \end{bmatrix} \begin{bmatrix} \mathbf{v}_1^H \\ \mathbf{v}_2^H \\ \cdot \\ \cdot \\ \mathbf{v}_n^H \end{bmatrix} \qquad (4)$$

$$\text{m} \times \text{n} \qquad\qquad \text{m} \times \text{n} \qquad\qquad \text{n} \times \text{n} \qquad \text{n} \times \text{n}$$

DETECTION BASED ON SIGNAL SPACE REPRESENTATION

The SSR based detection scheme is based on representing each tag transfer function or frequency signature $T_k(f)$ as a linear combination of a set of orthogonal signals. These signals serve as the fundamental building blocks of all the $T_k(f)$s, hence they are termed "basis functions." Detection of a noisy estimated tag signature is performed by calculating the amount of each basis function contained in it. This detection using basis functions is capable of filtering out a significant amount of noise affecting the estimated signatures which results in a good detection performance.

The first step is the derivation of a set of orthonormal basis functions which are capable of accurately describing any of the possible tag signatures. Given the tag is designed to represent b bits, there are $n = 2^b$ possible different tags. Let each tag signature $T_k(f)$, $k = 1, \ldots, 2^b$ be sampled in the frequency domain with m frequency samples which gives rise to an m dimensional vector $\mathbf{t}_k$. Generally $\mathbf{t}_k$ is a complex vector; i.e. $\mathbf{t}_k \in \mathbb{C}^m$, where $\mathbb{C}^m$ is the set of m dimensional complex vectors. These n independent tag signatures $\mathbf{t}_k$, $k = 1, \ldots, n$ each having m samples form a matrix $\mathbf{T}$,

$$\mathbf{T} = \begin{bmatrix} \mathbf{t}_1 & \mathbf{t}_2 & \cdots & \mathbf{t}_k & \cdots & \mathbf{t}_n \end{bmatrix} \qquad (2)$$

By performing singular value decomposition on $\mathbf{T}$ we obtain

$$\mathbf{T} = \mathbf{U}\boldsymbol{\Sigma}\mathbf{V}^H \qquad (3)$$

where $\mathbf{U}$ and $\mathbf{V}$ are unitary matrices composed of orthonormal column vectors $\{\mathbf{u}_i\}$ and $\{\mathbf{v}_i\}$, respectively and $\mathbf{A}^H$ is the Hermitian transpose of matrix $\mathbf{A}$. $\boldsymbol{\Sigma}$ is a diagonal and positive definite matrix which is called the singular value matrix containing singular values σ_i. If $\mathbf{T}$ has rank r it will contain r non-zero singular values. Since the $\mathbf{T}$ is composed of n independent tag signatures $r = n$. Therefore, (3) can be re-written as, (see Exhibit 1).

Since $\{\mathbf{u}_i\}$ is a set of orthogonal column vectors they serve as a good choice for a set of basis functions to represent the set of tag signatures $\{\mathbf{t}_k\}$. The amount of a particular $\mathbf{u}_i$ contained in a frequency signature $\mathbf{t}_k$ can be found by taking the inner product between them, i.e. $\langle \mathbf{t}_k, \mathbf{u}_i \rangle = \mathbf{t}_k^H \mathbf{u}_i$. Further simplification of (4) gives,

$$\begin{bmatrix} \mathbf{t}_1^H \\ \mathbf{t}_1^H \\ \vdots \\ \vdots \\ \mathbf{t}_n^H \end{bmatrix} \quad \mathbf{u}_i = \sigma_i \quad \mathbf{v}_i = \sigma_i \quad \begin{bmatrix} v_1^i \\ v_2^i \\ v_3^i \\ \vdots \\ v_n^i \end{bmatrix}$$

$$\text{n} \times \text{m} \quad \text{m} \times 1 \quad \text{n} \times 1 \quad \text{n} \times 1 \qquad (5)$$

where v_j^i is the *j*-th element in the column vector $\mathbf{v}_i$. Therefore, from (5) we see that the inner product between vectors $\mathbf{t}_k$ and $\mathbf{u}_i$ can be expressed as

$$\mathbf{t}_k^H \mathbf{u}_i = \sigma_i v_k^i \qquad (6)$$

Hence we can write,

$$\mathbf{t}_k = \sum_{i=1}^n \langle \mathbf{t}_k, \mathbf{u}_i \rangle \mathbf{u}_i = \sum_{i=1}^n \left(\sigma_i v_k^i \right) \mathbf{u}_i \qquad (7)$$

Since only a handful of σ_i are large and the rest are negligible each $\mathbf{t}_k$ can be approximated using a few $\mathbf{u}_i$ as expressed in (8). This approach is commonly used in many applications involving model simplification (Boyd, 2007; Strang, 2009).

$$\mathbf{t}_k \approx \sum_{i=1}^L \langle \mathbf{t}_k, \mathbf{u}_i \rangle \mathbf{u}_i = \sum_{i=1}^L \left(\sigma_i v_k^i \right) \mathbf{u}_i \qquad (8)$$

Hence, $\mathbf{u}_i, i = 1, \ldots, L$, serve as a basis for each tag signature $\mathbf{t}_k$. This will enable the construction of an L dimensional signal space where the 2^b tag signatures $\mathbf{t}_k$ are represented as 2^b signal points $\mathbf{s}_k \in \mathbb{C}^L$. Each $\mathbf{s}_k$ corresponding to a tag signature $\mathbf{t}_k$ will have the following coordinates in the signal space,

$$\mathbf{s}_k = \left[\langle \mathbf{t}_k, \mathbf{u}_1 \rangle, \langle \mathbf{t}_k, \mathbf{u}_2 \rangle, \ldots, \langle \mathbf{t}_k, \mathbf{u}_L \rangle \right] \qquad (9)$$

It is more suitable to call these signal points $\mathbf{s}_k$ as constellation points that form the signal constellation C which is a subset of the signal space. These 2^b constellation points are used in the detection of information bits from the measured estimates of the tag signature $T_k(f)$ through minimum distance detection.

Let $\hat{\mathbf{t}}_k$ be the sampled estimate of the tag signature (transfer function) $T_k(f)$ at the receiver which is obtained through (1). The detection process involves in first calculating the inner products between $\hat{\mathbf{t}}_k$ and $\mathbf{u}_i$, $i = 1, \ldots, L$. The inner product coefficients obtained give an idea of how much of each basis function $\mathbf{u}_i$ is contained in $\hat{\mathbf{t}}_k$. These inner product coefficients, $z_i = \langle \hat{\mathbf{t}}_k, \mathbf{u}_i \rangle$ form the noisy received signal point $\mathbf{z} = [z_1, \ldots, z_L]$ in $\mathbb{C}^L$. Next the distance from each of the constellation points $\mathbf{s}_k$ to the noisy received signal point $\mathbf{z}$ calculated. The constellation point which is closest to $\mathbf{z}$ is chosen as the estimated constellation point $\hat{\mathbf{s}}$ corresponding to the received signal point. Since each constellation point is associated with a unique $T_k(f)$ that carries a unique data, the data corresponding to the estimated tag signature $\hat{\mathbf{t}}_k$ can be retrieved. The above explained procedure can be summarized by the following equation,

$$\hat{\mathbf{s}} = \arg\left(\min_{\mathbf{s}_i \in C} \left\{ \|\mathbf{z} - \mathbf{s}_i\|^2 \right\} \right) \qquad (10)$$

VALIDATION OF THE DETECTION PROCESS

In order to validate the proposed detection method a retransmission based chipless RFID tag was design and fabricated. A low number of bits was considered in the design for the purpose of conducting a comprehensive and thorough analysis. Therefore a chipless tag capable of carrying 3 bits of data was designed. The tag essentially consists of two monopole antennas and a spiral resonator based passive microwave filter. All components of the tag are based on co-planar waveguide (CPW) theory. This is due to the fact that a co-planar design requires only a single conductive surface (S. Preradovic, Roy, & Karmakar, 2009) as oppose to a microstrip line based design (S. Preradovic & N. C. Karmakar, 2009), which would demand two conductive surfaces. Commercial production of tags would ultimately dictate a design constraint involving a single conductive surface in order lower the costs of tag fabrication.

Tag Transfer Function Design and Signal Space Representation

As explained in the previous sections the data is represented in features present in the transfer function $T_k(f)$ of the tag. This $T_k(f)$ is essentially based on passive microwave filter design. By the use of different resonating structures abrupt and sharp resonances or attenuations can be caused in $T_k(f)$. These resonances are used to represent logic "0" or "1" of the data bits carried by the chipless RFID tag. For the design of the 3-bit chipless RFID tag a spiral resonator based microwave filter was chosen. This filter consists of spiral resonators with each spiral resonating at a distinct frequency operating as a notch filter at the resonance. Eight different filters were designed using binary combinations of 3 different spiral resonators to produce the eight distinct $T_k(f)$s required to represent each of the 3 bit binary data combinations. Figure 3 shows four of the filters which are used in tags carrying data "000," "100," "110" and "011." Clearly, we can observe that by controlling the absence and presence of the spiral resonators the frequency characteristics of the $T_k(f)$ can be directly changed, where a resonance is used to represent a data bit "0" and its absence represents a data bit "1." The filters were designed and simulated using the full-wave electromagnetic simulation software ``Computer Simulation

Figure 3. Spiral resonator based filters carrying data "000," "100," "110" and "011." Each filter has the dimensions 2 cm x 5 cm and is fabricated on substrate Taconic TLX-0.

Technology (CST) Microwave Studio''. The simulation results showed that the resonances occur at the frequencies 2.42, 2.66 and 2.96 GHz. The simulated and experimentally measured tag transfer functions $T_k(f)$ are shown in Figure 4. The measurements were done using a vector network analyzer. The forward transmission scatter parameter S_{21} which is essentially equivalent to the filter transfer function $T_k(f)$ was measured using the instrument. The lowest and the highest resonance frequencies correspond to the most significant bit (MSB) and the least significant bit (LSB) respectively.

From the results shown in Figure 4 (a) it is clear that in the resonances corresponding to the different data carrying $T_k(f)$ accurately coincide to three distinct resonance frequencies, namely 2.42, 2.66 and 2.96 GHz. However, the same cannot be said for the experimentally measured transfer functions of the different filters. The resonances vary from one filter to another with a tolerance of around ±50 MHz. These discrepancies are due to the minute fabrication errors that have caused the physical dimensions of the resonating spiral structure and the depth of the substrate material to vary from one filter to another. These deviations are ultimately seen as a form of noise in the detection process and hence it will result in an increase in detection error probability. From the simulation and experimental measurements we can observe that the 3dB bandwidth of the resonances is around $150 - 200$ MHz.

Since the simulation results are close to the ideal theoretically expected results for $T_k(f)$ they were used as the reference transfer functions in calculating the set of basis functions discussed previously. Table 1 (a) shows the singular values resulting from performing singular value decomposition on the frequency signatures as shown in (3). Clearly, only the first 4 singular values are significant. Hence, all the $T_k(f)$s can be adequately described using only $L = 4$ basis functions $\mathbf{u}_i$, $i = 1, \ldots, L = 4$. Table 1 (b) lists the inner product coefficients, $\langle \mathbf{t}_k, \mathbf{u}_i \rangle$, between different tag signatures and basis functions $\mathbf{u}_i$, $i = 1, \ldots, L = 4$. The inner product $\langle \mathbf{t}_k, \mathbf{u}_i \rangle$ is a measure of the amount of basis function $\mathbf{u}_i$ in the signature $\mathbf{t}_k$. According to the table it is clear that $\mathbf{u}_1$ is the largest component in all the $\mathbf{t}_k$s. However, the relative variation of $\langle \mathbf{t}_k, \mathbf{u}_1 \rangle$ from one signature to another is small. Therefore, in identifying the individual $\mathbf{t}_k$s, $\mathbf{u}_1$ provides little information. As oppose to $\mathbf{u}_1$ the other basis functions show larger relative variations among different tag signatures. Therefore, the basis functions $\mathbf{u}_2$, $\mathbf{u}_3$ and $\mathbf{u}_4$ are sufficient to uniquely identify each signature. Hence, using these basis functions a three dimensional signal space was

constructed where all the 8 frequency signatures corresponding to data "000," "001," "010," …, "111" were plotted as signal points (constellations points) in it as shown in Figure 5. These 8 signal points can be considered a constellation against which measured unknown tag signatures will be matched in order to extract their data.

Figures 4 and 5 shows the clear contrast between the two representations (frequency domain and signal space) of $T_k(f)$. In the former the tag transfer functions are represented as functions of frequency where the resonances are clearly visible at certain frequencies where as in the latter the tag transfer functions are represented as points in a three dimensional space where the Euclidian distance separating them defines the uniqueness of each point or tag transfer function. In a way, the representation of transfer functions in frequency domain can also be thought of as a signal space representation where there exist an infinite number of basis functions in which the basis functions are the individual complex exponential frequencies. In such a representation each frequency in the frequency axis can be thought of as a basis function which is also a distinct dimension of the corresponding infinite dimensional signal space. Through the singular value decomposition based approach an alternative basis for representing these transfer functions is realized and of those only the most essential basis functions are chosen to represent the transfer functions. This gives rise to the proposed method where the $T_k(f)$s are represented using a smaller finite dimensional signal space.

In Figure 5 (b) the experimentally measured tag transfer functions are plotted (shown in black) against the constellation of tag transfer functions (shown in blue). The discrepancies observed in the resonance frequencies of the measured $T_k(f)$ seen in Figure 4 (b) are reflected as displacements from the expected locations (locations of the corresponding constellation points) in the signal space. As the error in the frequency domain rep-

Figure 4. (a) Simulated tag filter transfer functions (b) experimentally measured tag filter transfer functions

Figure 4. (a) Simulated tag filter transfer functions (b) experimentally measured tag filter transfer functions

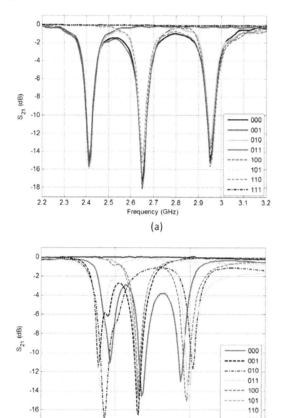

(a)

(b)

Figure 5. (a) Constellation points or the SSR of simulated $T_k(f)$s (blue) (b) SSR of the experimentally measured tag transfer functions (black)

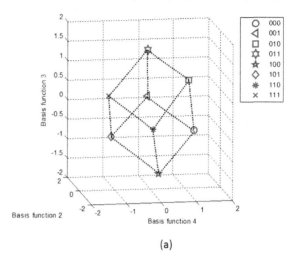

(a)

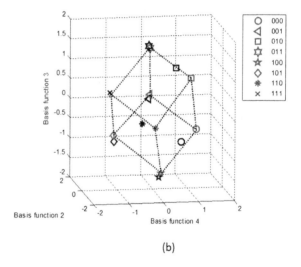

(b)

resentation gets larger the displacements observed in the signal space becomes greater. Therefore, we can see that minute fabrication errors can be interpreted as a noise source that causes the signal points to be displaced from their intended locations. Despite this noise the signal points corresponding to the measured tag transfer functions shown in Figure 5 (b) can be detected accurately. This is because the noise introduced by the fabrication error is not large enough for them to be displaced significantly and be mistaken for a wrong constellation point, i.e. the minimum distance defined by (10) is still achieved with the correct constellation point.

Detection Performance in AWGN

AWGN also causes the received signal point $\mathbf{z}$ to be displaced from the location of the correct constellation point $\mathbf{s}_k$ and cause detection errors. The amount of AWGN affecting the data carrying signal (in this case $T_k(f)$) is usually defined using a metric known as signal to noise ratio (SNR). However, in the context of chipless RFID tag signatures the definition of SNR cannot be directly used since the data is not contained in the

Table 1. (a) Singular values (b) Inner product coefficients

σ_1	σ_2	σ_3	σ_4	σ_5	σ_6	σ_7	σ_8
31.54	2.75	2.58	2.29	0.23	0.18	0.12	0.06

(a)

	$\langle \mathbf{t}_x, \mathbf{u}_1 \rangle$	$\langle \mathbf{t}_x, \mathbf{u}_2 \rangle$	$\langle \mathbf{t}_x, \mathbf{u}_3 \rangle$	$\langle \mathbf{t}_x, \mathbf{u}_4 \rangle$
$\mathbf{t}_1$, "000"	10.29	0.43	-0.59	1.33
$\mathbf{t}_2$, "001"	10.94	1.69	0.09	0.19
$\mathbf{t}_3$, "010"	10.76	-0.75	0.91	0.98
$\mathbf{t}_4$, "011"	11.37	0.34	1.51	-0.02
$\mathbf{t}_5$, "100"	10.83	-0.54	-1.52	0.17
$\mathbf{t}_6$, "101"	11.52	0.86	-0.78	-0.95
$\mathbf{t}_7$, "110"	11.34	-1.63	-0.19	-0.19
$\mathbf{t}_8$, "111"	12.05	-0.36	0.49	-1.23

(b)

received signal but in the transfer function of the tag filter. Therefore, in order to quantify the amount of noise affecting the tag signature the following quality factor γ was defined,

$$\gamma = \frac{d_0^2}{4\sigma^2} \tag{11}$$

where d_0 is the average length between adjacent constellation points (average length of the edges of the constellation) and σ^2 is the noise power spectral density. From (11) we can see that both the distance between adjacent constellation points and the noise affecting them are important in defining the overall performance. Figure 6 shows the effect of AWGN on the received signal point $\mathbf{z}$ The figure shows the results of a simulation on detecting the "110" data carrying tag in hundred trials for different γ We can clearly see that as the noise power increases the cloud formed by the hundred noisy received signal points grows centering around the "110" constellation point and also spreads close to other constellation points

which causes an increase in the probability of detection error (Haykin, 2009).

In order to evaluate the detection performance of the SSR based detection method Monte Carlo simulations were carried out for 100000 detection trials. The experiment involves in detecting a randomly chosen tag (out of the 8 possible tags) under only AWGN. Other sources of noise and interference were not considered for the sake of simplicity. Figure 7 shows the probability of detection error against SNR. The results shown here are in accordance with those shown in Figure 6. That is, for γ above 12 dB the detection error probability is very low (falls below 0.0001). In Figure 6 (a) and (b), where the γ is 20 dB and 15 dB respectively, by visual inspection of the signal space we can see that there is no detection error since the cloud of received signal points is concentrated only around the correct constellation point. However, when the γ is below 10 dB we can see from both Figures 6 and 7 that a considerable amount of detection errors occur.

Figure 6. Received signal points for (a) $\gamma = 20\ dB$ *(b)* $\gamma = 15\ dB$ *(c)* $\gamma = 10\ dB$ *(d)* $\gamma = 5\ dB$

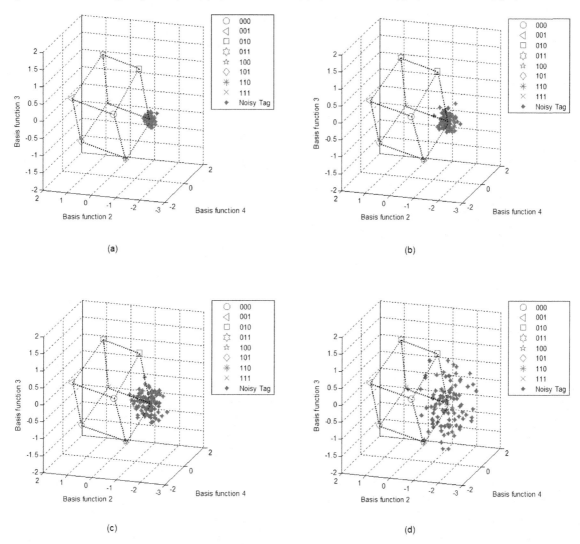

(a)

(b)

(c)

(d)

Detection was also performed using threshold levels on a filtered version of the noisy tag signature in order to compare with the performance of the SSR based new detection scheme. In this detection the signal is first filtered using moving average filtering to reduce the effects of noise on the signal. Afterwards, the filtered signal level near the resonance frequencies (±100 MHz) was compared against a predefined fixed threshold level in order to determine whether a "0" or "1" bit is present. If signal level is below the threshold "0" is detected if not "1" is detected. The detection

performance of both methods is shown in Figure 7. Clearly, we can see that SSR based detection has better performance over frequency domain fixed threshold based detection.

Experimental Results with the Complete Tag

The results presented so far are based on the assumption that the information carrying transfer function $T_k(f)$ can be accurately estimated. However, as expressed in (1) the accurate estima-

Figure 7. Tag detection error probability against γ

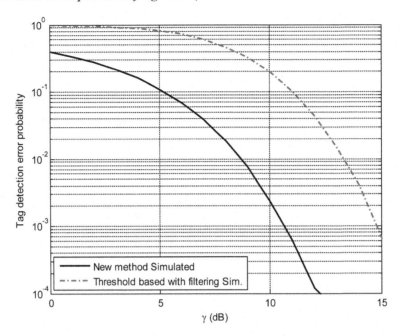

tion of $T_k(f)$ relies on successful removal or mitigation of the undesirable effects caused by the wireless channel and interference. In order to test the new detection under all these effects a complete chipless tag was constructed and experiments were performed. Figure 8 (a) shows the complete chipless RFID tag consisting of two monopole antennas and a spiral filter. The monopole antennas are oriented perpendicular to each other in order to create an orthogonal polarization between the received signal and the retransmitted signal which will reduce the interference between them. The experimental setup used for taking measurements is shown in Figure 8 (b). A two port vector network analyzer (VNA) was used as the reader where its ports were connected to two antennas for transmitting the interrogation signal and receiving the weak retransmitted signal. The transmitted interrogation signal is a repetitive frequency chirp signal starting from 2.2 GHz and ending at 3.2 GHz. Using the received weak signal the VNA computes the forward transmission scatter parameter, S_{21}, it observes for the two port network formed by the wireless channel

beyond the two ports. Using (1) and neglecting the effects of thermal noise this computed value of S_{21} can be approximated as,

$$S_{21} = \frac{R(f)}{S(f)} \approx H_F(f) T_k(f) H_R(f) + I(f)$$

(12)

Figure 9 (a) shows the measured S_{21} parameters or frequency signatures for tags measured at 1, 5 and 10 cm away from the reader antennas. From the results we can observe that only the measurement done at 1 cm shows a clear resemblance to the corresponding tag filter transfer function $T_5(f)$. In the measurement obtained at 5 cm the sharpness of the resonance occurring near 2.96 GHz has reduced. At 10 cm the observed tag frequency signature has become more distorted due to the effects of the channel, i.e. the resonance that is suppose to occur at 2.96 GHz appears to have shifted to 3 GHz. The signal space representation of these frequency domain tag signatures are shown in Figure 9 (b). Before plotting these

Figure 8. (a) The complete tag; (b) Experimental setup for acquiring measurements

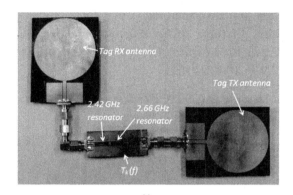

(a)

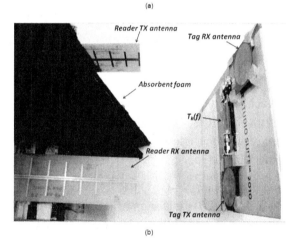

(b)

tag signatures in the signal space the effects of $H_F(f)$ and $H_R(f)$ was removed to some extent through the use of the gain measurements of the reader and tag antennas and by removing the path loss using the distance to the tag. From the SSR shown in Figure 9 (b) we can see that the estimates of $T_5(f)$ plotted in the signal space can be correctly detected as carrying data "100" for distances 1 and 5 cm. However, due to the presence of more distortion the estimate of $T_5(f)$ obtained from the measurement done at 10 cm is not correctly detected to be carrying data "100." It is erroneously detected to be the "000" data carrying constellation point hence causing a detection error.

Figure 9. (a) Measured S_{21} of tag carrying data "100" and the reference tag filter transfer function $T_5(f)$ (b) SSR of the frequency domain functions

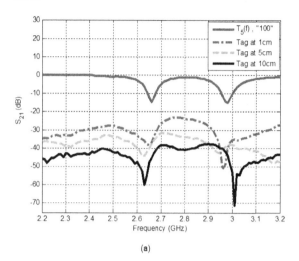

(a)

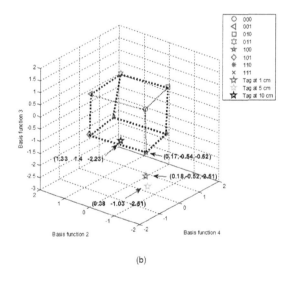

(b)

CONCLUSION

A novel method for detecting data in the frequency signatures of chipless RFID tags was introduced. Detection was based on the representation of tag frequency signatures as signal points in a signal space. Spiral resonator based chipless RFID tags were designed and fabricated in order to practically test the new approach. Experimental results show that the SSR based novel mathematical framework for modeling chipless RFIDs is useful in detecting information bits in the frequency signatures of chipless RFID tags. Through Monte Carlo simulations it was shown that the proposed detection method has better detection performance compared to threshold based detection of bits in frequency domain. Current research is focused on statistically modeling the noise sources and interference sources. This will enable the development of more efficient detection algorithms based on the presented mathematical model.

REFERENCES

Balbin, I., & Karmakar, N. C. (2009). Phase-encoded chipless RFID transponder for large-scale low-cost applications. *IEEE Microwave and Wireless Component Letters, 19*(8), 509–511. doi:10.1109/LMWC.2009.2024840

Blischak, A., & Manteghi, M. (2009, 1-5 June 2009). *Pole residue techniques for chipless RFID detection.* Paper presented at the Antennas and Propagation Society International Symposium (APSURSI), 2009 IEEE Charleston, South Carolina.

Bolic, M., Simplot-Ryl, D., & Stojmenovic, I. (2010). *RFID systems research trends and challenges.* John Wiley & Sons Ltd. doi:10.1002/9780470665251

Boyd, S. (2007). *Lecture 16: SVD appications.* Retrieved from http://see.stanford.edu/materials/lsoeldsee263/16-svd.pdf

Chamarti, A., & Varahramyan, K. (2006). Transmission delay line based ID generation circuit for RFID applications. *IEEE Microwave and Wireless Component Letters, 16*(11), 588–590. doi:10.1109/LMWC.2006.884897

Dullaert, W., & Reichardt, L., & H., R. (2011). Improved detection scheme for chipless RFIDs using prolate spheroidal wave function-based noise filtering. *IEEE Antennas and Wireless Propagation Letters, 10*, 472–475. doi:10.1109/LAWP.2011.2155023

Goldsmith, A. (2005). *Wireless communications.* Cambridge University Press.

Haykin, S. (2009). *Communication systems* (5th ed.). Wiley Publishing.

Hu, S., Law, C. L., & Dou, W. (2008). A balloon-shaped monopole antenna for passive UWB-RFID tag applications. *IEEE Antennas and Wireless Propagation Letters, 7*, 366–368. doi:10.1109/LAWP.2008.928462

IDTechEx. (2006). *Chipless RFID - The end game.* Retrieved 3/10/2011, from http://www.idtechex.com/research/articles/chipless_rfid_the_end_game_00000435.asp

Jang, H.-S., Lim, W.-G., Oh, K.-S., Moon, S.-M., & Yu, J.-W. (2010). Design of low-cost chipless system using printable chipless tag with electro-magnetic code. *IEEE Microwave and Wireless Component Letters, 20*(11), 640–642. doi:10.1109/LMWC.2010.2073692

Koswatta, R., & Karmakar, N. C. (2010, Dec. 2010). *Development of digital control section of RFID reader for multi-bit chipless RFID tag reading.* Paper presented at the International Conference on Electrical and Computer Engineering (ICECE), Dhaka, Bangladesh.

Lazaro, A., Ramos, A., & Girbau, D., & R., V. (2011). Chipless UWB RFID tag detection using continuous wavelet transform. *IEEE Antennas and Wireless Propagation Letters, 10*, 520–523. doi:10.1109/LAWP.2011.2157299

Manteghi, M. (2010, 11-17 July 2010). *A novel approach to improve noise reduction in the Matrix Pencil Algorithm for chipless RFID tag detection.* Paper presented at the Antennas and Propagation Society International Symposium (APSURSI), 2010 IEEE Toronto, Ontario.

Preradovic, S., Balbin, I., Karmakar, N. C., & Swiegers, G. (2008, April 16 - 17). *A novel chipless RFID system based on planar multiresonators for barcode replacement.* Paper presented at the IEEE International Conference on RFID, 2008, Las Vegas, Nevada.

Preradovic, S., & Karmakar, N. C. (2009, Sept.). *Design of fully printable planar chipless RFID transponder with 35-bit data capacity.* Paper presented at the European Microwave Conference, 2009. EuMC 2009., Rome.

Preradovic, S., & Karmakar, N. C. (2009, December). *Design of short range chipless RFID reader prototype.* Paper presented at the Intelligent Sensors, Sensor Networks and Information Processing (ISSNIP), Melbourne, Australia.

Preradovic, S., & Karmakar, N. C. (2010). Chipless RFID: Bar code of the future. *IEEE Microwave Magazine, 11*(7), 87–97. doi:10.1109/MMM.2010.938571

Preradovic, S., Roy, S., & Karmakar, N. (2009, 7-10 Dec. 2009). *Fully printable multi-bit chipless RFID transponder on flexible laminate.* Paper presented at the Microwave Conference, 2009. APMC 2009. Asia Pacific.

Shao, B., Chen, Q., Amin, Y., David, S. M., Liu, R., & Zheng, L. R. (2010, Nov). *An ultra-low-cost rfid tag with 1.67 gbps data rate by ink-jet printing on paper substrate.* Paper presented at the IEEE Asian Solid-State Circuits Conference (A-SSCC) Beijing

Strang, G. (2009). *Linear algebra and its applications* (4th ed.). Wellesley-Cambridge Press.

Weinstein, R. (2005). RFID: A technical overview and its application to the enterprise. *IT Professional, 7*(3), 27–33. doi:10.1109/MITP.2005.69

KEY TERMS AND DEFINITIONS

Backscatter: Reflections caused by an electromagnetic wave falling on a target.

Signal Space: A mathematical vector space consisting of the set of all signals

Signal Space Representation: A representation in which signals are treated as vectors in a vector space.

Signal Constellation: A diagram that visualizes distinct signals representing digital data as points in a signal space.

Inner Product: The scalar or dot product between two vectors in a vector space.

Transfer Function: A mathematical relationship that describes the behavior of the input and the output of a linear system.

Thermal Noise: Undesirable and random fluctuations observed in measured electrical signals due to the random motion of electrons caused by thermal energy.

Chapter 12
Object Analysis with Visual Sensors and RFID

Gour C. Karmakar
Monash University, Australia

Laurence S. Dooley
The Open University, UK

Nemai C. Karmakar
Monash University, Australia

Joarder Kamruzzaman
Monash University, Australia

ABSTRACT

Object analysis using visual sensors is one of the most important and challenging issues in computer vision research due principally to difficulties in object representation, segmentation, and recognition within a general framework. This has motivated researchers to investigate exploiting the potential identification capability of RFID (radio frequency identification) technology for object analysis. RFID however, has a number of fundamental limitations including a short sensing range, missing tag detection, not working for all objects, and some items being just too small to be tagged. This has meant applying RFID alone has not been entirely effective in computer vision applications. To address these restrictions, object analysis approaches based on a combination of visual sensors and RFID have recently been successfully introduced. This chapter presents a contemporary review on these object analysis techniques for localisation, tracking, and object and activity recognition, together with some future research directions in this burgeoning field.

INTRODUCTION

The innate desire for humans to have their computers and machines do exactly whatever they are able to do has, to date remained largely unfilled. While humans can effortlessly analyse and recognise an object and its activity or event, computer-based automatic systems still remain stubbornly in an embryonic state. This has lead to computer vision being one the most challenging research topic areas over the past few decades. The application of computer vision is now however, rapidly in-

DOI: 10.4018/978-1-4666-1616-5.ch012

creasing due to the advancement of software and hardware technologies, including RFID (*radio frequency identification*), digital image analysis, artificial intelligent and robotic vision systems.

Computer vision has been extensively applied in a wide range of diverse domains. These include, but are not limited to: robotic vision (Hsiao et al., 2008), autonomous car and mobile robot (Milanés et al., 2010), event detection, surveillance and tracking (Yilmaz et al., 2006; Lee et al., 2007), airport identification from aerial photographs, object-based image identification and retrieval (Veltkamp & Tanase, 2000), object and activity recognition (Smith et al., 2005; Wu et al., 2007; Bashir et al. 2007), behaviour prediction (Skinner, 1953; Pentland, 1999; Chen et al., 2004), object-based image and video coding (ISO/IEC 14496-2, 1999), criminal investigation using video footage analysis, computer graphic, and medical diagnoses (cancerous cell detection, segmentation of brain images, skin treatment, and intrathoracic airway trees) (Paul & Paul, 1993; Pham & Prince, 1999; Liu et al., 1997). An essential preprocessing step in all the aforementioned real-world examples is object detection and separation (segmentation), which both involve the core element of image analysis to retrieve and understand an event's information in real-time (Computer vision, 2011). Image analysis, typically includes image segmentation, feature extraction, and object classification (Baxes, 1994).

Image segmentation is the process of separating mutually exclusive homogeneous regions of interest (i.e., having similar pixel intensities) from other regions in an image. In reality, most natural objects are not homogeneous and comprise numerous objects in a hierarchical order. Furthermore, there are generally a huge number of objects and a myriad of variations amongst them. This makes object segmentation in an image an interesting if intractable task as it somewhat contradicts the above definition of object-based image segmentation. This is because there is no universally accepted standard definition of im-

age segmentation. The properties of an object to be segmented depend on its applications and human perception, so segmenting an object is in essence an ad hoc process, which depends on the emphasis given to the set of desired properties and a trade-off among them (Paul & Paul, 1993; and Karmakar et al., 2001). To reduce the complexity of detecting and separating an object and capturing its dynamic movement pattern (i.e., motion), computer vision researchers are progressively exploiting temporal information from visual sensors such as digital video cameras (Song & Fan, 2006; Ahmed et al., 2007).

Even if an object has been successfully separated, there still remains the question of *how can the object be automatically identified or recognized?* Automatic object identification is regarded as the most difficult problem in computer vision because of being unable to define a unified feature set to represent generic natural objects. Any identification system requires an accurate representation and supervised machine learning of an object, its interaction with other objects and a physical context. Many techniques (Lou et al., 2002; Yu et al. 2002) have evolved though their performance is limited. In addition, the development of a suitable training set requires considerable manual intervention and so is time consuming (Shirasaka et al. 2006). This has prompted the adoption of RFID technology in object analysis, which has been already proved as an effective approach in application domains like object tracking in automated assembling lines (Wang, 2007), workflow optimization (Faschinger et al., 2007) and inventory control (Goodrum et al., 2006; Ko, et al., 2007).

The main drawbacks of RFID are very short sensing range, missed detection, sensitivity to environmental noise and the inherent fact that the technology does not work for certain materials like metals and food items, while some items are simply too small to have a tag attached. For these reasons, RFID alone is often ineffectual, which motivated the investigation of alternative approaches based on combining RFID and computer vision techniques (Wu et al., 2007). This chapter presents an overview of contemporary approaches towards applying RFID in combination with computer vision for object analysis, involving location estimation, tracking and, object and activity recognition.

OBJECT ANALYSIS USING RFID

Object analysis applications have tended to evolve using either vision sensors or RFID tags separately rather than in combination. The latter is now gaining traction because as alluded above, the main benefit of RFID tags is they provide a general framework for object identification, which addresses the main shortcoming of visual sensor-based systems. This section presents an exploration of some contemporary research projects involving object analysis and which have exploited the potential of combining RFID and visual sensors.

Object Localisation

For many applications that require multiple targets detection such as, activity recognition and robotic vision in home environments to either monitor or track particular objects, localisation of the object is the important initial step. Based on the type of signal analysis, object localisation techniques can be classified into two categories: *i)* visual and *ii)* acoustic sensor. The latter based approaches are mainly applied to determine the location of a passive source such as the object producing a sound (Sheng et al., 2005; Rahman et al., 2005), and are normally sensitive to environmental noise. Conversely, object localisation using visual information is a classical problem for the computer vision research community due to its high accuracy requirement for location estimation, with much literature available (Park et al., 2006). Moreover, the number of approaches combining RFID with visual sensors (Kamol et al., 2007) is expanding due to the potential identification capability

of RFID, which addresses the key issue facing visual analysis techniques, namely approximate location estimation. A RFID tag strategy with both mobile and static RFID antennas together with some ceiling cameras has been proposed to determine object location in an indoor scenario (Kamol et al., 2007). The main processing steps involved are shown in Figure 1, where the location is estimated from the information derived by recognition. Colour histogram features of an object are stored in the tag attached to it. Initially RFID tags and RFID antennas are used to determine the area (approximate location) of an object which is nearby an RFID antenna. Since mobile robots attached with RFID antennas continuously roam the floor, they can detect those objects which are not detected by the static RFID antennas.

The features obtained from both the RFID tags and images captured by ceiling cameras are used ·to recognise an object by creating the back project of the colour histogram features, details of which are given in (Kamol et al., 2007). Following object recognition, its precise location or *location vector* is estimated by particle filtering, which used the multiple views obtained from the ceiling cameras. The authors conducted an experiment in an indoor environment comprising 10 different objects. The results revealed that both recognition accuracy and time overheads using RFID tags significantly improved compared to not using RFID tags, with a 34.5% improvement in recognition accuracy achieved allied with a reduction in computational time of around half an hour. They also showed the average localization error was around 0.96%, i.e., 79cm of the room diameter.

Object Tracking

Object tracking is another very challenging research problem, with numerous diverse real-world applications including the military, hospital, manufacturing lines and mining. Tracking an object of interest using the *global positioning*

Figure 1. Processing steps for localisation using the method articulated in (Kamol et al., 2007)

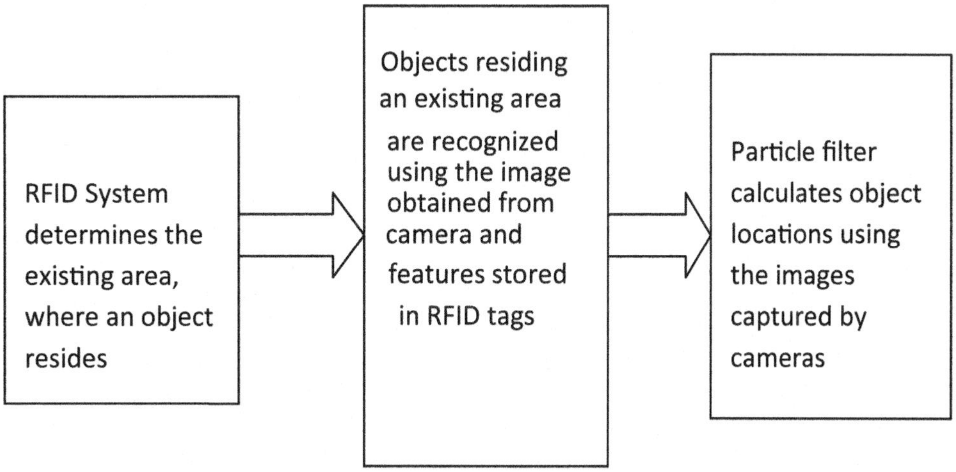

```
┌──────────────┐      ┌──────────────────┐      ┌──────────────────┐
│ RFID System  │      │ Objects residing │      │ Particle filter  │
│ determines the│ ═══▷ │ an existing area │ ═══▷ │ calculates object│
│ existing area,│      │ are recognized   │      │ locations using  │
│ where an object│     │ using the image  │      │ the images       │
│ resides       │      │ obtained from    │      │ captured by      │
│               │      │ camera and       │      │ cameras          │
│               │      │ features stored  │      │                  │
│               │      │ in RFID tags     │      │                  │
└──────────────┘      └──────────────────┘      └──────────────────┘
```

system (GPS) (Lee, 2009) is popular and widely adopted provided the object to be tracked is known *a priori*. Since GPS determines the object location using information obtained from satellites, it is highly sensitive to shadowing effects caused by surrounding structures such as buildings, bridges, hills and streets for example (Hosrt, 2007). Depending on the weather conditions, noise, radio signal, satellite position and shadowing effect, the location estimation error for GPS can be in the range of 1 to 10 metres[1], and for this reason, researchers are attempting instead to track objects using their attributes, like sound and appearance.

Tracking passive sources have been used in many applications. These applications primarily use acoustic sensors which are normally low cost and easy to deploy, though to determine the location of any object requires knowledge of the location of at least three sensors (Rahman et al., 2005), which can be affected by environmental noise so compromising the location estimate. Object tracking using only RFID tag is widely applied in automatic manufacturing lines (Wang et al., 2007) however this is very susceptible to the prevailing environment and often produces inaccurate object positions and movement paths.

To solve this problem, visual sensors have been proposed as they afford both reliable and consistent location estimation (Yilmaz et al., 2006). As mentioned previously, their main drawback however, is the requirement for either object identification using an appearance-based feature or coarse localisation. In order to provide coarse location information about an object, a new paradigm is emerging which jointly exploits RFID coverage and identification facility together with the reliability and location estimation accuracy of visual sensors. Lee et al. (2007) introduced one such model consisting of the following two steps:

1. Coarse location estimation with RFID tags
2. Refinement of the coarse location using visual sensors

These will now be respectively discussed.

Coarse Location Estimation with RFID Tags

While the accuracy and reliability of visual sensors-based location estimation is high (Lee et al., 2007), their disadvantage is requiring *a*

Figure 2. Overlapping regions ($S_1, ..., S_7$) amongst the coverage of RFID readers ($R_1, ..., R_3$)

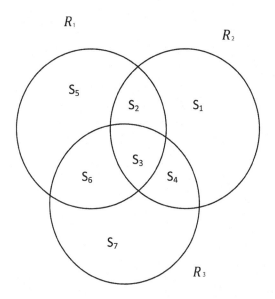

Figure 2. Overlapping regions ($S_1, ..., S_7$) amongst the coverage of RFID readers ($R_1, ..., R_3$)

priori information in terms of an approximate location estimate of an observed object. Coarse location estimation using RFID readers potentially bridges this hiatus. A viewing area is covered by a number of RFID readers, where the coverage of a RFID reader depends on its range. This inevitably leads to there being a number of overlapping and non-overlapping RFID coverage regions. To utilise these regions for coarse localization, the concept of *virtual sensing nodes* is introduced. As an example, three RFID readers (and their corresponding virtual sensing nodes (are displayed in Figure 2.

The maximum number of visual sensing nodes is represented by all possible combinations of RFID readers, which in this example is 7 $(^3c_1 + {}^3c_2 + {}^3c_3)$, with the linkage between RFID coverage and the virtual sensing node being shown in Figure 3. For instance, readers R_1, $R_2$and R_3 respectively cover the set of virtual sensing nodes: $\{S_2, S_3, S_5, S_6\}$, $\{S_1, S_2, S_3, S_4\}$, and $\{S_3, S_4, S_6, S_7\}$.

Each virtual sensing node is represented by a reference point, which is the centre of its rough sensing range. The reference point of each virtual sensing node in Figure 4 is shown by ©. The coarse localisation of an object is the reference point of a virtual sensing node covering that object. The actual object location (✪) at different time slots from t_1 to t_5 and its real movement path are indicated by the solid lines in Figure 4, while the estimated tracking trajectory with the coarse localisations are displayed by dotted lines.

Refinement of the Coarse Location Using Visual Sensors

As shown in Figure 4, the course localisation error depends on the sensing range of a visual sensing node with in the worst case, it being equal to the RFID reader coverage range. To reduce this error, two parallel projection cameras are used to take snapshots of the object to be tracked. The virtual viewable plane of each camera is the image frame generated by the parallel projection of an object. The object plane passes through the actual position of an object and is parallel to its corresponding virtual viewable plane i.e. P_y^o is parallel to the virtual viewable plane for the first camera, while P_x^o is for the second. Let P_y^c and P_x^c be the parallel planes passing through the estimated coarse location (x_c, y_c). The location (x_r, y_r) approximated with the visual sensors (two cameras) is then defined as:

$$x_r = x_c \pm \Delta x \tag{1}$$

$$y_r = y_c \pm \Delta y \tag{2}$$

where Δx and Δy are the distance between P_y^o and P_y^c, and P_x^o and P_x^c respectively. The sign $\pm$ reflects the relative positions of an object with respect to its estimated coarse location (x_c, y_c).

Figure 3. RFID reader coverage for virtual sensing node regions

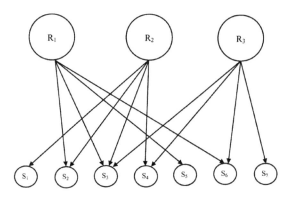

Figure 4. The reference points of virtual sensor nodes shown by ©. Dotted line - tracking trajectory based on coarse estimation; solid line - the actual moving path.

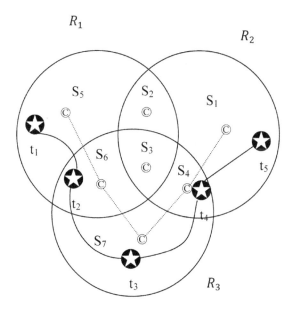

Simulation Results for Object Tracking

The authors conducted two experiments for a single object tracking inside a room and considered both one and two RFID readers, together with two visual sensors – Camera 1 (C_1) and Camera 2 (C_2) located on two walls of the room. The simulation setup is shown in Figure 5. An example of the object trajectory for two RFID readers, namely R_1 and R_2 is also presented in Figure 5 where the positions of the two RFID readers are represented by ©.

The simulation results showed that the usage of visual sensors improved the location estimation accuracy and does not depend on the number of RFID readers i.e., the tracking accuracy was almost the same for both one and two RFID readers, though the more RFID readers were used, the better the coarse localization accuracy achieved.

The proposed method is entirely dependent on parallel projection cameras which are generally not available (Chai & Shum, 2000). Moreover, it raises many questions over the effectiveness of this method in practical field deployments for both single and multiple objects tracking. For instance, how will the camera automatically detect an object from a cluttered and complex background scene? And, since RFID reader coverage for the passive tags is generally between $4cm$ and a few metres, how effective will this method be for location estimation in larger field deployments?

Activity Recognition

There are many vision-based techniques in the literature that can semantically interpret an object based on activity patterns analysis (Lou et al., 2002), perform behaviour-based similarity measurement using motion (Shechtman et al., 2007) and learn to recognize human action sequences using eye and head movements (Yu et al. 2002). All these methods are constrained however, to a specific action for a particular object (especially for humans) and only a fixed environment is ever considered. In most techniques, objects are represented by spatio-temporal video features such as colour/brightness distribution, height and width. However, these are insufficient to represent all object types since typical video sequences usu-

Figure 5. Object (✪) trajectory determined using two RFID readers

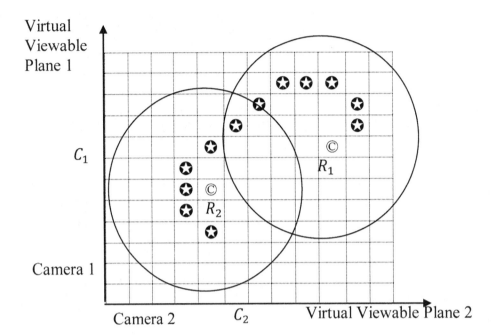

ally have a large number of perceptual objects and a multitude of variations amongst them. In most cases, temporal information is considered with either *Hidden Markov Model* (HMM) (Yu et al. 2002) or *Dynamic Bayesian Networks* (DBN) (Murphy, 2002), both of which use supervised machine learning techniques. To improve the recognition of action, the relationship between an object and other objects that exist in its context is exploited (Han et al., 2009). Even with the rapid advances in computer vision, activity recognition remains extremely difficult in terms of object separation and recognition and it is very much in this context that the alternative paradigm of using RFID technology for activity and object recognition has evolved.

RFID cannot alone produce a robust recognition system because of the many limitations previously delineated, such as the reader being too far away or wrongly sensing a tag if the reader is close to another object. In addition, there are many objects which are either so small that it is

not possible to attach tag on them or where RFID does not work such as for metallic and food objects. This provides the backdrop for exploiting visual information in conjunction with RFID and *common-sense* knowledge, to directly solve the problems created by incorrect and missing tag detections and losing objects without tags attached.

Any robust recognition system should be able to operate in different environments having varying degrees of complexity over a range of objects and activities. This led to the development of learning-based models, which ultimately restricts scalability, because in such models, objects used in an activity belong to a training set which needs to be identified, separated and manually labelled, which is extremely time consuming. It is clearly insuperable to do this for all images as a typical hour long video with 25fps, contains 90000 images.

To recognise an activity with a smaller number of related actions, Wu et al. (2007) introduced a scalable and robust activity recognition system, which uses object identification information obtained by sensing its RFID tag to automatically develop the learning model. RFID tags are attached on certain objects and the object (i.e., the person) which performs an action wears an RFID reader (bracelet on their hand). When the person performs an action, the RFID reader reads and wirelessly transmits the tag information associated with an object, to the activity processing server. The recognition system then uses this tag information to identify the object involved with that action. As a representative example of an experiment, the recognition of an activity within a kitchen environment is shown in Figure 6, where the person is holding and manipulating a water jug, with other objects also having RFID tags.

A DBN model is then designed which synergistically combines RFID detections, objects usage, common-sense knowledge and video sequence information. The complete model is shown in Figure 7(a), where common-sense knowledge is encoded along the edge between an activity and object, and A_t, O_t, R_t and I_t respectively denote activity, object, RFID and the frame at time t.

The various DBN parameters are defined as follows:

$P(A_t)$ is the *a priori* distribution;
$P(0_t|A1)$ and $P(0_{t+1}|0_t A_{t+1})$ are the observation probabilities;
$P(A_{t+1}|A_t)$ is the state transition probability; while $P(R_t|0_t)$ and $P(I_t|0_t)$ are the output probabilities.

$P(I_t|0_t)$ is the only parameter automatically learned via RFID detection during the DBN learning process. During this process, objects are separated from their backgrounds using a change detection technique and are represented by texture features, namely Scale Invariant Feature Transform (SIFT) (Lowe, 2004). The values of all the other parameters are heuristically determined from domain knowledge (Wu et al., 2007).

The authors conducted a series of experiments for each of the models shown in Figure 7(a) to 7(e), by considering 16 kitchen activities including boiling water, making tea, coffee and cereal, packed lunch, drinking juice, taking medicine, making salad which involved in total, 33 different objects. The experiments were conducted using three different video sequences, *Video_1*, *Video_2* and *Video_3*. The recognition results of various testing scenarios using the model learned with common-sense knowledge shown in Figure 7(a) for *Video_1* are presented in Table 1.

The test data were constructed i.e., the activities and objects to be predicted were labelled using their respective models represented in Table 1. The corresponding results reveal that due to the incorporation of common-sense knowledge, the combination RFID+visual sensor produced activity and object recognition rates of 80.67% and 72.36%, respectively, compared with 80.97% and 73.30% when only the visual sensors were applied. This vindicates that after learning the model, extra information from RFID detection of the test set does not provide any further useful information irrespective of the common-sense knowledge. In contrast, RFID only recognised 64.31% and 63% of activities and objects accurately, from which it can be concluded that even learned with RIFD detection, visual sensor information and common-sense knowledge, RFID alone in a test set produces inferior recognition. Similar observations can be made for the results for both *Video_2* and *Video_3*, though *Video_3* contained mainly texture-less objects that were not detected by visual sensor alone due to the fact that in DBN, objects are only represented by texture features, namely SIFT.

The validation test of the learned model was also performed using *Video_1* and *Video_2*, with the model learning with the former and tested by the latter sequence. RFID only had activity and object recognition rates of 80.33% and 66.54%,

Figure 6. A sample kitchen environment. The user wearing a RFID bracelet (blue rectangle) and holding water jug (red rectangle). Note some other objects with RFID tags (green rectangle) attached.

respectively compared with 73.37% and 71.02% produced by RFID+visual sensor. The lower activity recognition rate achieved by RFID+visual sensor was directly due to its inability to classify texture-less objects such as the kettle and lunch bag.

A further series of experiments were conducted to test the robustness of the model in regard to either missing tag detections or objects that do not have tag attached. Results plotted recognition rate versus number of missing tag for both activity and object. The recognition rates for RFID+visual sensor were significantly higher than those of RFID only for both activity and objects, though in both cases, the recognition rates decrease when the number of missing tag detections increased.

While this research project presented fundamental work in activity recognition and will help to develop many practical applications, its main limitation is the use of only texture features in object representation. Key primitive object features such as shape, which is one of the most perceptual characteristics of an object, were not considered (Ahmed et al., 2007). The actions of a particular object are represented by the movement patterns of different body parts (e.g., head and eye

movement are tightly coupled with actions (Yu et al., 2002)), gesture and interactions with other objects and/or physical context, so an elementary action is reflected by a pattern of underlying motion. Furthermore, an activity has a particular sequence of actions and the interactions among objects and their movement patterns that have not been explicitly considered in this project.

Training Set Construction for Object Detection

As mention previously, most vision sensor based applications exploit the object analysis techniques and thus require the detection and recognition of an object in advance. These techniques generally use supervised learning and need to develop training sets consisting of either geometric or appearance-based object features. Creating such training data sets is laborious and incurs considerable time expenditure. As it is difficult to observe which tag ID represents which object due to the inability of RFID readers to accurately detect all tags, this mandates that vision sensors and/or cameras are able to detect and recognise objects. However, their potential is limited and varies with the complex-

Figure 7. Bayesian network structures for activity and object recognition

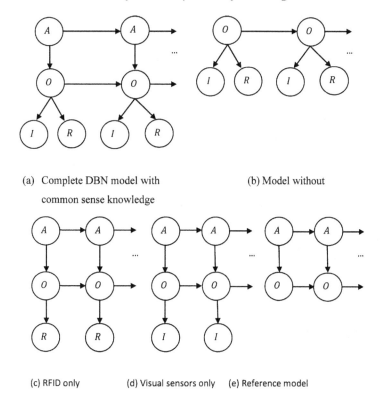

(a) Complete DBN model with common sense knowledge

(b) Model without

(c) RFID only   (d) Visual sensors only   (e) Reference model

ity of objects and their surrounding environment. This motivated Shirasaka et al. (2006) to introduce an approach to construct a training set using the appearance features of objects, where RFID tags are used as supervisors. The various processing steps involved in this approach are presented in Figure 8.

Obtaining Target Information from RFID Tags

All the objects to be detected in an environment have RFID tags attached to them, which contains identification information. The RFID readers shown in Figure 8 detect these tags and using the identification information to determine the relevant object. To apply supervised learning in this scenario, it is essential to derive two different types of information – the appearance-based features of an object and its corresponding target

information. This latter information can be derived directly from its RFID tag, while the former can be calculated using any suitable feature approximation technique such as in (Karmakar et al., 1999). In this approach however, spatial resolution reduction together with kernel *principal component analysis* (PCA) for dimensionality reduction (Sch¨olkopf et al., 1999) has been applied.

Object Detection and Separation

The most challenging part of this particular process is to detect and separate an object from the image captured by vision sensors. Since it is assumed the model is being deployed in a controlled indoor environment, the background complexity is generally limited and almost static. For this reason an image difference approach has been adopted in Figure 8 to separate the object from its background, though if the background changes are not insig-

Table 1. Recognition rates using different test scenarios

Common-sense knowledge used	Testing Scenarios	Activity (%)	Object (%)
Yes	RFID only (model in Figure 7 (c))	64.31	63.00
Yes	RFID+Visual Sensors (model in Figure 7 (a)) Visual Sensor only (model in Figure 7 (d))	80.67 80.97	72.36 73.30
No	RFID+Visual Sensor (model in Figure 7 (b))	60.84	74.68
No	Visual Sensors only	62.76	74.72

nificant, the image difference approach becomes ineffectual. While the performance is satisfactory for a single object, if the image contains multiple objects, it is then necessary to determine which RFID tag corresponds to which object. How this can be achieved is not presented in (Shirasaka et al., 2006).

Storing Object Images and Corresponding ID in a Database

The RGB colour images containing the objects are obtained by the differential technique described above and then spatially reduced to 30 *pixels* square. All background pixels, except the object pixels are set equal to zero to eliminate this information and then these object images and their corresponding RFID tags used to construct a training set which is stored in a database (see Figure 8).

Feature Dimensionality Reduction

The reduction of feature set dimensionality is a major issue as identifying the optimal set of attributes of a feature set, which perform equally well for all datasets, is extremely intractable. If the dimension of a feature set exceeds a certain limit, the feature set loses its distinguishing capability and hence compromises the detection accuracy. In addition, the higher the dimension, the greater the computational complexity, so to make the system suitable for real-time applications, judiciously applying dimensionality reduction is essential.

Figure 8. The processing steps for training set construction (Shirasaka et al., 2006)

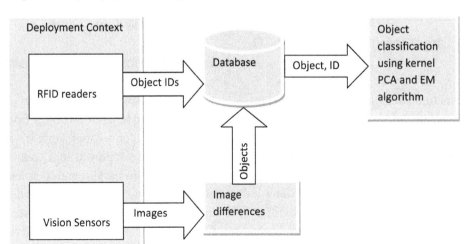

While there are many techniques available for feature dimensionality reduction (Fodor, 2002), the authors chose the most widely adopted, namely kernel PCA (Sch"olkopf et al., 1999). This decreases the original dimensions of the feature set, from 30 x 30 x 3 to *n* dimensions, where the feature set derived by kernel PCA is represented by $x = \{x_1, x_2, \ldots, x_n\}$.

Object Classification Using the EM Algorithm

The overall objective is to design an object classification model using the constructed training set, i.e., to determine the *a posteriori* probabilistic function $p(id/x)$ from a training set, which can automatically detect any object (i.e., estimate the object id) from a given appearance-based feature set *x* Depending on the location and contextual object, RFID readers may fail to detect the tag of an object. The *expectation maximisation* (EM) algorithm is often used in the Gaussian mixture model to estimate the maximum likelihood of missing data and concomitantly improve the computational complexity. The EM algorithm seeks to maximise the *likelihood function* $l\left(x \, / \, \theta_{id}\right)$ where, θ_{id} represents the set of parameters (the mean and covariance matrix) of a normal distribution function for an arbitrary object referenced by its id.

Experiments have been conducted with three scenarios, which had one, two and three objects respectively. All objects and a RFID reader were placed both on and under a table, respectively. Vision sensors focused only upon the objects on the table. The mean classification accuracy, which is the ratio between the number of features for which the classifier classified accurately and the total number of feature used in the experiment, was a very commendable 98.7%.

The fundamental concept introduced in (Shirasaka et al., 2006) is the development of a semi-automatically learning model which avoids significant manual processing. It advances many supervised learning-based research questions for numerous computer vision applications including tracking and surveillance. The classification accuracy of this method has been shown to be high because the experimental environment was restricted and simple. If the deployment context is more complex such as a room consisting of many different static and mobile background objects, and the objects are moving from one place to another, it is anticipated the classification accuracy of this method will decrease significantly. This is mainly due to the poor method adopted to separate objects from their background and not in considering multiple object movements, which may force a RFID reader to detect multiple objects. This confusion will lead the recognition system to match which object relates to which RFID tag. This can be resolved by storing the main object features into the RFID tags. What type of features, how many of them are required for accurate object identification; and how can they be seamlessly added to a RFID tag, remains a major research question? Some other future research directions are considered in the following section.

FUTURE RESEARCH DIRECTIONS

The potential to store colour histogram features of an object into a RFID tag to improve recognition has already been considered (Kamol et al., 2007), however colour information alone is not sufficient to recognise all possible natural objects as it cannot differentiate between objects with similar colour distributions. This mandates object representations based more on structural representations such as shape and texture (Karmakar et al, 2003). It is probable that in the future a large number of cutting edge applications will require RFID tags to store various types of object features. A myriad of natural objects and huge variations among them have made the representation of an object in a generic form very complicated. Moreover, the perception and definition of an object is application dependent so it will be a challenging task to determine what type and which set of features should be stored in RFID tags to effectively recognise a wide variety of objects for a particular application context. This firstly requires designing RFID tags which are capable of storing essential features along with their tag id.

An activity is represented by a number of actions, and the objects and context that are involved in that action, most literature has mainly considered isolated actions. However, without identification of an object, and interpretation and interaction with other objects and context, the activity recognition performance will be limited as the same action may represent a number of different activities for different objects and contexts. For example, consider a chef's cutting action i.e., vegetable, fruits, fish and meat, with a knife in a kitchen. This can represent the cooking activity, while for a doctor in an operation theatre, performing exactly the same action is interpreted as surgical activity. There are a myriad number of activities and an activity comprises a number of hierarchically organised elementary actions distributed in different levels that make the activity recognition challenging. The various temporal sequences of the same action set may represent different activities, which makes activity recognition even more complicated. Without a proper understanding of the hierarchical relationships and temporal sequences amongst actions in conjunction with object interactions with domain or context, the potential of any activity recognition system will remain limited.

To understand and semantically interpret the behaviour of a particular object of interest from a video sequences remains a major research problem to both the artificial intelligence and multimedia technology research communities, as it covers a very wide range of applications from national security to medical diagnostics. In all these cases, the users know what types of objects they need to detect and can then predict the object's behaviour. This raises a question on how to detect and predict the behaviour of a moving object from its given descriptor. The answer has defiantly remained elusive due to the huge number of objects and the myriad of variations amongst them. This makes the efficacy of the existing detection techniques very limited and difficult to segment any object based solely on pixel properties. There is no existing approach to both detect and separate an object for a given descriptor except a rather rudimentary technique proposed in (Ahmed et al., 2007). The descriptor could be developed automatically using RFID technology and then applied to a method similar to (Wu et al., 2007) for improving recognition rates. The activity and objects of interest can then be recognised using such descriptor and RFID technologies.

Contextual relationships are essential in identifying particular events such as a car-bomb attack. This can for example, involve the making and distributing of the bomb, identification of the bomb maker, abnormal behaviour of a person, psychological profiling, and unattended bags in an airport. If we can understand and consider the social context in which an event is initiated, developed, and finally occurs, it will add another valuable modality to the decision making. For example, which events or acts or contexts make a child hyperactive or a mentally ill patient violent? The inclusion of this information means a better likelihood that the prediction of an object's behaviour will be more accurate, and where it is currently impossible to detect suspicious events, it may make them at least feasible. This will open a new avenue of research in for example, child growth and mentally ill patient analysis based on applying RFID technology within a controlled environment. McFate (2005) analysed the social environment in which *improvised explosive devices* originated, and were subsequently produced, distributed and applied. He discovered that social network analysis provided a valuable tool to locate insurgents within the tribal networks of Iraq, which ultimately led to the capture of Saddam Hussain.

Chen et al. (2004) have analysed social interaction patterns and provided reports of activities and behaviour of patients in nursing home environments. In their research, the labelling of doors, walls, ceilings and floors were performed manually, so this is not suitable for dynamic environments consisting of complex objects, especially for arbitrarily shaped objects and where their locations are not *a priori* known. To address this issue, the identification objects (e.g., human, contextual information - doors, walls, ceiling and floors) could each have RFID tags attached to them and their various features stored to give impetus to the computer vision field in managing things automatically, just like we humans can do in any given context.

CONCLUSION

Object analysis is a fundamental processing step in many leading-edge applications involving location estimation, tracking and monitoring, activity recognition, event detection, and behaviour analysis and prediction. Either vision sensors or RFID technology alone have been shown to be inadequate in terms of their performance which has promoted the introduction and investigation by the computer vision community of hybrid approaches based on a combination of vision sensors and RFID. This chapter has presented a contemporary review of these evolving hybrid approaches. The advantages and drawbacks of each approach have been highlighted along with some insightful future research directions. This has revealed key aspects, such as the interactions with other objects, deployment contexts and social interactions which are deemed essential to consider in the analysis and prediction of event behaviour. This chapter has also offered some perceptive insights into the future research scope that may help to improve the human condition.

REFERENCES

Ahmed, R., Karmakar, G. C., & Dooley, L. S. (2007). Probabilistic spatio-temporal video object segmentation using a priori shape descriptor. *IEEE International Conference on Acoustics, Speech and Signal Processing (ICASSP-07), 1*, (pp. 15-20).

Bashir, F. I., Khokhar, A. A., & Schonfeld, D. (2007). Object trajectory-based activity classification and recognition using hidden Markov models. *IEEE Transactions on Image Processing, 16*(7), 1912–1919. doi:10.1109/TIP.2007.898960

Baxes, G. A. (1994). *Digital image processing: Principles and applications.* New York, NY: John Wiley & Sons, Inc.

Chai, J., & Shum, H. (2000). Parallel projections for stereo reconstruction. *IEEE Conference on Computer Vision and Pattern Recognition, 2*, (pp. 493-500).

Chen, D., Yang, J., & Wactlar, H. D. (2004). Towards automatic analysis of social interaction patterns in a nursing home environment from video. *ACM SIGMM International Workshop on Multimedia Information Retrieval*, (pp. 283-290).

Faschinger, M., Sastry, C. R., Patel, A. H., & Tas, N. C. (2007). An RFID and wireless sensor network-based implementation of workflow optimization. *IEEE International Symposium on a World of Wireless, Mobile and Multimedia Networks (WoWMoM07)*, (pp. 1-8).

Fodor, I. K. (2002). *A survey of dimension reduction techniques.* Technical Report UCRL-ID-148494, Lawrence Livermore National Laboratory, Center for Applied Scientific Computing, USA.

Goodrum, P., McLaren, M., & Durfee, A. (2006). The application of active radio frequency identification technology for tool tracking on construction job sites. *Automation in Construction, 15*(3), 292–302. doi:10.1016/j.autcon.2005.06.004

Han, D., Bo, L., & Sminchisescu, C. (2009). Selection and context for action recognition. *IEEE 12th International Conference on Computer Vision (ICCV09)*, (pp. 1933-1940).

Hosrt (2007, August 19). *GPS accuracy – Myth and truth – Part 1 – DGPS/Shadowing effect*. Sensors-GPS, http://autocom.wordpress.com/2007/08/19/genaues-gps-mythos-und-wahrheit-teil-1-dgps-abschattung/

Hsiao, K., Vosoughi, S., Tellex, S., Kubat, R., & Roy, D. (2008). Object schemas for responsive robotic language use. In the *Proceedings of the 3rd ACM/IEEE International Conference on Human Robot Interaction*, (pp. 233-240).

ISO/IEC 14496-2. (1999). *Information technology—Coding of audio-visual objects. Part 2: Visual.*

Kamol, P., Nikolaidis, S., Ueda, R., & Arai, T. (2007). RFID based object localization system using ceiling cameras with particle filter. In the *Proceedings of the Future Generation Communication and Networking (FGCN07), 2*, (pp. 37–42).

Karmakar, G. C. (2003). *An integrated fuzzy rule based image segmentation framework*. Australia: Gippsland School of Information Technology, Monash University.

Karmakar, G. C., Dooley, L., & Rahman, S. M. (2001). Review on fuzzy image segmentation technique. In Rahman, S. M. (Ed.), *Design and management of multimedia information systems: Opportunities and Challenges* (pp. 282–313). Hershey, PA: Idea Group Publishing. doi:10.4018/978-1-930708-00-6.ch014

Karmakar, G. C., Rahman, S. M., & Bignall, R. J. (1999). Composite features extraction and object ranking. In the *Proceedings of the International Conference on Computational Intelligence for Modeling, Control and Automation (CIMCA'99)*, Vienna, Austria, (pp. 134 – 139).

Ko, Y., Roy, S., Smith, J. R., Lee, H., & Cho, C. (2007). An enhanced RFID multiple access protocol for fast inventory. *IEEE Global Telecommunications Conference (GLOBECOM07)*, (pp. 4575 – 4579).

Lee, J. (2009). Global positioning/GPS. In Kitchen, R., & Thrift, N. (Eds.), *International encyclopedia of human geography* (pp. 548–555). doi:10.1016/B978-008044910-4.00035-3

Lee, J., Park, K., Hong, S., & Cho, W. (2007). Object tracking based on RFID coverage visual compensation in wireless sensor network. *IEEE International Symposium on Circuits and Systems (ISCAS)*, (pp. 1597 – 1600).

Liu, J., Bowyer, K., Goldgof, D., & Sarkar, S. (1997). *A comparative study of texture measures for human skin treatment*. Presented at International Conference on Information, Communications and Signal Processing (ICICS '97), Singapore.

Lou, J., Liu, Q., Tan, T., & Hu, W. (2002). Semantic interpretation of object activities in a surveillance system. *International Conference on Pattern Recognition, 3*, 777 – 780.

Lowe, D. G. (2004). Distinctive image features from scale-invariant keypoints. *International Journal of Computer Vision, 60*(2), 91–110. doi:10.1023/B:VISI.0000029664.99615.94

McFate, M. (2005). Iraq: The social context of IEDs. *Military Review*, (May-June): 2005.

Milanés, V., Llorca, D. F., Vinagre, B. M., González, C., & Sotelo, M. A. (2010). Clavileño: Evolution of an autonomous car. *International IEEE Annual Conference on Intelligent Transportation Systems,* Madeira Island, Portugal, (pp. 1129-1134).

Murphy, K. P. (2002). *Dynamic Bayesian networks: Representation, inference and learning*. Berkeley, USA: University of California.

Pal, N. R., & Pal, S. K. (1993). A review on image segmentation techniques. *Pattern Recognition, 26,* 1277–1294. doi:10.1016/0031-3203(93)90135-J

Park, S., Kim, K., Park, S., & Park, M. (2006). Object entity-based global localization in indoor environment with stereo camera. *Proceedings of SICE,* (pp. 2681-2686).

Pentland, A. (1999). Modelling and prediction of human behaviour. *Neural Computation, 11,* 229–242. doi:10.1162/089976699300016890

Pham, D. L., & Prince, J. L. (1999). An adaptive fuzzy c-means algorithm for image segmentation in the presence of intensity inhomogeneities. *Pattern Recognition Letters, 20,* 57–68. doi:10.1016/S0167-8655(98)00121-4

Rahman, M. Z., Karmakar, G. C., & Dooley, L. S. (2005). Passive source localization using power spectral analysis and decision fusion in wireless distributed sensor networks. *International Conference on Information Technology: Coding and Computing* (ITCC-05), Las Vegas, USA, (pp. 260–264).

Scholkopf, B., Smola, A., & M¨uller, K. (1999). Kernel principal component analysis. In Scholkopf, B., Burges, C. J. C., & Smola, A. J. (Eds.), *Advances in kernel methods–Support vector learning* (pp. 327–352). Cambridge, MA: MIT Press.

Shechtman, E., & Irani, M. (2007). Space-time behaviour based correlation or How to tell if two underlying motion fields are similar without computing them? *IEEE Transactions on Pattern Analysis and Machine Intelligence, 29*(11), 2045–2056. doi:10.1109/TPAMI.2007.1119

Sheng, X., & Yu-Hen Hu, Y. (2005). Maximum likelihood multiple-source localization using acoustic energy measurements with wireless sensor networks. *IEEE Transactions on Image Processing, 53*(1), 44–53.

Shirasaka, Y., Yairi, T., Kanazaki, H., Junichi Shibata, J., & Machida, K. (2006). Supervised learning for object classification from image and FDID data. *SICE-ICASE International Joint Conference,* Bexco, Busan, Korea, (pp. 5940-5944).

Skinner, B. F. (1953). *Science and human behavior.* New York, NY: The Free Press.

Smith, J. R., Fishkin, K. P., Jiang, W. B., Philipose, M., Rea, A. D., Roy, S., & Rajan, K. S. (2005). RFID-based techniques for human-activity detection. *Communications of the ACM, 48*(9), 39–44. doi:10.1145/1081992.1082018

Song, X., & Fan, G. (2006). Joint key-frame extraction and object segmentation for content-based video analysis. *IEEE Transactions on CSVT, 16*(7), 904–914.

Veltkamp, R. C., & Tanase, M. (2000). *Content-based image retrieval systems: A survey.* Technical Report UU-CS-2000-34, Department of Computing Science, Utrecht University.

Wang, J., Luo, Z., Wong, E. C., & Tan, C. J. (2007). RFID assisted object tracking for automating manufacturing assembly lines. *IEEE International Conference on e-Business Engineering,* (pp. 48-53).

Wikipedia. (2011). *Computer vision.* Retrieved October 1, 2011, http://en.wikipedia.org/wiki/Computer_vision

Wu, J., Osuntogun, A., Choudhury, T., Philipose, M., & Rehg, J. M. (2007). A scalable approach to activity recognition based on object use. *IEEE 11th International Conference on Computer Vision,* Rio de Janeiro, Brazil, (pp. 1–8).

Yilmaz, A., Javed, O., & Shah, M. (2006). Object tracking: A survey. *ACM Computing Surveys, 38*(4), 1–45. doi:10.1145/1177352.1177355

Yu, C., & Ballard, D. H. (2002). Learning to recognize human action sequences. *International Conference on Development and Learning*, (pp. 28-33).

KEY TERMS AND DEFINITIONS

Computer Vision: A field of computer science that uses image analysis for retrieving and understanding the information of an event in real-time.

GPS: The global positioning system which determines the object location using information obtained from satellites.

Image Analysis: An image processing technique which includes image segmentation, feature extraction, and object classification.

Object Localisation: The process of locating an object.

Recognition: To identify an object or event.

RFID: Radio Frequency Identification technology which uses radio waves to read the data stored in an electronic tag.

Segmentation: The process of separating mutually exclusive homogeneous regions or objects of interest.

SIFT (Scale Invariant Feature Transform): A feature approximation technique which is invariant to scale, rotation, and translation.

Training Set: A data set used for learning of a machine.

Visual Sensor: A digital electronic device which is used to capture images

ENDNOTE

[1] http://www.maps-gps-info.com/gps-accuracy.html

Chapter 13
Wireless Sensor Network Protocols Applicable to RFID System

A. K. M. Azad
Monash University, Australia

Joarder Kamruzzaman
Monash University, Australia

Nemai C. Karmakar
Monash University, Australia

ABSTRACT

Radio Frequency Identification (RFID) systems and Wireless Sensor Networks (WSNs) are believed to be the two most important technologies in realizing the ubiquitous computing vision of Future Internet. RFID technology provides much cheaper solution for object identification and tracking based on radio wave. On the other hand, data on various parameters about the physical environment can be acquired using WSNs. Integration of the advantages of both RFID systems and WSNs would benefit many application domains. In RFID system, either an active RFID tag itself or an RFID reader (reading passive or semi-passive tags) consisting of an RF transceiver poses communication capability similar to that for nodes in WSNs. Therefore, instead of using single hop RFID protocol, RFID networks can take advantage of WSN-like multihop communication, and in this regard a number of WSN protocols can be useful for such RFID systems. In this chapter we present possible scenario of the integration of RFID system and WSNs and study a number of wireless sensor network protocols suitable to use in RFID system.

INTRODUCTION AND BACKGROUND

Ubiquitous computing is a vision of the next generation computing environment where future computing devices will become invisibly embedded in the world around us and accessed through simple and intelligent interfaces whenever needed (ITU 2005; Abowd et al. 2005). Imagine a ubiquitous computing world where while snowboarding a person would be notified with the information about the slope ahead, ice thickness and softness, and the location of surrounding skiers for safe running, or a washing machine would automatically read tags embedded in our clothes and adjust

DOI: 10.4018/978-1-4666-1616-5.ch013

its wash cycle accordingly. Increasingly then, connections are not just people-people or people-computers, but also between people-things, and most strikingly, between things-things. This is what the International Telecommunication Union (ITU) calls the "*internet of things*" (ITU 2005). Such a vision would be transformational and have profound implications on how we live, work, interact and learn, and promises to revolutionize a wide range of application domains (ITU 2005; Abowd et al. 2005). By embedding information storing, processing and communication capabilities, *RFID system* and *Wireless Sensor Networks* (WSNs) can be used as a tool to bridge the real world with the virtual computation world, and such a schematic illustration is shown in Figure 1.

A RFID system is made up of RFID tags, RFID readers and data processing system, where the reader reads the tag information and sent to the data processing system where tag information is decoded and used for object identification and tracking. However, a WSN is made up of many tiny and low cost sensor nodes each having a sensing module, a processing module and a wireless transceiver module, a central host node (often called sink), and optionally a number of intermediate fusion nodes, namely, the cluster heads. Sensor nodes sense the surrounding environment and send the data generated thereby to the sink, which further process the sensor data to generate a meaningful scenario. RFID technology provides much cheaper solution for object identification and tracking wirelessly based on radio wave. On the other hand, data on various parameters about the physical environment can be acquired using WSNs. Integration of the advantages of both RFID system and WSNs would benefit many application domains.

The rest of the chapter is organized as follows: Section 2.1 and 2.2 introduce the RFID system and wireless sensor networks, respectively, and then their possible integration scenarios are discussed in Section 2.3. Throughout Section 3 a number of WSNs protocols applicable to RFID systems are discussed, and finally, Section 4 concludes the chapter.

Figure 1. Bridging the physical world and virtual world with RFID/sensor networks

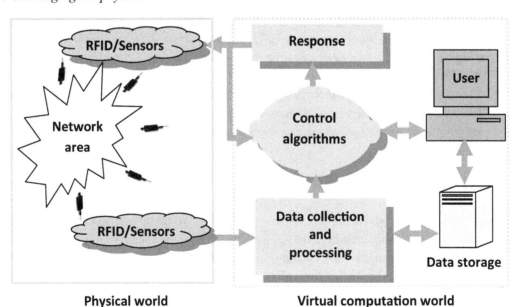

RFID System and Tags

With the recent advances in Micro-Electro Mechanical Systems (MEMS) and communication technologies, devices are becoming smaller and more intelligent. This advance will proliferate ubiquitous computing in daily lives. One of the key technologies for ubiquitous computing is radio frequency identification (RFID), which enables an object (e.g., a product, animal, or person) to be identified at a short distance without manual intervention. Radio Frequency Identification (RFID) is a technology that transmits the identity of an object, for example a series of numbers, through wireless radio

communication to an electronic reader. Today, RFID is applied widely in supply-chain tracking, retail stock management, parking access control, library book tracking, marathon races, airline luggage tracking, electronic security keys, toll collection, theft prevention, and healthcare. A generic RFID system, as shown in Figure 2, usually consists of three main components: tags, readers and data processing system. A tag stores a unique identification number (ID) and may have memory that stores additional data such as manufacturer name, product type, and environmental factors including temperature, humidity, and so on. In some RFID tags, the reader can read and/or write data to tags through wireless transmissions. In a typical RFID application, tags are attached or embedded in objects that must be identified or tracked. By reading nearby tag IDs and then consulting a background database that provides mapping between IDs and objects, the reader can monitor the existence of the corresponding objects.

Existing RFID tags can be classified based on different grounds such as the power management, the frequency of operation, the tag's reading range and the presence or absence of a microchip (Bhuiyan 2011), and the classification of RFID tags is shown in Figure 3. Based on the power management, RFID tags can be of three types: passive, semi-passive and active tags. Passive tag has no

Figure 2. Generic structure of a RFID system

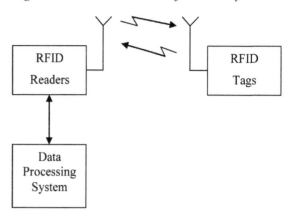

own power supply, but the power required to turn on the tag IC is collected from the reader/interrogator signal impinge on the tag antenna (Chawla & Sam 2007). Passive tags are cheaper but operate successfully in a very short read distance e.g. a few meters only. Semi-passive tags are similar to passive tags apart from the fact that they have a small power supply (battery) (Wenyi et al. 2009) to keep the tag IC constantly powered up. However, they use the power supply only to modify the signal transmitted by the interrogator to pass the information contained in their IC, rather than broadcasting the tag information, as active tags do. Active tags have their own internal power supply and they use that power supply to turn on the microchip as well as to transmit the outgoing signal containing the tag information (Karmakar et al. 2006). They have longer operating ranges compared to passive tags. Based on operating frequency, RFID tags may be classified into low frequency (LF) tags operating from 125-134KHz, high frequency (HF) tags operating frequency centered at 13.56MHz, ultra-high frequency (UHF) tags with operating frequency limited to 433 MHz and 860 MHz to 960 MHz, and Microwave Frequencies tags with operating frequencies 2.4 GHz and 5.8 GHz (Nikitin, Rao & Lazar 2007; Sanming et al. 2010). Some reported RFID tags (Sanming et al. 2010) also use the license free ultra wide-band (UWB) frequency band (3.1 GHz to 10.6 GHz). Based on reading range, RFID tags can be of two types (Nikitin, Rao & Lazar 2007; Nikitin & Rao 2008): near field and far filed tags. The near field coupled tag utilizes either inductive coupling (using coils) or capacitive coupling (using antennas) to establish communication between the tag and the reader. These tags only operate at LF and HF carrier frequencies. Far field coupled tag utilizes the far field radiation properties of both tag and reader antennas to transfer data. These tags can operate at microwave frequencies (2.4 GHz, 5.8 GHz) and hence, their required antenna size is small. These tags have longer reading range compared to near field tags (Chawla & Sam 2007; Nikitin & Rao 2008). Based on the presence of chips, RFID tags are of two kinds: Chipless and Chip-based tags. Chip based tag uses a microchip to store the identification information of the tags (Karmakar et al. 2006). These tags can be either read only or have read/ write capability. Chipless tags do not use any memory chip to store data bits; rather it uses the resonance (Balbin & Karmakar 2009) or reflection (Linlin et al. 2008) properties of microwave circuits to encode bits. To date, chipless tags are read only and once

fabricated, the information contained in the tag cannot be rewritten wirelessly.

Wireless Sensor Networks (WSNs)

A wireless sensor network (WSN) is made up of many tiny and low-cost sensor nodes each include three basic components: a sensing module for data acquisition from the surrounding environment, a processing module for local data processing, and a wireless communication module for data transmission and reception. Usually, a wireless sensor network is spatially deployed in an ad hoc fashion that performs distributed sensing tasks in a collaborative manner without relying on any underlying infrastructure support (Culler, Estrin & Srivastava 2004; Tseng, Pan & Tsai 2006; Heinzelman, Chandrakasan & Balakrishnan 2002; Liu, Wan & Jia 2007). Sensor nodes are generally deployed randomly, and once deployed they organize themselves as a network through radio communication. Sensor nodes generate data from the surrounding environment and send to the *sink*, which then generates a meaningful scenario about the phenomena of interest and responds accordingly. The sink node is usually more powerful in

Figure 3. Classification of RFID tags

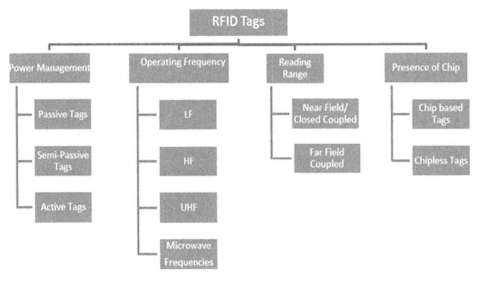

terms of resources and processing capabilities compared to the sensor nodes and can be located internal or external to the sensor field. The decrease in the size and cost of sensors and an envisaged flexibility and ease of deployment has fuelled interest in the possible use of large set of disposable unattended sensors in many application domains (Tseng, Pan & Tsai 2006; Heinzelman, Chandrakasan & Balakrishnan 2002; Liu, Wan & Jia 2007). Networking such unattended sensor nodes in an ad hoc manner would have significant impact on the efficiency of many military and civil applications, such as environmental monitoring, disaster management system, asset tracking, logistic and inventory control, target detection and tracking, health care systems, emergency navigation, traffic management, etc.

From an applications' perspective, based on the data delivery model WSNs can be classified as: *data gathering* WSN and *query-based* WSN. Data gathering WSNs can further be categorized into *periodic* and *event-based* networks. In a periodic data gathering WSN, the sensors send their data to the sink periodically at a predetermined rate (Heinzelman, Chandrakasan & Balakrishnan 2002; Bhardwaj, Garnett & Chandrakasan 2001). In an event-based model, the users are interested in the occurrence of a certain phenomenon (or set of phenomena) and the sensors report information only if an event of interest occurs (Tseng, Pan & Tsai 2006; Wang et al. 2008). In query-based WSNs, the sensors only report their results in response to explicit request from the users (Demirbas, Xuming & Singla 2009; Lee, Lee & Kim 2009; Yao & Gehrke 2002). Moreover, based on the types of nodes present in the network, WSNs can be categorized into i) *homogeneous* WSNs (Heinzelman, Chandrakasan & Balakrishnan 2002; Bhardwaj, Garnett & Chandrakasan 2001; Efthymiou, Nikoletseas & Rolim 2002), where all nodes (except the sink) are of the same type, and ii) *heterogeneous* WSNs (Mhatre & Rosenberg 2004; Zhang, Ma & Yang 2004), where there are a number of different classes of nodes that differ

in terms of parameters like sensing range, data generation rate, energy storage, transmission range, etc. Based on applications' requirement, the scale of a WSN may vary from small to very large coverage areas and the number of nodes deployed can be low to very high. To accommodate such variations, WSNs topology can be single-tier (Bhardwaj, Garnett & Chandrakasan 2001; Efthymiou, Nikoletseas & Rolim 2002) or hierarchical with each level of the network divided into a number of clusters (Mhatre & Rosenberg 2004; Zhang, Ma & Yang 2004). Figure 4 shows examples of single tier and hierarchical clustered WSNs.

Usually, an embedded power source, often consisting of a limited-power battery, supplies the energy needed by the sensor node to perform sensing, data processing and wireless communication tasks. Typically, in WSNs, the energy expenditure in data processing is much less compared to data communication. For example, in a typical sensor node, the energy cost of transmitting a single bit of information is comparable to that needed for processing a thousand data processing instructions (Heinzelman, Chandrakasan & Balakrishnan 2002; Mahfoudh & Minet 2008). Moreover, the energy consumption for sensing operation is application specific and depends on the sensor hardware. Of the three major sources of energy usage, a sensor spends the majority of its energy in data communication (Heinzelman, Chandrakasan & Balakrishnan 2002; Mahfoudh & Minet 2008) and, therefore, a significant portion of research on WSNs has pursued energy-efficient designs at the various layers of the communication protocol stack.

Integration of RFID and Wireless Sensor Networks

As can be seen from Table 1 that RFID networks and WSNs represent two complementary technologies, and there are a number of advantages to merging them. RFID tags are much cheaper than

Figure 4. (a) A single-tier flat wireless sensor network with many sensor nodes and a sink, and (b) a 4-tier hierarchical clustered wireless sensor network

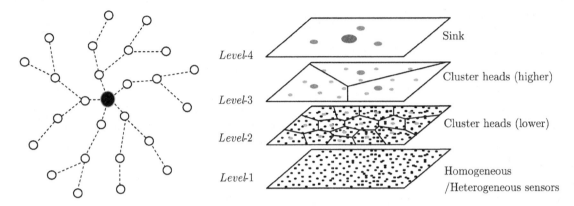

sensor nodes. It is economical to use RFID tags to replace some of the sensor nodes in WSNs. Moreover, because an object that is embedded with an RFID tag is traceable, RFID technology provides a reasonable addition to wireless sensor networks in tracking objects that otherwise are difficult to sense. On the other hand, WSNs offer a number of advantages over traditional RFID systems. Sensors can provide various sensing capabilities to RFID tags, push logic into nodes to enable RFID readers and tags to have intelligence, and provide an RFID system with the capability of operating in multihop fashion, which potentially can extend the applications of RFIDs. In the following we discuss four possible scenarios for the integration of RFID and WSNs: (1) Adding sensing capability to RFID tags,(2) Adding RFID tags with WSN nodes, (3) Integrating RFID readers and sensor nodes, and (4) Mix of RFID and sensors.

Adding Sensing Capability to RFID Tags

Integration of this class enables the adding of sensing capabilities to RFID systems. The RFID tags with sensors (sensor tags) use the same RFID protocols and mechanisms for reading tag IDs, as well as for collecting sensed data. Because integrated sensors inside RFID tags are used only for sensing

purposes, current protocols of these RFID tags rely on single hop communication. Passive tags with integrated sensors operate without batteries and gather power from the RF signal of readers. They belong to the class of applications in which sensing is required only at specific time instants. Sensing is performed when the sensor tag is in the reading range of the reader, and the sensing and reading is triggered by the reader. Passive tags with integrated sensors are used in several applications including temperature sensing and monitoring, photo detection, and movement detection (Cho et al. 2005; Kitayoshi & Sawaya 2005; Philipose et al. 2005). In comparison with passive sensor tags that can perform measurements only when interrogated by the readers, semi-passive sensor tags can perform independent measurements even when they are not in the proximity of readers. This requires a greater amount of memory for storing measured values. But, sensed data is transmitted using the power from reader's interrogating signal. Semi-passive tags with integrated sensors could be used in temperature sensing and monitoring, location recording, vehicle and asset tracking, and access control. Active sensor tags have more memory and improved range and functionality in comparison with passive and semi-passive sensor tags, and therefore can be used in a wide range of applications such as remote healthcare, asset

Table 1. WSNs vs RFID system (Liu et al. 2008)

Feature	WSNs	RFID systems
Purpose	Sensing surrounding environment or providing information on the condition of the attached object	Detect presence of tagged object
Component	Sensor nodes, relay nodes (optional), cluster heads (optional) and sinks	Tags, readers and data processing system
Communication	Multihop	Singlehop
Protocols	Zigbee, WiFi	RFID standard
Mobility	Sensor nodes are usually static	Tags moved with attached objects
Deployment	Random or fixed	Fixed, usually require careful placement
Power supply	Battery powered	Tags are batter powered or passive
Programmability	Sensor nodes are programmable	Usually closed system
Design goal	WSNs are general purpose	Tags are optimized to perform a particular task
Price	Sensor node-> medium Sink -> expensive	Tag -> cheap Reader -> expensive

tracking, habitat monitoring, etc. Here, sensors embedded with the tag sample environment data periodically and independently, and the sensing data is transmitted to the reader proactively by the tag. Active sensor tags may be compliant with existing RFID standards, or they can rely on existing wireless protocols such as wireless local area network (WLAN) or ZigBee. Tags that rely on non-RFID wireless protocols usually use a medium access control (MAC) address as a unique number that identifies the tag. The advantage is that the tags can be integrated easily into the existing wireless infrastructure, and they do not require expensive readers.

Adding RFID Tags with WSN Nodes

Sensor tags obtained by integrating sensing capability with RFID tags have limited communication capabilities. In high-end applications, it is possible to integrate RFID tags with WSN nodes and wireless devices, such that the integrated tagged node can communicate with many wireless devices, not just the RFID readers. Based on the complexity and functionality of the integrated node, there could be three possible integration scenarios: 1) WSN nodes with RFID capability, 2) Advanced WSN nodes

with RFID and some other protocols, and 3) WSN nodes with attached RFID for wake up control. In Case-1, we consider WSN nodes attached with RFID tags in that they have a unique ID number that can be used in identifying objects or people. Unlike RFID sensor tag (discussed above in Section 2.3.1), these integrated WSN nodes can have multiple sensors and additional communication capabilities including communication between the nodes to form a multihop network, forming self-organized networks, and using power-efficient routing and MAC protocols to exchange information. Integrated WSN nodes mainly rely on existing wireless standards such as Zigbee and WLAN. Case-2 includes advanced smart wireless sensor nodes with RFID capabilities. A wireless smart sensor platform uses wireless technologies such as Wi-Fi, Bluetooth, and RFID for communications in a point-to-point topology (Ramamurthy et al. 2007). The platform uses external sensors that are equipped with a smart sensor interface (SSI). The interface extracts data from sensors and provides a data communication interface to the central control unit, which then coordinates timely storage and transmission of data through RF transceiver. Case-3 is the combination of RFID and WSNs in which RFID can be used to wake up the WSN node. Each sensor node is integrated with an RFID tag and is also provided with an RFID reader capability. The integrated tag listens to the RFID reader radio of neighboring nodes. If channel activity is detected, the tag awakens the sensor to listen to the channel and then receives data through the RF sensor radio. Otherwise, the sensor node can stay in sleep mode. Because an RFID radio uses much less energy than an RF sensor radio, the RFID-impulse technique can reduce energy consumption significantly while providing short end-to-end delay (Liu et al. 2008).

Integrating RFID Readers and Sensor Nodes

Another type of integration of RFID and WSNs is the combination of RFID readers with WSN nodes. The integrated readers can sense environmental conditions, communicate with each other in wireless fashion, read identification numbers from tagged objects, and transmit effectively this information to the sink. The three possible ways for integration of RFID reader and sensor nodes are: 1) RFID readers with attached sensors, 2) sensor nodes attached with RFID reader, and 3) RFID readers integrated with sensor nodes using multifunctional intermediate device. In Case-1, sensors attached to the RFID reader performs sensing tasks and generates data. The RFID reader reads tags and these readings are added with the sensed data and sent to the host. The reader might also do some preprocessing before sending the tag and sensed information to the host. Unlike Case-1 where the RFID reader is part of a RFID network, in Case-2 the RFID reader is attached with the WSN node and become part of WSNs. Here, the RFID reader reads the tags in its vicinity and this information is used by the sensor nodes together with its sensed data. Sensor nodes send the (reader and sensor) data generated thereby to the sink using WSNs' multihop routing and MAC protocols. In Case-3, RFID readers and sensors are combined with multi-functional devices, such

as cell phones, personal digital assistants (PDAs) and laptops. The basic idea is that by combining RFID and sensors in cellular networks or the Internet, consumers will be able to read any RFID tag and sensor data in almost any application. For example, a doctor's PDA can read physiological data from a number of body sensors attached to a patient's body and also read RFID tag attached as his/her wrist belt to identify the patient.

Mix of RFID and Sensors

Unlike the previous cases, RFID tags/readers and sensors in this class are physically separated. An RFID system and a WSN both exist in the application, and they work independently. However, there is an integration of RFID and WSN at the software layer when data from both the RFID tags and the WSN nodes are forwarded to the common control center. In such scenarios, successful operation of either the RFID system or the WSN may require assistance from others. For example, the RFID system provides identification for the WSN to find specific objects, and the WSN provides additional information, such as locations and environmental conditions, for the RFID system. The advantage of the mix of RFID and sensors is that there is no need to design new integrated nodes, and all operations and the collaboration between the RFID and the WSN can be done at the software layer.

Wireless Sensor Network Protocols for RFID System

Discussion throughout Sections 2.1 to 2.3 reveal that in RFID system, either an active RFID tag itself or a RFID reader (reading passive or semi-passive tags) consisting of an RF transceiver poses similar communication capability like nodes in WSNs. On the other hand, if RFID tags and/or reader are part of sensor nodes in WSNs then the resulting RFID system is embedded into the wireless sensor networks. Therefore, instead of using single hop RFID protocol, RFID network can take advantage of WSN-like multihop communication, and in this regard a number of WSN protocols can be useful for such RFID systems. In the rest of this chapter we studied a number of wireless sensor network protocols. Although some of the works presented here are not directly fit for using in RFID/sensor networks, but they present the pioneering idea of that genre of work, and based on those concept many WSN protocols were developed. In Section 3 we represent basic communication protocols like *node activation cycling protocols*, *MAC protocols* and *routing protocols*. Then in Section 4 we studied a number of *transmission policies* that regulate the transmission range/power of nodes in WSNs. Finally, in Section 5, protocols related to the *scalability and node heterogeneity* of WSNs are addressed. Hereafter, we refer by node either a sensor node or its equivalent (*i.e.*, active tag, RFID reader, RFID tag attached to sensor node or to any other wireless devices, etc.) in the RFID network which has RF capability.

BASIC COMMUNICATION PROTOCOLS

Node Activation Cycling Protocols

A node's radio transceiver can be in one of the following four states: *transmitting, receiving, idle* (*listening*) or *sleep*. Each state corresponds to a different energy consumption level which drops significantly in the sleep state compared to the others (Mahfoudh & Minet 2008). As a consequence, the "radio always on" solution is unacceptable. On the contrary, radio should be switched off whenever communication is not required, and be resumed as soon as needed. In this way the nodes alternate between the *active* (transmitting, receiving or idle) and sleep states depending on the network activity. Throughout we refer to this behavior as the *node activation cycling*. Node activation cycling can be achieved through

two different and complementary approaches. The first approach exploits the node redundancy, and adaptively selects only a subset of nodes to remain active for maintaining the required level of network coverage and connectivity. The nodes that are not required can go to sleep and thus save energy. We refer to this approach as the *subset protocol*. On the other hand, active nodes (*i.e.*, nodes belonging to the currently active subset) do not need to keep their radio continuously on. They can switch off radio in absence of network activity. We refer to this second approach of activation cycling operating on the active nodes as *sleep/wakeup protocol*. In the following, we review the illustrative research on node activation cycling, breaking it down into the subset protocols and the sleep/wakeup protocols.

Node activation cycling (subset or sleep/wakeup protocol) can be implemented at the active tag or reader of an RFID system. Reducing the number of 'ON' tags/readers at a given instance using subset protocol or reducing the effective duty cycle (wakeup and sleep time ratio) of tags/readers using sleep/wakeup protocol would reduce the collision during tag reading and/or reader to base station communications, and thereby would increase the overall system throughput. Moreover, node activation cycling would reduce the energy consumption at the tags and/or readers significantly. In the following we study a number of pioneering subset and sleep/wakeup protocols at Section 3.1.1 and Section 3.1.2, respectively.

The Subset Protocols

Several criteria can be used to decide which nodes to activate/deactivate and when. In this regard, subset protocols can be broadly classified into two categories: *location driven* (Casari et al. 2005; Zorzi & Rao 2003) and *connectivity driven* (Chen et al. 2002; Cerpa & Estrin 2004) protocols.

- **Location-driven subset protocols**: These protocols assume that the nodes are location-aware and decide which node to turn on and when based on their location information. Two popular location-driven subset protocols are Geographical Adaptive Fidelity (GAF) (Casari et al. 2005) and Geographic Random Forwarding (GeRaF) (Zorzi & Rao 2003). In GAF, the region is divided into small *virtual grids* as shown in Figure 5(*a*) and at any instance, one node denoted as the *leader* is kept active in a grid while the others sleep. In GAF load balancing is achieved through a periodic selection of the leader in the virtual grid using a ranked-based election algorithm. In GeRaF, the portion of the radio coverage of the source which is closer to the intended destination is split into a number of regions as shown in Figure 5(*b*). Each region is assigned a priority with the region closest to the destination having the highest priority. Here, the nodes follow a given duty cycle to switch between the active and sleep states. Once a sender broadcasts a packet, the active nodes in the highest priority region contend for forwarding. If all nodes in that region are sleeping, in the next transmission attempt, the forwarder is chosen among nodes in the second highest priority region, and so on. Location-driven protocols are usually independent of routing protocol and extend the network lifetime in proportion to the node density.

- **Connectivity-driven subset protocols**: These protocols dynamically activate and deactivate nodes so that network connectivity is fulfilled. Examples of such protocols are SPAN (Chen et al. 2002) and ASCENT (Cerpa & Estrin 2004). SPAN adaptively elects a number of nodes from all nodes in the network, denoted as *co-ordinators*, following a coordinator eligibility rule. The coordinators stay awake continuously and perform multi hop routing, while the other nodes stay in sleeping

mode and periodically check if they need to wake up and become coordinators. In SPAN, the coordinator election algorithm requires knowledge of the neighbor and connectivity information which is provided by the routing protocol; hence SPAN depends on it and may require modification in the routing lookup process. Unlike SPAN, ASCENT decides whether a node should join the backbone or sleep based on the local connectivity and packet loss information and is independent of the routing protocol. The basic idea of ASCENT is that initially only some nodes are *active* while the others are *passive*. If the number of active nodes is not large enough, the sink node may experience a high message loss from the sources. The sink then starts sending *help messages* to solicit neighboring passive nodes to join the network by changing their state to active. A passive node moves into a *sleep* state if it does not get a help message from any active node within a certain period. In connection-oriented subset protocols, the energy saving does not increase proportionally with the node density, rather depends on the active-sleep duty cycle.

The Sleep/Wakeup Protocols

As mentioned earlier, another approach to save power is to switch off an active node when it is not involved in any activity. The sleep/wakeup protocol can be implemented either as an independent running protocol on the top of a MAC protocol, or strictly integrated with the MAC protocol itself. MAC protocols will be discussed later; here we review the independent sleep/wakeup protocols. Such schemes offer a great flexibility as they can be tailored to any MAC protocol. Independent sleep/wakeup protocols can further be divided into three main categories: *on-demand* (Schurgers, Tsiatsis & Srivastava 2002; Yang & Vaidya

Figure 5. (a) The virtual grids in GAF (grid side is u and radio range t_x) and (b) the priority regions in GeRaF (with increasing priority from A_3 to A_1)

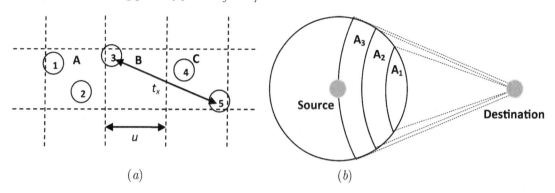

(a)   (b)

2004), *scheduled rendezvous* (Faizulkhakov 2007; Cao 2005; Hohlt, Doherty & Brewer 2004), and *asynchronous* (Zheng, Hou & Sha 2003; Paruchuri et al. 2004) schemes.

• **On-demand protocols**: The basic idea behind on-demand sleep/wakeup protocols, such as STEM (Schurgers, Tsiatsis & Srivastava 2002) and PTW (Yang & Vaidya 2004) is that a node wakes up only when another node wants to communicate with it. Here, the main problem is to inform the sleeping node that some other node wants to communicate. To this end, on-demand schemes typically use separate radios for signaling and data communication. Each node periodically turns on its *wakeup radio* for a certain period to detect possible communication requests. When a source node has to communicate with a neighboring target node, it sends periodic control signals on the wakeup channel. As soon as the target node receives the control signal, it sends back a wakeup acknowledgement and turns on its *data radio*. In STEM, the wake up process is followed by data transmission and thus, in low bit-rate networks STEM suffers from very large wakeup latency. But in PTW, the wakeup procedure is pipelined with the packet transmission

so as to reduce the wakeup latency. In both STEM and PTW, a side effect of using a second radio for wakeup channel is the additional power consumption.

• **Scheduled rendezvous protocols**: Here nodes wake up according to a wakeup schedule to check for potential communications with their neighbors. Different scheduled rendezvous protocols differ in the way the nodes sleep and wake up during their lifetime. The simplest way is using a *fully synchronized pattern* (Faizulkhakov 2007) where all nodes in the network wake up at the same time according to a periodic schedule. The *staggered wakeup pattern* (Cao 2005) shown in Figure 6(a) takes advantage of the internal network organization and nodes located at different levels of the data-gathering tree wake up at different times. Moreover, since nodes located at different levels manage different amounts of data, the active periods should be different. Obviously, the active parts of nodes belonging to adjacent levels are partially overlapped to allow child nodes to communicate with parent nodes. Finally, the *adaptive wakeup pattern* (Hohlt, Doherty & Brewer 2004) sets

Figure 6. (a) The staggered sleep/wakeup pattern and (b) an example of asynchronous sleep/wakeup pattern based on a symmetric-(7, 3, 1) schedule function

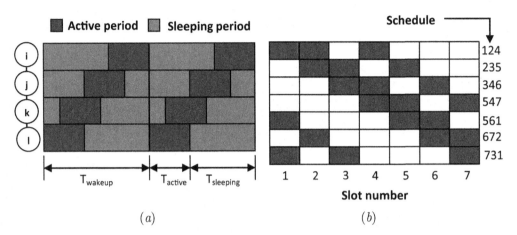

(a)

(b)

the length of the active period to its minimum value adaptively based on the current network traffic. This scheme not only minimizes the energy consumption but also provides lower average packet latency with respect to the fixed staggered scheme.

- **Asynchronous protocols**: In asynchronous protocols, each node wakes up independently of the others and is still able to communicate with its neighbors. This goal is achieved by properties implied in the sleep/wakeup scheme and thus no explicit information exchange is needed among the nodes. AWP (Zheng, Hou & Sha 2003) is an asynchronous sleep/wakeup scheme that can detect neighboring nodes in a finite time without requiring slot alignment. The idea behind AWP is illustrated in Figure 6(*b*) for a set of seven neighboring nodes based on a *symmetric*-(7, 3, 1) wakeup schedule function. As shown in the figure, by following its own schedule, each node is able to communicate with any other neighboring node. However, the packet latency introduced may be large, especially in multi-hop networks. Alternatively, in Random Asynchronous Wakeup (RAW) (Paruchuri et al. 2004), each node wakes

up randomly once in a fixed time interval (*T*), remains active for a predefined time T_a ($T_a \leq T$) and then sleeps again. RAW is extremely simple and relies only on local decisions which makes it well suited for networks with frequent topology changes. With RAW, however, it is not guaranteed that a node can find another active neighbor upon wakeup and so it is not suitable for sparse networks.

The MAC Protocols

Medium access control (MAC) protocol provides the functionality of accessing the shared medium to the nodes in a network. In an RFID system, MAC protocols can be implemented at active tag and reader for shared channel access between tag and reader during tag reading and at reader and base station for shared channel access between them. This would reduce the major sources of energy wastage during tag reading and RFID reader to base station communication by avoiding *collision, overhearing, idle listening, control-packet overhead*, and *overemitting* [44]. In the following, we review some of the most common MAC protocols by classifying them into *TDMA-based* (Rajendran, Obraczka & Garcia-Luna-Aceves

2006; Rajendran, Garcia-Luna-Aceves & Obraczka 2005), *contention-based* (Polastre, Hill & Culler 2004; Heidemann & Estrin 2004; Dam & Langendoen 2003; Lin, Qiao & Wang 2004), and *hybrid* (Rhee et al. 2008) protocols.

The TDMA-Based MAC Protocols

In TDMA-based MAC protocols, time is divided into periodic frames where each frame consists of a number of time slots. Every node is assigned to one or more slots per frame according to a certain scheduling algorithm, and uses such slots for transmitting/receiving packets to/from other nodes. TRAMA (Rajendran, Obraczka & Garcia-Luna-Aceves 2006), a TDMA-based MAC protocol, divides time in two portions, a random-access period and a scheduled access period. The random access period is devoted to slot reservation where each node uses a (hash) priority function to compute its winning slots. The scheduled access period, consisting of one or more slots assigned to an individual node, is used for packet transmission. FLAMA (Rajendran, Garcia-Luna-Aceves & Obraczka 2005) is another TDMA protocol derived from TRAMA which is optimized for periodic monitoring applications and has reduced messaging overheads.

TDMA-based protocols are inherently energy efficient as the nodes turn on their radio only during their assigned slots and sleep the rest of the time. In addition, TDMA-based MAC protocols can avoid problems like packet collision and overemitting, as it is possible to schedule the transmissions of neighboring nodes to occur at different times, which improves energy-efficiency significantly. However, TDMA-based protocols have several drawbacks as well. Firstly, they have limited flexibility and scalability. Secondly, they need a tight synchronization and are very sensitive to interference. Finally, TDMA-based protocols are energy inefficient in low traffic conditions due to the high messaging overhead.

The Contention-Based MAC Protocols

Contention-based protocols perform duty cycling by tightly integrating channel access functionalities with a sleep/wakeup scheme similar to those described in Section 3.1.2. But, in this case the sleep/wakeup algorithm is not independent of the MAC protocol but is tightly coupled with it. The B-MAC (Polastre, Hill & Culler 2004) is a carrier-sense multiple access (CSMA) protocol that achieves a low duty cycle by using asynchronous wakeup/sleep scheme based on periodic listening. Nodes periodically wakeup to check the channel availability and the interval between two consecutive wakeups can be specified by the application. Once awake, a node remains active for a fixed period. Another well-known MAC protocol is S-MAC (Heidemann & Estrin 2004) which adopts scheduled rendezvous scheme with static active and sleep periods. During the active period, the channel access time is split into two parts: in the *listen period*, nodes exchange SYNC packets with neighboring nodes for collision avoidance and in the remaining period, the actual data transfer takes place. The two main deficiencies of S-MAC are the high latency and the insensitivity to varying traffic loads due to its fixed duty cycle. T-MAC (Dam & Langendoen 2003) and DS-MAC (Lin, Qiao & Wang 2004) further extend S-MAC and attempt to solve these problems. They use the same scheduling scheme as S-MAC but instead of a fixed active and sleep duty cycle, the active window is varied adaptively with the traffic load.

Contention-based MAC protocols are robust, scalable, introduce a lower delay than TDMA-based protocols, and can easily adapt to traffic conditions. But, their energy expenditure is higher than TDMA MAC protocols because of the contention and collisions.

The Hybrid MAC Protocols

The motivation behind the hybrid MAC protocol is to combine the respective strength of the TDMA and CSMA-based protocols while offsetting their weaknesses. Unlike TDMA, hybrid MAC protocols achieve a high channel utilization and low latency, and unlike CSMA-based protocols, they reduce data collisions. One of the most interesting hybrid protocols is Z-MAC (Rhee et al. 2008) which uses CSMA as the baseline MAC scheme and a TDMA schedule to enhance the contention resolution. During the neighbor discovery process, each node builds a list of two-hop neighbors and then a distributed slot assignment scheme is applied to ensure that no two nodes in the list are assigned to the same slot. Unlike TDMA, a node in Z-MAC may transmit during any time slot if it is not being used by its owner. A node with data to send first performs carrier sensing and then transmits its data when the channel is clear. By mixing CSMA and TDMA, Z-MAC reduces collisions and hence reduces energy wastage and data transmission latency.

The Routing Protocols

In existing RFID system, the communication between tag and reader is single hop, i.e., readers interrogate passive tags and the tags respond directly to the reader, and active tags directly send their data to the readers. Although the reading process for passive tag shall remain single hop, but active tag composing of RF transceiver can adopt multi hop communication during tag to reader communication. This would increase the effective reading zone for a reader. Moreover, the single hop communication from the reader to base station might not be feasible or highly inefficient due to their mutual physical distance, and therefore multi hop communication would improve the scalability of the overall system. For multi hop communication, at the network layer, the main aim is to find ways for energy-efficient route setup and reliable relaying of data from the tags to reader and from the readers to the base station. In this section, we review the most representative routing approaches classifying them into *data-centric*, *location-based*, and QoS-*aware* routing protocols.

The Data-Centric Routing Protocols

In flat networks, each node typically plays the same role and the nodes collaborate to perform the routing task. Due to the large number of such nodes, it is not feasible to assign an IP-like global address to each node. This consideration leads to data-centric routing where attribute-based naming is used to specify the properties of the data, and *interest* is disseminated throughout the network. There are two approaches that have been used for interest dissemination in data-centric routing. The first is the SPIN-type protocols (Akkaya & Younis 2005; Heinzelman, Kulik & Balakrishnan 1999; Kulik, Heinzelman & Balakrishnan 2002) in which the nodes broadcast the advertisement for the data and wait for requests from the interested nodes. The second is the Directed Diffusion-type protocols in which the sink broadcasts the interest to the nodes located in a specific area. Directed Diffusion (Intanagonwiwat, Govindan & Estrin 2000) is a breakthrough which has inspired the design of many other data-centric routing protocols, such as Rumor (Braginsky & Estrin 2002), Gradient Based, COUGAR (Yao & Gehrke 2002), ACQUIRE (Schurgers & Srivastava 2001), etc.

Directed Diffusion (Intanagonwiwat, Govindan & Estrin 2000) is an on-demand protocol where data is sent by the nodes in response to a query. In order to issue a query, sink broadcasts an *interest* consisting of a list of *attribute-value pairs* to the nodes. Upon receiving an interest, nodes send back their *gradient*, characterized by the data rate, duration, expiration time, etc., to the sink. Based on the gradient values, the sink then selects one or more paths among the discovered paths for query routing. Directed Diffusion is highly energy efficient as it is on-demand, nodes can do in-network processing and all communications are neighbor-to-neighbor, hence there is no need to maintain a global network topology.

REEP (Zabin et al. 2008) is a recently proposed data-centric, energy efficient and reliable routing protocol for WSNs inspired by the Directed Diffusion protocol. In REEP, the sink initiates a *sense event*, a kind of query, and all nodes acquire information to support the sense event. Nodes satisfying the query respond by generating the *information event* which specifies the detected object type and the location information of the source node. After receiving information from a number of source nodes, the sink generates *request events* which are used to retrieve the actual data from the selected nodes. Each node in REEP uses an *energy threshold value* by checking which node agrees or denies participating in any further activities. Moreover, each node uses a *request priority queue* (RPQ) to track over the sequence of information event reception from different neighbors and to select a neighbor with the highest priority in order to request for a path setup.

The Location-Based Routing Protocols

The location information of nodes can be made available by either using GPS or through distributed localization techniques based on measuring the incoming signal strengths from the neighboring nodes. In location-based routing protocols, the node location information is used to route data. A significant number of location-based routing protocols have been proposed, and in the following, we review two representative energy-aware and location-based routing protocols.

Geographic and Energy Aware Routing (GEAR) (Yu, Estrin & Govindan 2001) uses an energy-aware and geographically informed neighbor selection heuristic to route a packet towards the destination region. The key idea is to restrict the number of interests in directed diffusion by considering only certain regions rather than

sending the interests to the whole network. By doing this, GEAR can conserve more energy than directed diffusion. In GEAR, each node keeps an *estimated cost* and a *learning cost* of reaching the destination through its neighbors. The estimated cost is a combination of the residual energy and the distance to destination. The learned cost is a refinement of the estimated cost which depends on the presence of the *network hole*. Compared to similar non-energy-aware routing protocols like GPSR (Karp & Kung 2000), GEAR can deliver 70-80% more packets for same amount of node energy.

In conventional location-based routing schemes, maintaining the neighbor information causes a high messaging overhead. Moreover, the neighbor information can quickly get outdated, which in turn leads to frequent packet drops. Recently, beaconless location-based routing protocols (Zhang & Shen 2011; Chawla et al. 2006) have been proposed in which nodes can forward packets without the help of beacons or maintaining the neighbor information. Energy-efficient Beaconless Geographic Routing (EBGR) (Zhang & Shen 2011) is a beaconless geographic routing for dynamic wireless networks. Without any prior knowledge of neighbors, EBGR works as follows: each node first calculates its ideal next-hop relay position based on the optimal forwarding distance in terms of minimizing the total energy consumption for delivering a packet to the sink. Then, when a node has a packet to transmit, it first broadcasts a RTS message to detect its best next-hop relay. Each candidate that receives this RTS message sets a delay for broadcasting a corresponding CTS message based on a discrete delay function, which guarantees that the neighbor closest to the optimal relay position has the shortest delay. The neighbor that has the minimum delay broadcasts its CTS message first, and the other candidates notice that a node has already responded the request and quit the contention process. Finally, the packet is unicasted to the established next-hop relay.

The QoS-Aware Routing Protocols

There has been an increasing interest in applications that require certain *quality-of-service* (QoS) guarantees, such as, end-to-end delay, reliability, fault-tolerance, etc. Such QoS-aware data routing balances between energy consumption and service quality while delivering data from the nodes to the sink. Several QoS provisioning protocols have been proposed for wireless ad hoc networks. However, they are based on the end-to-end path discovery and resource reservation, which renders their application impractical for large scale dynamic networks.

Sequential Assignment Routing (SAR) (Sohrabi & Pottie 2000) is one of the first routing protocols to introduce the notion of QoS into the routing decisions in WSNs. It is a table-driven multi-path approach aimed to achieve energy efficiency and fault tolerance. The SAR protocol creates trees rooted at the one-hop neighbours of the sink considering the QoS metrics, the energy resources on each path, and the priority level of each packet. By using created trees, multiple paths from the sink to the nodes are formed, and one or more paths are selected according to the energy resources and QoS on the path. Although multiple paths from the nodes to the sink ensure fault-tolerance and easy recovery, the protocol suffers from the overhead of maintaining the tables and states at each node, especially when the number of nodes is huge.

The Multi-Path and Multi-SPEED (MMSPEED) (Felemban, Lee & Ekici 2006) routing protocol is a fairly new QoS-aware routing protocol for WSNs. Here the QoS provisioning is performed in two quality domains, namely, timeliness and reliability. Multiple QoS levels are provided in the timeliness domain by guaranteeing multiple packet delivery speed options, so that various traffic types can dynamically choose the proper speed options for their packets depending on their end-to-end deadlines. In the reliability domain, various reliability requirements are supported by probabilistic multipath forwarding to control the number of packet delivery paths depending on the required end-to-end reaching probability. These mechanisms for QoS provisioning are realized in a localized way without global network information by employing localized geographic packet forwarding augmented with dynamic compensation, which compensates for local decision inaccuracies as a packet travels towards its destination.

THE TRANSMISSION POLICIES

As discussed throughout Section 3, activation cycling selects a subset of nodes to become active in a given instance, the MAC protocol controls the shared medium access among neighboring active nodes, and the routing protocol establishes paths to send data from the sensors to the sink. A common approach taken in the design and operation of these protocols is the use of a common-range transmission power for nodes (Heinzelman, Chandrakasan & Balakrishnan 2002; Bhardwaj, Garnett & Chandrakasan 2001; Efthymiou, Nikoletseas & Rolim 2002). Although the energy-awareness is incorporated in their decision making process, the inherent many-to-one data flow from the sensor nodes to the sink in WSNs leads to an unbalanced energy usage among nodes located at various distances from the sink, and few nodes in the critical zone deplete their energy early becoming the bottleneck for the overall network performance.

As discussed in Section 3.3, adopting multi hop communication from (active) tags to readers would improve the effective reading zone of RFID readers and from readers to base station would enhance the scalability of the overall system. But then, similar to the WSNs, the RFID network will also suffer from the aforementioned unbalanced traffic problem. In the following we study how this issue has been addressed in WSNs which can be used in RFID system as well.

Several strategies have been proposed as the additional aids to the basic communication protocols in extending the overall network lifetime through balancing the energy usage among nodes across the network. For example, lifetime extending node deployment strategy based on a non-uniform node density was proposed in (Liu 2006) where the deployment of extra relay nodes around the sink helps solve the unbalanced energy usage problem. A similar non-uniform energy node deployment strategy considers installing nodes with increasing initial energy towards the sink to avoid early collapse of the close-by nodes due to heavy incoming traffic (Maleki & Pedram 2005). But in many applications, the aforementioned deterministic node deployment strategies are not feasible. Alternatively, there have been a number of studies focused on issues such as using multiple sinks or sink mobility for balancing the energy usage among nodes. In the sink mobility scheme (Pan et al. 2005), the sink moves inside the network region and collects data from the nodes as it passes by. Such studies generally attempt to prolong the network lifetime by exploiting the mobility pattern of the sink. But, this approach is not realistic for large-scale networks since it causes excessive messaging overheads for route updates on each sink location change. Another class of protocols uses multiple stationary sinks and determines their optimal placement (Azad & Chockalingam 2006). Then each node is optimally assigned to a sink (Hou, Shi & Sherali 2006) so that the overall network performances like the lifetime, data latency etc. are optimized. The problem of unbalanced energy usage among nodes assigned to a particular sink still persists as long as a common-range is used by the nodes for data transmission to the sink.

To address the unbalanced traffic problem, recently, a number of studies have shown that choosing variable transmission ranges for transporting data to the sink have a profound impact on the traffic and energy usage distribution among nodes across the network and can play an effective role in extending the overall network performance. For example, increasing the transmission range of the nodes results in less relay traffic on the nodes located close to the sink and hence reduces their energy consumption, while nodes located further from the sink consumes more energy due to the higher transmission range. Moreover, during multi hop transmission, the relay load on a node depends on its distance from the sink, hence allowing the use of different transmission ranges by nodes located at various distances also improves the traffic distribution among nodes across the network. Throughout we refer *transmission policy* as the means to regulate the transmission ranges and their respective duty cycles used by the nodes to achieve a more uniform traffic distribution among nodes across the network. Transmission policy has no conflicts with the other protocols, but it is an additional boost and can be used together with them. In the next section, we review a number of notable energy-efficient transmission policies presented in the literature.

We define a *transmission parameter* as a set of values that determines a number of transmission ranges and their duty cycles which would be used by the nodes for data transmission. Transmission policy regulates the transmission parameters used by various nodes over the lifetime. The performance and complexity of transmission policies vary with the degree of freedom in choosing transmission parameters by the nodes. At one extreme, transmission parameter is node-wise distinct and at the other end, same transmission parameter is used by all nodes in the network. In the middle, a number of transmission parameters are used where each parameter is followed by the nodes located in a particular region. From this observation, in the following, we review some prominent transmission policies dividing them into *network based* (Bhardwaj, Garnett & Chandrakasan 2001; Mhatre & Rosenberg 2004), *region based* (Efthymiou, Nikoletseas & Rolim 2002; Gao et al.2006; Azad & Kamruzzaman 2011), and *node-by-node based* (Cardei, Pervaiz & Cardei 2006; Kirousis et al. 2000) approaches.

The Network Based Transmission Policies

Here, the same transmission parameter is used by all nodes in the network to regulate their data transmission. Usually, in network based transmission policy, with the aim to maximize the overall network lifetime the nodes determine the transmission parameter in a distributed manner and thus this policy is easy to implement and incurs a low messaging overhead (Bhardwaj, Garnett & Chandrakasan 2001; Mhatre & Rosenberg 2004). But, this approach is not efficient since all nodes use the same parameter irrespective of their position which fails to consolidate the unbalanced traffic flow across the network. However, in the following we review two notable works in this category.

Characteristic Distance for Transmission

In (Bhardwaj, Garnett & Chandrakasan 2001), Bhardwaj *et al.* studied the upper bound on the lifetime of a network by finding an optimal hop distance, namely *characteristic distance*, to send a packet from the node to the sink to minimize the total energy usage along the path. Here, a data link between a radio transmitter and a receiver separated by *d* meters is divided into *k* sub paths by introducing *k*-1 intervening relay nodes as shown in Figure 7. The authors have shown that, for given distance *d* and number of hops *k*, the overall energy dissipation along the path is minimum when the length of all the sub paths are made equal to *d/k* and the optimal number of hops is given by

$$k_{opt} = \lfloor d / d_{char} \rfloor \quad \text{or} \quad \lceil d / d_{char} \rceil \quad (1)$$

where the distance d_{char}, called the *characteristic distance*, is independent of *d* and is given by

$$d_{char} = \left[\frac{2\alpha_s}{\beta_s(\gamma - 1)} \right]^{\frac{1}{\gamma}} \quad (2)$$

where $2 \leq \gamma \leq 4$ is the path loss factor, α_s is the energy/bit needed to run the transceiver circuit of the node and β_s is the energy consumed in the amplifier circuit to transmit one bit of data over unit distance (Heinzelman, Chandrakasan & Balakrishnan 2002; Mhatre & Rosenberg 2004).

The above study focuses on one source-destination pair at a time and assumes uniform relay traffic on all the intermediate nodes along the multi hop path without taking into account the non-uniform relay traffic pattern. Although this study does not fit for WSNs/RFID-networks, it is one of the pioneer works of its kind that attempted to find an optimal transmission distance with the aim to maximize network lifetime.

Single Hop, Multi Hop, and their Hybrid Transmission Policy

In (Mhatre & Rosenberg 2004), Mhatre and Rosenberg studied single hop (SH) and multi hop (MH) transmissions and obtained an optimal transmission distance for MH communication. A hybrid of SH and MH transmissions was also proposed and the optimal ratio of the number of data cycles operated in these two modes over the network lifetime was obtained. The authors assumed a circular region with the sink at the centre. The whole region is divided into co-centric rings (with ring thickness *w*) around the sink and the nodes send their data towards the sink using SH or MH transmission or combination of them as shown in Figure 8. In their analysis, the authors adopted the energy model as in (Heinzelman, Chandrakasan & Balakrishnan 2002) where the energy usage by a node to transmit (to a distance *d*) and receive one bit of data, respectively, are defined as

Figure 7. k-1 relay nodes between source and sink to reduce energy needed to transmit a bit

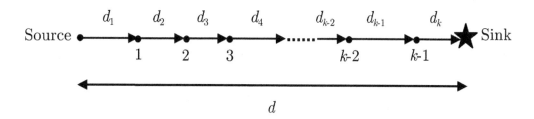

$$\varphi_t = \alpha_s + \beta_s d^\gamma \text{ and} \tag{3}$$

$$\varphi_r = \alpha_s. \tag{4}$$

In SH transmission, a node sends data directly to the sink without using any intermediate relay nodes. A node located in the i-th ring ($C_{i,w}$) needs to send data to a distance of iw on average to reach the sink, hence using (3) the energy consumption to transmit one bit of data is

$$\Omega_s(i, SH) = \alpha_s + \beta_s (iw)^\gamma. \tag{5}$$

On the other hand, during MH communication, data is relayed through a number of intermediate nodes on the way from the source node to the sink with one ring forward at each hop. Using the energy consumption model as in (3) and (4), when each node generates unit amount of data (one bit), the energy usage by a node in ring $C_{i,w}$ to transmit its self generated data and relay incoming traffic to the next ring is given by

$$\Omega_s(i, MH) = (\alpha_s + \beta_s w^\gamma) + (2\alpha_s + \beta_s w^\gamma) \frac{\tilde{N}_s(C_{i,w})}{N_s(C_{i,w})} \tag{6}$$

where $N_s(C_{i,w})$ is the number of nodes present in ring C_i and $\tilde{N}_s(C_{i,w})$ is the number of nodes located outside of ring $C_{i,w}$ (i.e., in rings $C_{i+1,w}$ to $C_{l,w}$). The authors defined the *network lifetime*

as the total number of completed data cycles before the first node dies and obtained the optimal transmission distance $\hat{w}_{MH}$ (optimal value for the ring thickness) that maximizes the network lifetime for MH transmission as

$$\hat{w}_{MH} = \left[\frac{4\alpha_s}{\beta_s(\gamma - 2)} \right]^{\frac{1}{\gamma}}, \gamma > 2. \tag{7}$$

In SH transmission, the nodes located at the farthest ring $C_{l,w}$ consume energy at the highest rate compared to the nodes residing in the inner rings (i.e., $C_{1,w}$ to $C_{l-1,w}$). But in MH transmission, the nodes closer to the sink need to relay higher incoming traffic than the nodes in the outer rings and thus consume energy at a higher rate. Comparing these two communication modes the authors show that, SH transmission performs better than MH transmission in terms of network lifetime for $\gamma \leq 2$ and vice versa for $\gamma > 2$. Moreover, having observed the two opposite energy decay characteristics, a hybrid of SH and MH transmission policies was proposed and a unique ratio of the number of data cycles in single hop, τ_{SH} and multi hop, τ_{MH} as in (8) was obtained that would be followed by all nodes.

$$\tau_{SH} : \tau_{MH} = \Delta\Omega_{MH} : \Delta\Omega_{SH} \tag{8}$$

where $\Delta\Omega_{MH} = \Omega_s(1, MH) - \Omega_s(l, MH)$ and $\Delta\Omega_{SH} = \Omega_s(l, SH) - \Omega_s(1, SH)$.

Figure 8. Single hop (SH) and multi hop (MH) transmissions

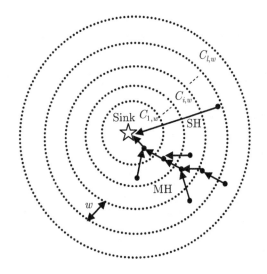

The Region Based Transmission Policies

In this approach, instead of using the same transmission parameters for all nodes, the network is divided into a number of regions with distinct transmission parameters followed by the nodes located in each region (Efthymiou, Nikoletseas & Rolim 2002; Gao et al.2006). Region based transmission policy has great potential in extending the network lifetime through achieving a highly uniform energy usage distribution among nodes across the network. Although it appears promising, its full potentials and benefits are yet to be explored. However, in the following we review two such notable studies.

Transmission Range Adjustment

In (Gao et al.2006), Gao *et al.* deduce the relationship between the optimal radio range and traffic generation rate for a linear network. This work uses the characteristic distance obtained in (Bhardwaj, Garnett & Chandrakasan 2001) and enhances it by incorporating the traffic generation rate and idle state energy consumption of the nodes. Similar to

(Bhardwaj, Garnett & Chandrakasan 2001), the linear network of length d is divided into equal length k grids and only one node wakes up in each grid as the relay node. Initially, it is assumed that the intermediate relay nodes do not generate any data and that all nodes use a uniform transmission range of at least twice the grid length irrespective of their location. The authors obtained an optimal transmission distance for this scenario that minimizes the total energy consumption along the path of the linear network and is given by

$$R_{opt} = \frac{r_{opt}}{2} = \left[\frac{2\alpha_s A + c\alpha_s(1 - 2A)}{2^{\gamma}(\gamma - 1)\beta_s A} \right]^{\frac{1}{\gamma}} \qquad (9)$$

where R_{opt} is the optimal value of the grid length, r_{opt} is the optimal radio range, A is the rate of data generation by a node and c is a constant such that $c\alpha_s$ represents the rate of energy consumption during the idle state which is close to the energy consumption in the receiving state and thus c is close to 1. Here we have rewritten the Equation (9) of (Bhardwaj, Garnett & Chandrakasan 2001) for the energy model defined in (3) and (4).

The above result is then used to extend the analysis considering the traffic generation at the intermediate nodes. In this scenario, the nodes far from the sink have less data to forward and have longer idle times; therefore, they should use a longer radio range while the nodes close to the sink relay more traffic and should use a shorter radio range. This leads to a non-uniform grid covering the network as shown in Figure 9.

Now, the traffic forwarded by nodes at different distances from the sink varies and for a node at distance x from the sink, it is parameterized as

$$A(x) = (d - x)\rho_s a \qquad (10)$$

where ρ_s is the node density and a is the traffic generation rate by a node. Finally, the authors presented two heuristic algorithms based on relationship (9). In the first algorithm, the grid sizes are calculated iteratively as follows

$$R_i = R_{opt}\left(x = \sum_{j=1}^{i} R_j\right) \qquad (11)$$

where R_{opt} is parameterized by x which means that in (9), $A(x)$ as determined by (10) will be used in place of A. The second heuristic algorithm saves a little more energy and the grid sizes are determined as follows

$$R_{i-1} + R_i = r_{opt}\left(x = \sum_{j=1}^{i} R_j\right) \qquad (12)$$

where the first grid size R_1 is adopted as $R_1 = r_{opt}(x = R_1 + R_2)\,/\,2$. Once the grid lengths are known, the transmission ranges that would be used by the nodes located at various grids are calculated as

$$r_i = R_i + R_{i-1}, i > 1 \text{ and } r_1 = R_1. \qquad (13)$$

The network model considered in the above work does not fit to the traffic pattern of a RFID/sensor network. Here, a linear network scenario where intermediate nodes on a multi hop path receive relay load from only one predecessor node. But in actual RFID/sensor network, during multi hop communication, a node close to the sink may receive traffic from a number of neighboring nodes (Figure 8). Moreover, to maximize the overall network lifetime, instead of minimizing the total energy consumption along the path, the optimization objective should focus on minimizing the energy consumption by the highest energy consuming nodes which ultimately become the bottleneck for network operation. However, this study pioneered the idea of using region based variable transmission ranges for balancing energy usage across the network.

Energy Balanced Data Propagation with Hybrid Transmission

Energy balanced data propagation based on the hybrid of SH and MH transmissions were studied in (Efthymiou, Nikoletseas & Rolim 2002) where the average per node energy dissipation over the entire lifetime is the same for all the nodes in the network. Here, a node sends data either one hop closer to the sink or directly to the sink with a certain probability where this probability varies with the nodes location relative to the sink. Here the authors assume the network area as a circle sector with the sink at the center. The sector area is then divided into a number of ring sectors or 'slices' with each slice having thickness w as shown in Figure 10. For balancing load and spreading energy dissipation evenly among the nodes, a node in ring sector $C_{i,w}$ forwards data to $C_{i-1,w}$ (*i.e.* the next sector towards the sink) with probability p_i, while with probability 1-p_i it transmits data directly to the sink. There is a tradeoff for choosing p_i: i) if p_i increases then transmissions tend to happen locally, thus the energy consumption per transmission is low; however, nodes close to the sink tend to be overused since more data pass through them and ii) on the other hand, if p_i decreases, distant transmission by a node increases resulting in a higher energy drainage by the node per transmission. The authors Equationed a recursive relation to determine such probabilities for each ring sector that guarantees energy balanced data propagation among the nodes. Moreover, the authors also presented a close form estimate of p_i considering the energy consumption for the transmission operation only (no receiving energy) which is given by

$$p_i = 1 - \frac{3x}{(i+1)(i-1)}, \ 3 \le i \le m \qquad (14)$$

Figure 9. Linear network divided by virtual grids of different size

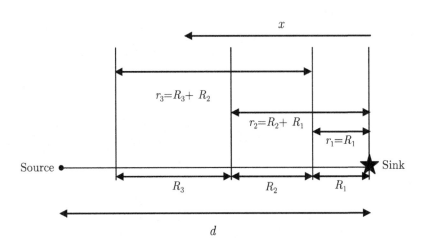

where $p_2 = x \in (0,1)$ is a free parameter and $p_1 = 0$. Although the scheme attains balanced energy usage among nodes, the use of only two options, either direct transmission to the sink or one ring forward towards the sink, results in an overall energy inefficiency. Instead, better energy efficiency is expected when a number of transmission ranges are used. Moreover, the authors do not mention about the optimal transmission range for multi hop transmission.

Energy Balanced Transmission Policy with Variable Range Transmission

In this section we present our work on energy balanced transmission policy for homogeneous WSNs (Azad & Kamruzzaman 2011). Similar to (Efthymiou, Nikoletseas & Rolim 2002; Mhatre & Rosenberg 2004) we divide the region by equally spaced co-centric arcs (see Figure 11) and refer the area enclosed between two consecutive arcs as *ring segment*. But our approach differs in a number of ways: i) we decompose the transmission distance of traditional MH scheme into two parts: *ring thickness* and *hop size*. We refer ring thickness as the width of a ring segment, *i.e.*, the minimum distance between two consecutive arcs, and hop size represents how many ring segments a sensor

forwards its data toward the cluster head (CH) in a single transmission. Instead of forwarding data only one ring segment during MH or directly to the sink in SH, in our proposed scheme hop size is varied to a number of feasible values, ii) varying hop size over lifetime would result in more uniform energy drainage among sensors, and iii) ring thickness and duty cycles for various hop sizes are determined optimally to enhance the overall sensor lifetime.

For ring thickness w and hop size η, a node in ring segment $C_{i,w}$ sends self generated and incoming traffic to another node in ring segment $C_{i-\eta,w}$ if $i > \eta$; otherwise directly to the sink. The average relay load $I_s(i,w,\eta)$ and energy usage $e_s(i,w,\eta)$ per data cycle for a node located in the ring segment $C_{i,w}$ is obtained as eq. (14) and eq. (16) in (Azad & Kamruzzaman 2011), respectively, which we use throughout this paper, and are given as in (15) and (16).

$$I_s(i,w,\eta) = \begin{cases} \dfrac{l^2 + \eta l - l}{\eta(2i-1)}\lambda_s - \lambda_s, & 1 \le i \le \eta \\ \dfrac{l^2 - i^2 + \eta l - i\eta - l + i}{\eta(2i-1)}\lambda_s, & \eta < i \le l-\eta \\ 0, & i > l-\eta \end{cases}$$

(15)

Figure 10. The hybrid of multi hop (with probability p_i) and single hop (with probability $1-p_i$) transmissions

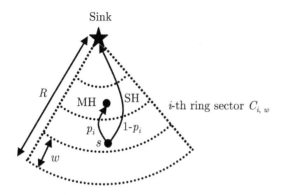

Figure 11. Traffic flow for hop size $\eta=1$ and $\eta=2$ with a fixed ring thickness w

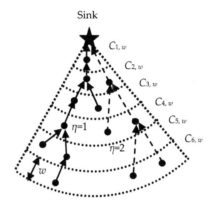

$$e_s(i,w,\eta) = \{\alpha_s + \beta_s x^\gamma(i,w,\eta)\}\lambda_s + \{2\alpha_s + \beta_s x^\gamma(i,w,\eta)\}I_s(i,w,\eta)$$

$$(16)$$

where for $i \geq \eta$, $x(i,w,\eta) = \eta w$, otherwise $x(i,w,\eta) = iw$, and I is the index of the farthest ring segment.

In (Azad & Kamruzzaman 2011), *critical ring segment* refers to the ring segment whose nodes have the highest energy usage rate and *critical energy* as the energy consumption per data cycle by a node located in the critical ring segment. From (Azad & Kamruzzaman 2011),

Theorem 1. *Given a ring thickness w and hop size η (>1): (a) for $\dfrac{\eta^\gamma - 2\eta + 1}{\eta - 1} \leq \dfrac{4\alpha_s}{\beta_s w^\gamma}$, the critical ring segment is $C_{1,\,w}$; otherwise it is $C_{\eta,\,w}$. Both $C_{1,\,w}$ and $C_{\eta,\,w}$ become critical ring segments when the equality holds. (b) for $\dfrac{\eta^\gamma - 2\eta + 1}{\eta - 1} = \dfrac{4\alpha_s}{\beta_s w^\gamma}$ and $\eta>2$, there exists $1<i<\eta$ so that $e_s(1,w,\eta) > ... > e_s(i,w,\eta)$ and $e_s(i,w,\eta) < ... < e_s(l,w,\eta)$ and 'i' is obtained as the ceiling of the root of the equation $(\gamma - 1)i^\gamma - \dfrac{\gamma}{2}i^{\gamma-1} - \dfrac{2\alpha_s}{\beta_s w^\gamma} = 0$.*

Through further analysis we also find the optimal relation between ring thickness w and hop size η that minimizes the critical energy and is given by (17). Then, following Theorem 1, for optimally related ring thickness and hop size pair both $C_{1,\,w}$ and $C_{\eta,\,w}$ are critical.

$$w = \left[\frac{4\alpha_s(\eta-1)}{\beta_s(\eta^\gamma - 2\eta + 1)}\right]^{\frac{1}{\gamma}}, \quad \eta > 1 \qquad (17)$$

In (Azad & Kamruzzaman 2011), an optimal ring thickness and hop size pair $(\hat{w}, \hat{\eta})$ is determined, which is related as per (17) and gives the minimum value of critical energy among all feasible ring thickness and hop size pairs related by the above equation. In that sense any (w,η) related by (17) represents local minima while $(\hat{w}, \hat{\eta})$ represents global minima in terms of the critical energy value in w,η space. Hereafter, w_η denotes the optimal ring thickness for hop size η, determined as per (11) for $\eta > 1$, and as per (10) for $\eta = 1$ and $\gamma > 2$, and $w_\eta = a(X)$ for $\eta = 1$ and $\gamma = 2$.

To attain balanced energy usages and extend network lifetime, in *synchronous variable hop size* (SVHS) transmission scheme proposed in (Azad & Kamruzzaman 2011), optimal ring thickness $(\hat{w})$ is used and hop size is varied over to a number of different feasible values over lifetime and optimal duty cycles are determined for each feasible hop size.

The Node-by-Node Based Transmission Policies

The location and relay load on a node is unique and thus a distinct transmission parameter should be used for each individual node to obtain ideal result in terms of balancing energy usage and extending network lifetime. Ideally, in node-by-node transmission policy, each node should use a number of transmission ranges with their associated frequencies during lifetime and these values would be different for different nodes. In this case, the nodes would adjust their transmission ranges and associated duty cycles based on the locally available information. Although it seems attractive, practically, it is a very hard problem. A simplified problem, namely, *range assignment problem*, is studied where a distinct transmission range is assigned to individual node so that the resulting communication graph is strongly connected and the energy cost is minimized (Cardei, Pervaiz & Cardei 2006). The computational complexity of range assignment problem has been analyzed in Kirousis *et al.* (Kirousis et al. 2000). The problem is solvable in polynomial time (more specifically in $O(n^4)$) for the one-dimensional linear network and it is shown to be NP-hard in the case of two and three-dimensional networks. Thus, computing the optimal range assignment for RFID/sensor networks having a large number of nodes is an impossible task. Moreover, node-by-node transmission policy is much more complicated since each node may use a number of transmission ranges over the lifetime. So, for limited energy and processing power nodes, node-by-node basis transmission policy is practically unrealistic.

SCALABILITY AND HETEROGENEITY IN WSNS

Many applications entail the deployment of a large population of nodes covering a vast area. Designing and operating such a large network would require scalable architecture and data transmission strategies. Early studies considered flat network architecture, where all nodes send their data to a single sink which is suitable for small networks. Such flat networks suffer from poor scalability and their performances (*e.g.*, the network lifetime, energy usage, throughput, resource utilization, delay, etc.) degrade very quickly as the size of the network increases. Instead, grouping nodes into clusters has been of wide interest to the research community to achieve network scalability and improve network performances. In clustered networks, the whole network is divided into a number of non-overlapped cluster regions where each cluster has a leader called the *cluster head*. Each node is assigned to a cluster head to which it sends its data. Each cluster head aggregates the data received from nodes in its vicinity and send the aggregated data to the sink. Throughout, we refer to the data transmission from the nodes to cluster heads as *intra-cluster transmission* and the data transmission from the cluster heads to the sink as *inter-cluster transmission*. In RFID system, we may consider the tag to reader communication as the intra-cluster transmission where the RFID readers represent the cluster head and the communication between RFID readers and base station as the inter-cluster transmission.

In addition to scalability, clustering has numerous advantages. It can localize the route setup within the cluster and thus reduce the size of the routing table stored at the individual nodes. Clustering enables bandwidth reuse and thus increases the system capacity. It also reduces the communication overheads, thereby reducing the energy consumption and interferences among the nodes. Furthermore, data aggregation in the cluster head reduces the overall traffic in the network and hence saves valuable energy and bandwidth. However, there are also some disadvantages associated with clustering, such as additional overheads during cluster head selection, cluster construction and re-construction, and nodes assignment.

Based on the node and cluster head characteristics, the clustering techniques proposed in WSNs literature can be divided broadly into two categories: *homogeneous node clustering*, where the cluster heads are selected from the nodes, and *heterogeneous node clustering*, where some designated nodes, equipped with significantly more resources (*e.g.*, energy, computation power, communication range, etc.) compared to the nodes, are used as cluster heads. In the following, we review some well-known clustering techniques from both categories.

The Homogeneous Node Clustering

This clustering technique is applied to networks in which all the nodes are identical and a number of nodes are selected as the cluster heads. But, when a node functions as a cluster head it consumes significantly more energy than a non-cluster head node. Therefore, in homogeneous networks, a node can't act as a cluster head throughout its lifetime, instead a periodic role rotation among the nodes as cluster heads should be done to achieve balanced energy usage among the nodes. This requires all nodes to be capable of working as a cluster head which imposes an extra hardware and software complexity to perform MAC and routing coordination, data aggregation, and the long range transmission to the distant sink. In addition, a significant communication overhead is needed for the periodic election of cluster heads and change in data routes accordingly. In the following, we review two illustrative homogeneous node clustering techniques.

Low Energy Adaptive Clustering Hierarchy (LEACH)

LEACH (Heinzelman, Chandrakasan & Balakrishnan 2002) is a pioneer clustering algorithm for homogeneous networks. The operation of LEACH is divided into *rounds*. Each round begins with a setup phase when the cluster heads are selected and the clusters are organized, followed by a steady-state phase when data is transferred from the nodes to the cluster heads and on to the sink. During the setup phase, LEACH selects cluster heads by using a distributed algorithm in which the nodes make autonomous decisions without any centralized control and ensure that at each round there are at least a certain number of nodes that elect themselves as cluster heads. In order to achieve load and energy usage balancing, the cluster heads are periodically rotated among the nodes. If k number of nodes is selected as cluster heads among the N nodes at each round, then each node needs to be cluster head once in N/k rounds on average. To attain such cluster head election, nodes use a probabilistic measure to become a cluster head in a particular round based on its role in earlier rounds. A node i elects itself to be a cluster head at the beginning of round $r+1$ with probability $p_i(r+1)$ based on an indicator function $\sigma_i(r)$ which determines whether or not node i became a cluster head at the most recent $r \bmod (N/k)$ rounds; $\sigma_i(r) = 0$ if node i has been a cluster head or 1 otherwise. Then the probability $p_i(r+1)$ is given by

$$p_i(r+1) = \begin{cases} \dfrac{k}{N - k \times \left(r \bmod \dfrac{N}{k} \right)}, & \sigma_i(r) = 1 \\ 0, & \sigma_i(r) = 0. \end{cases}$$

$$(18)$$

The above process ensures that there will be k cluster heads at each round among the N nodes and each node becomes cluster head approximately the same number of times over the network lifetime. Once the cluster heads have been organized in the setup phase, the cluster heads collect data from their one hop neighbors, aggregate the gathered data and send directly to the sink during the steady-state phase. For a network with N nodes uniformly distributed over a region having area $M \times M$, the authors obtained the optimal number of cluster heads that minimizes the total energy consumption over the network in a round which is given by

$$k_{opt} = \sqrt{\frac{N}{2\pi}} \sqrt{\frac{\beta_s}{\alpha_s}} \frac{M}{d_{toSink}^2} \qquad (19)$$

where d_{toSink} is the distance from the cluster head to the sink. In LEACH, the cluster heads send data directly to the sink, which is either infeasible for large networks or leads to a very high energy usage by the cluster heads located far from the sink. Moreover, LEACH assumes perfect data aggregation, *i.e.*, the data coming from all nodes assigned to a cluster head is always compressed to a single message, irrespective of the number of neighboring nodes, which is not realistic.

Power-Efficient and Adaptive Clustering Hierarchy (PEACH)

The main idea of PEACH (Yi et al. 2007) is an adaptive cluster formation using the overheard information by each node. In PEACH, a node overhears and recognizes the source and destination of packets transmitted by its neighboring nodes. Based on this overheard information, PEACH forms clusters without any additional messaging overhead, and hence avoids the energy wastage due to advertisement, joining and scheduling message overheads that incur in clustering protocols like LEACH (Heinzelman, Chandrakasan & Balakrishnan 2002) and HEED (Younis & Fahmy 2004).

Figure 12 shows how PEACH forms clusters. *NodeSet*(A, B) is the set of all nodes that can overhear a packet sent by node A to node B with a transmission range equals to the distance between A and B. *ClusterSet*(A, B) is the set of all nodes that can overhear transmission from A to B and their distance from the sink is more than the distance between B and the sink. For example, in Figure 12, the *NodeSet*(A, B) is $\{A, B, C, D, E, F, G\}$ and the *ClusterSet*(A, B) is $\{A, C, D, E\}$. Then node B becomes the cluster head of the *ClusterSet*(A, B). Cluster head B sets a timer with duration T_{delay} to gather multiple packets from the nodes in the *ClusterSet*(A, B), aggregates the received data and sends the aggregated data to the next hop.

The Heterogeneous Node Clustering

In this approach, a network consists of two types of nodes: i) a small number of nodes possessing higher resources and become the natural candidate to act as the cluster heads and ii) a large number of resource-constraint nodes. Since there is no need for the cluster head selection and role rotation, here the problem of clustering reduces to finding an optimal assignment of nodes to the cluster heads. This reduces the communication overhead significantly compared to homogeneous node clustering. Moreover, in this approach, with the expense of few costly cluster heads, the extra hardware and software complexity can be reduced for large number of nodes. In the following, we review two notable works on heterogeneous node clustering.

Energy-Efficient Cluster Head Assignment Scheme

The study in (Wang 2008) addresses the problem of minimizing the total energy consumption across the network by adequately assigning cluster heads in a hierarchical network. This technique finds how many cluster heads are needed and where they should be positioned. The authors assume

Figure 12. An example of packet transmission and overhearing in PEACH

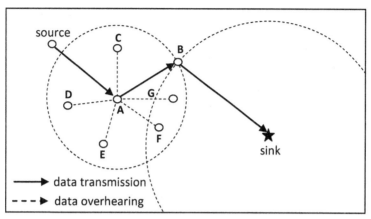

that the whole network is divided into square grids with each cell containing a node, the nodes are organized into equal-sized square-shaped clusters and the sink is located at the centre of the network as shown in Figure 13. Here, a node can only communicate with its four neighboring nodes while a cluster head can communicate with any nodes in the cluster and only guarantees reliable communication with its four neighboring cluster heads. For example, in Figure 13 node 's' can communicate with nodes s_1, s_2, s_3, and s_4 while cluster head 'h' can communicate with cluster heads h_1, h_2, h_3, and h_4. In their analysis, the authors used a simplified abstraction, where the data flow is considered from the sink to all the nodes. The authors obtained an optimal cluster size that minimizes the total energy consumption of all nodes in the network for the proposed data model. For an $N \times N$ cell network, the network is divided into $\left(\dfrac{N}{a}\right)^2$ number of clusters of dimension $a \times a$ and the system's total energy is minimized if

$$a = \sqrt[3]{2N}. \tag{20}$$

Figure 13. A clustered hierarchical sensor network

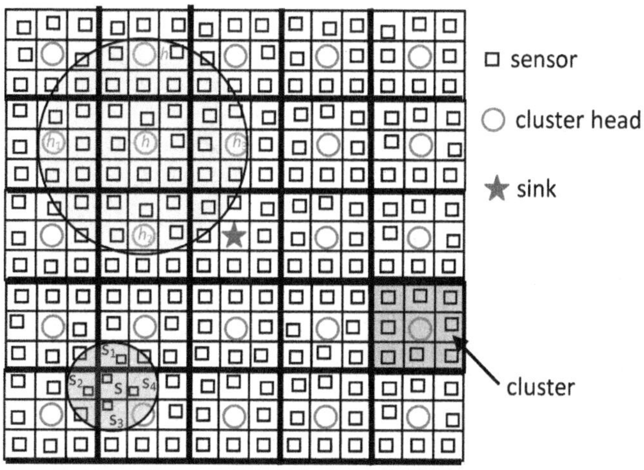

This work minimizes the total energy consumption across the network in one round in which the sink transmits a unit amount of data to all nodes. In a heterogeneous clustered WSN, minimizing the total energy does not guarantee network lifetime maximization, since during multi hop intra-cluster and inter-cluster data transmissions a few nodes near each cluster head and a few cluster heads near the sink consume energy at the highest rate and become the bottlenecks for the overall network lifetime. To maximize the network lifetime, optimization should focus on minimizing the energy consumption by those critical nodes. In addition, the authors assume a very rigid network structure where all clusters are of square-shapes and equal sizes, the cluster heads are at the center of the clusters, and the sink is at the center of the network region, which does not reflect the real application scenario. Finally, the data flow considered in this work from sink to all nodes is exactly the opposite of the real data gathering.

Minimum Cost Heterogeneous Sensor Network

In (Mahtre et al. 2005), Mhatre *et al.* studied a heterogeneous network consisting of two types of nodes: type-1 cluster heads and type-0 rfid/sensor nodes. The whole network is divided into Voronoi cell like polygonal clusters around the cluster heads and each cluster is approximated by a circular region with the cluster head at its centre and the cluster sizes are considered uniform. A typical cluster is shown in Figure 14. The type-0 nodes use multi hop communication to send their data to the cluster head. The incoming traffic from the outer nodes is assumed to be uniformly distributed among the nodes located one hop away from the cluster head. The authors consider that, a surveillance aircraft (sink) flying at an altitude of H sweeps the area periodically and triggers a data cycle during which each node sends its data to the nearest cluster head, the cluster heads then ag-

gregate the received data and send it to the aircraft using direct transmission. In their Equationtion the authors assume perfect data aggregation, *i.e.*, at each cycle a cluster head performs data fusion of the packets received from all nodes in its vicinity and generates only a single packet to transmit to the sink. The authors Equationte a model for the overall hardware cost for network deployment as a function of the node (node and cluster head) intensities and energies. Through mathematical analysis, the authors found the following relations as in (21) and (22) among the node intensities and node energies that minimize the network deployment cost and guarantee a minimum lifetime of T data cycles.

$$\rho_h = \frac{t_s^{\frac{\gamma-2}{2}}\sqrt{\rho_s}}{\sqrt{\frac{\pi}{4}\left(H^\gamma + \frac{c_1 - c_0}{cT\beta_s}\right)}} \qquad (21)$$

$$E_s : E_h = \frac{\alpha_s + \beta_s t_s^\gamma + (2\alpha_s + \beta_s r^\gamma)\left(\frac{e^{-\rho_h \pi t_s^2}}{1 - e^{-\rho_h \pi t_s^2}}\right)}{\alpha_s + \beta_s H^\gamma + (\alpha_s + \xi_a)\left(\frac{\rho_s}{\rho_h}\right)} \qquad (22)$$

Figure 14. A typical cluster and its approximation by a circular region

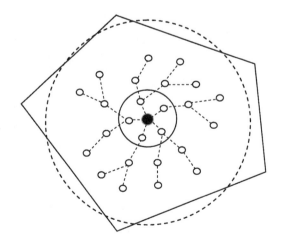

where ρ_s and ρ_h are the deployment intensities, E_s and E_h are the initial energy storage for node and cluster head, respectively, t_s is the transmission range of node, c_i is the hardware cost for a type-i node, c is the proportionality constant for the battery cost, α_s is the per packet energy spent at the transceiver circuit, β_s is the energy spent in the RF amplifier to transmit data and ξ_a is the energy consumption per packet of data for processing and compressing during data aggregation.

CONCLUSION

Radio Frequency Identification (RFID) systems and Wireless Sensor Networks (WSNs) are two emerging technologies. RFID technology provides cheaper solution for object identification and tracking wirelessly based on radio wave and using WSNs data on various parameters about the physical environment can be acquired. In this chapter, firstly, we discuss the fundamental principles behind the RFID systems and wireless sensor networks (WSNs) and thereby compare their advantages and disadvantages. Secondly, we present possible ways of integration of the RFID and sensors so that strong aspects of both the systems can be converged. Then we discuss how active RFID tags or reader (reading passive or semi-passive tags) are similar to that of sensor nodes in WSNs and therefore many WSNs protocols can be used in RFID systems. Finally, we present WSNs protocols such as node activation cycling, medium access control (MAC), routing, transmission policy and clustering techniques, which can be used in RFID systems.

Node activation cycling would reduce the number of 'ON' tags/readers at a given instance and the effective duty cycle of tags/readers, and hence the collision during tag reading and/or reader to base station communications and the energy consumption at the tags and/or readers would be reduced. MAC protocol provides the functionality of accessing the shared medium to the nodes in a network. In an RFID system, implementing MAC protocol at active tags and reader for shared channel access during tag reading and at readers and base station for shared channel access between them would avoid collision, overhearing, idle listening, control-packet overhead, and overemitting significantly. In RFID system, active tags directly communicate with the readers. While readers interrogate passive tags and the tags respond directly to the reader. Although the reading process for passive tags shall remain single hop, multi hop communication is possible during reading active tags and for RFID readers to base station communications. Adoption of multi hop routing protocols at the network layer of RFID system would increase the effective reading zone for RFID reader and would improve the scalability of the overall system.

Transmission policy would regulate the transmission ranges of active tags and readers to address the unbalanced traffic flow near the readers and base station due to the adoption of multi hop routing in RFID network. For large application, instead of considering as a flat network, dividing the whole network into a number of clusters improves the scalability of the overall system. In RFID system, RFID readers would be considered as the cluster head and the whole RFID network would be divided into a number of non-overlapped cluster regions. In addition to scalability, clustering has numerous advantages. It can localize the route setup within the cluster and thus reduce the size of the routing table. Clustering enables bandwidth reuse and thus increases the system capacity. It also reduces the communication overheads, thereby reducing the energy consumption and interferences among the nodes.

Despite distinct characteristics of RFID system and wireless sensor networks (WSNs), in this chapter we present how a number of WSNs protocols can be adopted in RFID systems and the advantages of such espousal. In our future study, we will focus on identifying the challenges of adopting various WSNs protocols in RFID systems and their solutions in detail.

REFERENCES

Abowd, G. D. (2005). Prototypes and paratypes: Designing mobile and ubiquitous computing applications. *IEEE Pervasive Computing / IEEE Computer Society and IEEE Communications Society, 4*(4), 67–73. doi:10.1109/MPRV.2005.83

Akkaya, K., & Younis, M. (2005). A survey on routing protocols for wireless sensor networks. *Ad Hoc Networks, 3*, 325–349. doi:10.1016/j.adhoc.2003.09.010

Azad, A. K. M., & Kamruzzaman, J. (2011). Energy balanced transmission policies for wireless sensor networks. *IEEE Transactions on Mobile Computing, 10*(7). doi:10.1109/TMC.2010.238

Azad, A. P., & Chockalingam, A. (2006). Mobile base stations placement and energy aware routing in wireless sensor networks. *Proceedings of the IEEE WCNC, 2006*, 264–268.

Balbin, I., & Karmakar, N. (2009). RFID tag for conveyor belt tracking using multi-resonant dipole antenna, *European Microwave Conference*, Rome, Italy, 2009, 1109-1112.

Bhardwaj, M., Garnett, T., & Chandrakasan, A. P. (2001). Upper bounds on lifetime of sensor networks. *Proceedings of the IEEE, ICC*, 785–790.

Bhuiyan, M. S. (2011). *PhD confirmation report: Electrical and Computer Systems Engineering*. Australia: Monash University.

Braginsky, D., & Estrin, D. (2002). Rumor routing algorithm for sensor networks. *Proceedings of the Sensor Networks and Apps, 2002*, 22–31.

Cao, Q. (2005). Toward optimal sleep scheduling in sensor networks for rare event detection. *Proceedings of the IPSN, 2005*, 21–27.

Cardei, M., Pervaiz, M. O., & Cardei, I. (2006). Energy-efficient range assignment in heterogeneous wireless sensor networks. *Proceedings of the International Conference on Wireless and Mobile Communications, 2006*, (pp. 11-16).

Casari, P. (2005). A detailed simulation study of geographic random forwarding (GeRaF) in wireless sensor networks. *Proceedings of the MILCOM, 2005*, 59–68.

Cerpa, A., & Estrin, D. (2004). Ascent: adaptive self-configuring sensor network topologies. *IEEE Transactions on Mobile Computing, 3*(3), 272–285. doi:10.1109/TMC.2004.16

Chawla, M. (2006). Beaconless position based routing with guaranteed delivery for wireless ad-hoc and sensor networks. *FIP International Federation for Information Processing, 2006*, 61–70. doi:10.1007/978-0-387-34738-7_5

Chawla, V., & Sam, H. D. (2007). An overview of passive RFID. *IEEE Communications Magazine, 45*, 11–17. doi:10.1109/MCOM.2007.4342873

Chen, B. (2002). Span: An energy efficient coordination algorithm for topology maintenance in ad hoc wireless networks. *ACM Wireless Networks, 8*(5), 481–494. doi:10.1023/A:1016542229220

Cho, N., et al. (2005). *A 5.1-uW UHF RFID tag chip integrated with sensors for wireless environmental monitoring.* European Solid-State Circuits Conf. (ESSCIRC), Grenoble, France, Sept. 2005.

Culler, D., Estrin, D., & Srivastava, M. (2004). Overview of sensor networks. *IEEE Computer, 37*(8), 41–49. doi:10.1109/MC.2004.93

Dam, T. V., & Langendoen, K. (2003). An adaptive energy-efficient MAC protocol for wireless sensor networks. *Proceedings of SenSys, 2003*, 171–180.

Demirbas, M., Xuming, L., & Singla, P. (2009). An in-network querying framework for wireless sensor networks. *IEEE Transactions on Parallel and Distributed Systems, 20*(8), 1202–1215. doi:10.1109/TPDS.2008.217

Efthymiou, C., Nikoletseas, S., & Rolim, J. (2002). The cougar approach to innetwork query processing in sensor networks. *SIGMOD, 31*(3), 9–18. doi:10.1145/601858.601861

Faizulkhakov, Y. R. (2007). Time synchronization methods for wireless sensor networks: A survey. *Programming and Computer Software, 33*(4), 214–226. doi:10.1134/S0361768807040044

Felemban, E., Lee, C.-G., & Ekici, E. (2006). MMSPEED: Multipath multi-SPEED protocol for QoS guarantee of reliability and timeliness in wireless sensor networks. *IEEE Transactions on Mobile Computing, 5*(6), 738–754. doi:10.1109/TMC.2006.79

Gao, Q. (2006). Radio range adjustment for energy efficient wireless sensor networks. *Ad Hoc Networks, 4*, 75–82. doi:10.1016/j.adhoc.2004.04.007

Heidemann, W., Ye, J., & Estrin, D. (2004). Medium access control with coordinated adaptive sleeping for wireless sensor networks. *IEEE/ACM Transactions on Networking, 12*(3), 493–506. doi:10.1109/TNET.2004.828953

Heinzelman, W., Kulik, J., & Balakrishnan, H. (1999). Adaptive protocols for information dissemination in wireless sensor networks. *Proceedings 5th ACM/IEEE Mobicom*, 1999, (pp. 174–85).

Heinzelman, W. R., Chandrakasan, A., & Balakrishnan, H. (2002). An application-specific protocol architecture for wireless microsensor networks. *IEEE Transactions on Wireless Communications, 1*(4), 660–670. doi:10.1109/TWC.2002.804190

Hohlt, B., Doherty, L., & Brewer, E. (2004). Flexible power scheduling for sensor networks. *Proceedings of the ISPN, 2004*, 205–214.

Hou, Y. T., Shi, Y., & Sherali, H. D. (2006). Optimal base station selection for anycast routing in wireless sensor networks. *IEEE Transactions on Vehicular Technology, 55*(3), 813–821. doi:10.1109/TVT.2006.873822

Intanagonwiwat, C., Govindan, R., & Estrin, D. (2000). Directed diffusion: A scalable and robust communication paradigm for sensor networks. *Proceedings of the ACM MobiCom*, 2000, (pp. 56–67).

ITU. (2005). The Internet of things. *ITU Internet Reports*, 2005. Retrieved from http://www.itu.int/osg/spu/publications/internetofthings/

Karmakar, N. C., et al. (2006). Development of low-cost active RFID tag at 2.4 GHz. *36th European Microwave Conference*, 2006, (pp. 1602-1605).

Karp, B., & Kung, H. T. (2000). GPSR: Greedy perimeter stateless routing for wireless sensor networks. *Proceedings of the MobiCom, 2000*, 243–254. doi:10.1145/345910.345953

Kirousis, L. (2000). Power consumption in packet radio networks. *Theoretical Computer Science, 243*(1-2), 289–305. doi:10.1016/S0304-3975(98)00223-0

Kitayoshi, H., & Sawaya, K. (2005). *Long range passive RFID-tag for sensor networks*. IEEE Vehicular Technology Conference.

Kulik, J., Heinzelman, W. R., & Balakrishnan, H. (2002). Negotiation-based protocols for disseminating information in wireless sensor networks. *Wireless Networks, 8*, 169–185. doi:10.1023/A:1013715909417

Lee, Y. W., Lee, K. Y., & Kim, M. H. (2009). Energy-efficient multiple query optimization for wireless sensor networks. *Proceedings of the SENSORCOMM, 2009*, 531–538.

Lin, P., Qiao, C., & Wang, X. (2004). Medium access control with a dynamic duty cycle for sensor networks. *Proceedings of the WCNC, 2004*, 1534–1539.

Linlin, Z., et al. (2008). Design and implementation of a fully reconfigurable chipless RFID tag using Inkjet printing technology. *IEEE International Symposium on Circuits and Systems*, USA, 2008, (pp. 1524-1527).

Liu, H. (2008). Taxonomy and challenges of the integration of RFID and wireless sensor networks. *IEEE Network, 22*(6), 26–35. doi:10.1109/MNET.2008.4694171

Liu, H., Wan, P., & Jia, X. (2007). Maximal lifetime scheduling for sensor surveillance systems with K sensors to one target. *IEEE Transactions on Parallel and Distributed Systems, 15*(2), 334–345.

Liu, S. (2006). A lifetime-extending deployment strategy for multi-hop wireless sensor networks. *IEEE Proceedings of the Communication Networks and Services Research Conference*, May 2006, (pp. 1-8).

Mahfoudh, S., & Minet, P. (2008).Survey of energy efficient strategies in wireless ad hoc and sensor networks. *Proceedings of the International Conference on Networking*, 2008, (pp. 1-7).

Mahtre, V. (2005). A minimum cost heterogeneous sensor network with a lifetime constraint. *IEEE Transactions on Mobile Computing, 4*(1), 4–15. doi:10.1109/TMC.2005.2

Maleki, M., & Pedram, M. (2005). QoM and lifetime-constrained random deployment of sensor networks for minimum energy consumption. *IEEE Proceedings of Information Processing in Sensor Networks, 2005*, 293–300.

Mhatre, V., & Rosenberg, C. (2004). Homogeneous vs heterogeneous clustered sensor networks: A comparative study. *IEEE ICC*, 2004, (pp. 646-651).

Mhatre, V., & Rosenberg, C. (2004). Design guidelines for wireless sensor networks: communication, clustering and aggregation. *Ad Hoc Networks, 2*, 45–63. doi:10.1016/S1570-8705(03)00047-7

Nikitin, P. V., & Rao, K. V. S. (2008). Antennas and propagation in UHF RFID systems. *IEEE International Conference on RFID*, Las Vegas, NV, 2008, (pp. 277-288).

Nikitin, P. V., Rao, K. V. S., & Lazar, S. (2007). An overview of near field UHF RFID. *IEEE International Conference on RFID*, 2007, (pp. 167-174).

Pan, J. (2005). Optimal base-station locations in two-tiered wireless sensor networks. *IEEE Transactions on Mobile Computing, 4*(5), 458–473. doi:10.1109/TMC.2005.68

Paruchuri, V., et al. (2004). Random asynchronous wakeup protocol for sensor networks. *Proceedings of the IEEE International Conference on Broadband Networks*, 2004, (pp. 710-717).

Philipose, M. (2005). Battery-free wireless identification and sensing. *IEEE Pervasive Computing / IEEE Computer Society and IEEE Communications Society, 4*(1), 37–45. doi:10.1109/MPRV.2005.7

Polastre, J., Hill, J., & Culler, D. (2004). Versatile low power media access for wireless sensor networks. *Proceedings of the Sensys, 2004*, 95–107. doi:10.1145/1031495.1031508

Rajendran, V., Garcia-Luna-Aceves, J., & Obraczka, K. (2005). Efficient application-aware medium access for sensor networks. *Proceedings of the MASS, 2005*, 630–637.

Rajendran, V., Obraczka, K., & Garcia-Luna-Aceves, J. J. (2006). Energy efficient, collision-free medium access control for wireless sensor networks. *Wireless Networks, 12*(1), 63–78. doi:10.1007/s11276-006-6151-z

Ramamurthy, H. (2007). Wireless industrial monitoring and control using a smart sensor platform. *IEEE Sensors Journal, 7*(5), 611–618. doi:10.1109/JSEN.2007.894135

Rhee, I. (2008). Z-MAC: A hybrid MAC for wireless sensor networks. *IEEE/ACM Transactions on Networking, 16*(3), 511–524. doi:10.1109/TNET.2007.900704

Sanming, H. (2010). Study of a uniplanar monopole antenna for passive chipless UWB-RFID localization system. *IEEE Transactions on Antennas and Propagation, 58*, 271–278. doi:10.1109/TAP.2009.2037760

Schurgers, C., & Srivastava, M. B. (2001). Energy efficient routing in wireless sensor networks. *Proceedings of the IEEE MILCOM, 2001*, 357–361.

Schurgers, C., Tsiatsis, V., & Srivastava, M. B. (2002). STEM: topology management for energy efficient sensor networks. *IEEE Aerospace Conference, 2002*, (pp. 1099-1108).

Sohrabi, K., & Pottie, J. (2000). Protocols for self-organization of a wireless sensor network. *IEEE Personal Communication, 7*(5), 16–27.

Tseng, Y. C., Pan, M. S., & Tsai, Y. Y. (2006). Wireless sensor networks for emergency navigation. *IEEE Computer, 39*(7), 55–62. doi:10.1109/MC.2006.248

Wang, D. (2008). An energy-efficient clusterhead assignment scheme for hierarchical wireless sensor networks. *International Journal of Wireless Information Networks, 15*(2), 61–71. doi:10.1007/s10776-008-0079-4

Wang, Q. (2008). Lightning: a hard real-time, fast, and lightweight low-end wireless sensor election protocol for acoustic event localization. *IEEE Transactions on Mobile Computing, 7*(5), 570–584. doi:10.1109/TMC.2007.70752

Wenyi, C., et al. (2009). Analysis, design and implementation of semi-passive Gen2 tag. *IEEE International Conference on RFID, 2009*, (pp. 15-19).

Yang, X., & Vaidya, N. (2004). A wakeup scheme for sensor networks: achieving balance between energy saving and end-to-end delay. *Proceedings of the IEEE Real-Time and Embedded Technology and Applications Symposium, 2004*, (pp. 19-26).

Yao, Y., & Gehrke, J. (2002). The cougar approach to in-network query processing in sensor networks. *SIGMOD, 31*(3), 9–18. doi:10.1145/601858.601861

Yi, S. (2007). PEACH: Power-efficient and adaptive clustering hierarchy protocol for wireless sensor networks. *Computer Communications, 30*(14-15), 2841–2852. doi:10.1016/j.comcom.2007.05.034

Younis, O., & Fahmy, S. (2004). HEED: A hybrid, energy-efficient distributed clustering approach for ad hoc sensor networks. *IEEE Transactions on Mobile Computing, 3*(4), 366–379. doi:10.1109/TMC.2004.41

Yu, Y., Estrin, D., & Govindan, R. (2001). *Geographical and energy-aware routing: A recursive data dissemination protocol for wireless sensor networks.* UCLA Comp. Sci. Dept. tech. rep., 2001.

Zabin, F. (2008). REEP: data-centric, energy-efficient and reliable routing protocol for wireless sensor networks. *IET Communication, 2008*, 995–1008. doi:10.1049/iet-com:20070424

Zhang, H., & Shen, H. (2011). Energy-efficient beaconless geographic routing in wireless sensor networks. *IEEE Transactions on Parallel and Distributed Systems, 21*(6), 881–896. doi:10.1109/TPDS.2009.98

Zhang, Z., Ma, M., & Yang, Y. (2004). Energy-Efficient multihop polling in clusters of two-layered heterogeneous sensor networks. *IEEE Transactions on Computers, 57*(2), 231–245. doi:10.1109/TC.2007.70774

Zheng, R., Hou, J., & Sha, L. (2003). Asynchronous wakeup for ad hoc networks. *Proc. of the ACM MobiHoc*, 2003, (pp. 35–45).

Zorzi, M., & Rao, R. R. (2003). Geographic random forwarding (GeRaF) for ad hoc and sensor networks: multihop performance. *IEEE Transactions on Mobile Computing, 2*(4), 337–348. doi:10.1109/TMC.2003.1255648

KEY TERMS AND DEFINITIONS

Chipless RFID: Chipless RFID tags do not use any memory chip to store data bits; rather it uses the resonance or reflection properties of microwave signals to encode bits.

Clustering: Dividing the whole WSN into a number of non-overlapped regions, namely, clusters, where each cluster has a leader called the *cluster head*. Each sensor is assigned to a cluster head to which it sends its data. Each cluster head aggregates the data received from nodes in its vicinity and send the aggregated data to the sink.

Heterogeneous Clustering: Here some designated nodes, equipped with significantly more resources (*e.g.*, energy, computation power, communication range, etc.) compared to the sensor nodes are used as cluster heads.

Heterogeneous WSNs: There are a number of different classes of sensor nodes which differ in terms of parameters like sensing range, data generation rate, energy storage, transmission range, etc.

Homogeneous Clustering: This clustering technique is applied to networks in which all the nodes are identical and a number of nodes are selected as the cluster heads. Moreover, in homogeneous clustering a periodic role rotation among the nodes as cluster heads is done to achieve balanced energy usage among the nodes.

Homogeneous WSNs: WSNs with all nodes (except the sink) are of the same type.

RFID: Radio Frequency Identification (RFID) is a technology that transmits the identity of an object at a short distance without manual intervention through wireless radio communication to an electronic reader.

Scalability: The property of repetitively extending architecture from smaller to larger application scopes.

Sensor: Small micro-electromechanical device to sense some physical parameters from the surrounding environment and send back to a central processing station. It consists of three basic components: a sensing module for data acquisition from the surrounding environment, a processing module for local data processing, and a wireless communication module for data transmission and reception.

Sensor-Tag: The RFID tags with sensors that use the RFID protocols and mechanisms for reading tag IDs, as well as for collecting sensed data.

Wireless Sensor Network (WSN): WSN is a collection of many tiny and low-cost sensors, usually, deployed in an ad hoc fashion that performs distributed sensing tasks in a collaborative manner without relying on any underlying infrastructure support.

Compilation of References

Abbak, M., & Tekin, I. (2009). RFID coverage extension using microstrip patch antenna array. *IEEE Antennas and Propagation Magazine*, *51*(1), 185–191. doi:10.1109/MAP.2009.4939065

Abboud, F., Damiano, J. P., & Papiernik, A. (1988). Simple model for the input impedance of coax-fed rectangular microstrip patch antenna for CAD. *IEE Proceedings. Microwaves, Antennas and Propagation*, *135*(5), 323–326. doi:10.1049/ip-h-2.1988.0066

Abo-Elnaga, T. G., Abdallah, E. A. F., & El-Hennawy, H. (2010a). *Universal UHF RFID rose reader antenna.* China: PIERS Proceedings.

Abo-Elnaga, T. G., Abdallah, E. A. F., & El-Hennawy, H. (2010b). *UWB circular polarization RFID reader antenna for 2.4GHz Band.* China: PIERS Proceedings.

Abou-Rjeily, C., & Fawaz, W. (2008). Space-time codes for MIMO ultra-wideband communications and MIMO free-space optical communications with PPM. *IEEE Journal on Selected Areas in Communications*, *26*(6). doi:10.1109/JSAC.2008.080810

Abowd, G. D. (2005). Prototypes and paratypes: Designing mobile and ubiquitous computing applications. *IEEE Pervasive Computing / IEEE Computer Society and IEEE Communications Society*, *4*(4), 67–73. doi:10.1109/MPRV.2005.83

Adachi, F. (2001). Wireless past and future – Evolving mobile communications systems. *IEICE Transactions in Fundamentals. E (Norwalk, Conn.)*, *84-A*(1).

Ahmed, R., Karmakar, G. C., & Dooley, L. S. (2007). Probabilistic spatio-temporal video object segmentation using a priori shape descriptor. *IEEE International Conference on Acoustics, Speech and Signal Processing (ICASSP-07)*, 1, (pp. 15-20).

Akkaya, K., & Younis, M. (2005). A survey on routing protocols for wireless sensor networks. *Ad Hoc Networks*, *3*, 325–349. doi:10.1016/j.adhoc.2003.09.010

Alamouti, S. M. (1998). A simple transmit diversity technique for wireless communications. *IEEE Journal on Selected Areas in Communications*, *16*, 1451–1458. doi:10.1109/49.730453

Ali, J. K. (2008). A new compact size microstrip patch antenna with irregular slots for handheld GPS application. *Engineering & Technology*, *26*, 1241–1246.

Along, K., Chenrui, Z., Luo, Z., Xiaozheng, L., & Tao, H. (2010, 17-19 June 2010). *SAW RFID enabled multifunctional sensors for food safety applications.* In The IEEE International Conference on RFID-Technology and Applications (RFID-TA), 2010.

Angerer, C., & Markus, R. (2009). Advanced synchronisation and decoding in RFID reader receivers. *IEEE Radio and Wireless Symposium, RWS '09*, (pp. 59-62).

Angerer, C., Langwieser, R., & Rupp, M. (2010). Experimental performance evaluation of dual antenna diversity receivers for RFID readers. *Proceedings of the 3rd International EURASIP Workshop on RFID Technology.*

Angerer, C., Langwieser, R., Maier, G., & Rupp, M. (2009). Maximal ratio combining receivers for dual antenna RFID readers. In *Proceedings of International Microwave Workshop on Wireless Sensing, Local Positioning, and RFID.*

Avila-Navarro, E., Cayuelas, C., & Reig, C. (2010). Dual-band printed dipole antenna for Wi-Fi 802.11n applications. *Electronics Letters*, *46*(21), 1421–1422. doi:10.1049/el.2010.2000

Azad, A. K. M., & Kamruzzaman, J. (2011). Energy balanced transmission policies for wireless sensor networks. *IEEE Transactions on Mobile Computing*, *10*(7). doi:10.1109/TMC.2010.238

Azad, A. P., & Chockalingam, A. (2006). Mobile base stations placement and energy aware routing in wireless sensor networks. *Proceedings of the IEEE WCNC*, *2006*, 264–268.

Baki, A. K. M., Shinohara, N., Matsumoto, H., Hashimoto, K., & Mitani, T. (2007, April). Study of isosceles trapezoidal edge tapered phased array antenna for solar power station/satellite. *IEICE Transactions in Communication. E (Norwalk, Conn.)*, *90-B*(4), 968–977.

Baki, A. K. M., Shinohara, N., Matsumoto, H., Hashimoto, K., & Mitani, T. (2008, February). Isosceles-trapezoidal-distribution edge tapered array antenna with unequal element spacing for solar power station/satellite. *IEICE Transactions in Communication. E (Norwalk, Conn.)*, *91-B*(2), 527–535.

Balanis, C. A. (1988). *Antennas: Theory, design and analysis* (2nd ed.). New York, NY: Wiley.

Balanis, C. A. (2005). *Antenna theory* (3rd ed.). Hoboken, NJ: John Wiley.

Balanis, C. A. (Ed.). (2008). *Modern antenna handbook*. New York, NY: John Wiley & Sons. doi:10.1002/9780470294154

Balbin, I., & Karmakar, N. (2009). RFID tag for conveyor belt tracking using multi-resonant dipole antenna, *European Microwave Conference*, Rome, Italy, 2009, 1109-1112.

Balbin, I., & Karmakar, N. C. (2009, August). Phase-encoded chipless RFID transponder for large-scale low-cost applications. *IEEE Microwave and Wireless Components Letters*, *19*(8).

Balbin, I., & Karmakar, N. C. (2009, September). *Novel chipless RFID tag for conveyer belt tracking using multi-resonant dipole antenna*. In The 39th European Microwave Conference, Rome, Italy.

Balbin, I., & Karmakar, N. C. (2009). Phase-encoded chipless RFID transponder for large-scale low-cost applications. *IEEE Microwave and Wireless Component Letters*, *19*(8), 509–511. doi:10.1109/LMWC.2009.2024840

Barbin, M. V., & Barbin, S. E. (2010). Antenna design for a portable RFID reader. *PIERS Proceedings*, USA.

Baro, S., Bauch, G., & Hansmann, A. (2000). Improved codes for space-time trellis coded modulation. *IEEE Communications Letters*, *4*(1), 20–22. doi:10.1109/4234.823537

Barthel, H. (2009). *Regulatory status for RFID in the UHF spectrum*. Belgium: EPC Global.

Bashir, F. I., Khokhar, A. A., & Schonfeld, D. (2007). Object trajectory-based activity classification and recognition using hidden Markov models. *IEEE Transactions on Image Processing*, *16*(7), 1912–1919. doi:10.1109/TIP.2007.898960

Baxes, G. A. (1994). *Digital image processing: Principles and applications*. New York, NY: John Wiley & Sons, Inc.

Bechevet, D., Tan-Phu, V., & Tedjini, S. (2005, 3-8 July). *Design and measurements of antennas for RFID, made by conductive ink on plastics*. In The IEEE Antennas and Propagation Society International Symposium, 2005.

Behzad, A. R., Shi, Z. M., Anand, S. B., Li, L., Carter, K. A., & Kappes, M. S. (2003). A 5-GHz direct-conversion CMOS transceiver utilizing automatic frequency control for the IEEE 802.11a wireless LAN standard. *IEEE Journal of Solid-state Circuits*, *38*(12), 2209–2220. doi:10.1109/JSSC.2003.819085

Bhagat, M., McFiggins, J., & Venkataraman, J. (2002, October). *Chip package co-design of the RF front end with an integrated antenna on multilayered organic material*. Paper presented at the IEEE Conference on Electrical Performance of Electronic Packaging and Systems, Monterey, Calif.

Bhardwaj, M., Garnett, T., & Chandrakasan, A. P. (2001). Upper bounds on lifetime of sensor networks. *Proceedings of the IEEE*, *ICC*, 785–790.

Bhuiyan, M. S. (2011). *PhD confirmation report: Electrical and Computer Systems Engineering*. Australia: Monash University.

Bialkowski, M. E., & Karmakar, N. C. (1999). Design of compact L-band 180° phase shifters. *Microwave and Optical Technology Letters, 22*(2), 144–148. doi:10.1002/(SICI)1098-2760(19990720)22:2<144::AID-MOP19>3.0.CO;2-D

Blayo, A., & Pineaux, B. (2005). Printing processes and their potential for RFID printing. In The *Proceedings of the 2005 Joint Conference on Smart Objects and Ambient Intelligence: Innovative Context-Aware Services: Usages and Technologies.*

Blischak, A., & Manteghi, M. (2009, June). *Pole residue techniques for chipless RFID detection.* IEEE Antennas and Propagation Society International Symposium, Charleston.

Bolic, M., Simplot-Ryl, D., & Stojmenovic, I. (2010). *RFID systems research trends and challenges.* John Wiley & Sons Ltd. doi:10.1002/9780470665251

Borowiec, R. (2010, 14-16 June 2010). *A printed RFID Tag antenna for metallic objects operating in UHF band.* In The 18th International Conference on Microwave Radar and Wireless Communications (MIKON), 2010. *Cabot Corp.* Retrieved 24 August, 2011, from http://www.cabot-corp.com/New-Product-Development/Printed-Electronics/Products

Boyd, S. (2007). *Lecture 16: SVD appications.* Retrieved from http://see.stanford.edu/materials/lsoeldsee263/16-svd.pdf

Braginsky, D., & Estrin, D. (2002). Rumor routing algorithm for sensor networks. *Proceedings of the Sensor Networks and Apps, 2002*, 22–31.

Brandl, M., Schuster, S., Scheiblhofer, S., & Stelzer, A. (2008). *A new anti-collision method for SAW tags using linear block codes.* IEEE International Frequency Control Symposium.

Brebels, S., Ryckaert, J., Come, B., Donnay, S., Walter De, R., & Beyne, E. (2004). SOP integration and codesign of antennas. *IEEE Transactions on Advanced Packaging, 27*(2), 341–351. doi:10.1109/TADVP.2004.828822

Brown, W. C., & Eugene, E. E. (1992, June). Beamed microwave power transmission and its application to space. *IEEE Transactions on Microwave Theory and Techniques, 40*(6), 1239–1250. doi:10.1109/22.141357

Cangialosi, A., Monaly, J. E., & Yang, S. C. (2007). Leveraging RFID in hospitals: Patient life cycle and mobility perspectives. *IEEE Communications Magazine, 45*, 18–23. doi:10.1109/MCOM.2007.4342874

Cao, Q. (2005). Toward optimal sleep scheduling in sensor networks for rare event detection. *Proceedings of the IPSN, 2005*, 21–27.

Cardei, M., Pervaiz, M. O., & Cardei, I. (2006). Energy-efficient range assignment in heterogeneous wireless sensor networks. *Proceedings of the International Conference on Wireless and Mobile Communications, 2006*, (pp. 11-16).

Casari, P. (2005). A detailed simulation study of geographic random forwarding (GeRaF) in wireless sensor networks. *Proceedings of the MILCOM, 2005*, 59–68.

Cerpa, A., & Estrin, D. (2004). Ascent: adaptive self-configuring sensor network topologies. *IEEE Transactions on Mobile Computing, 3*(3), 272–285. doi:10.1109/TMC.2004.16

Chai, J., & Shum, H. (2000). Parallel projections for stereo reconstruction. *IEEE Conference on Computer Vision and Pattern Recognition, 2*, (pp. 493-500).

Chamarti, A., & Varahramyan, K. (2006). Transmission delay line based ID generation circuit for RFID applications. *IEEE Microwave and Wireless Component Letters, 16*(11), 588–590. doi:10.1109/LMWC.2006.884897

Chamarti, A., & Varahramyan, K. (2006). Transmission delay line based ID generation circuit for RFID applications. *IEEE Microwave and Wireless Components Letters, 16*(11), 588–590. doi:10.1109/LMWC.2006.884897

Chan, K. T., Chin, A., Chen, Y. B., Lin, Y. D., Duh, T. S., & Lin, W. J. (2001, December). *Integrated antennas on Si, proton-implanted Si and Si-on-quartz.* Paper presented at the IEEE International Electron Devices Meeting, Washington, DC.

Chawla, M. (2006). Beaconless position based routing with guaranteed delivery for wireless ad-hoc and sensor networks. *FIP International Federation for Information Processing, 2006*, 61–70. doi:10.1007/978-0-387-34738-7_5

Chawla, V., & Sam, H. D. (2007). An overview of passive RFID. *IEEE Communications Magazine, 45*, 11–17. doi:10.1109/MCOM.2007.4342873

Chen, D., Yang, J., & Wactlar, H. D. (2004). Towards automatic analysis of social interaction patterns in a nursing home environment from video. *ACM SIGMM International Workshop on Multimedia Information Retrieval*, (pp. 283-290).

Chen, H., & Wang, Z. J. (2010). Gains by a space-time-code based signaling scheme for multiple-antenna RFID tags. *23rd Canadian Conference on Electrical and Computer Engineering (CCECE)*, (pp. 1-4).

Chen, Z. N. (Ed.) (2007). *Antennas for portable devices.* New York, NY: John Wiley & sons

Chen, B. (2002). Span: An energy efficient coordination algorithm for topology maintenance in ad hoc wireless networks. *ACM Wireless Networks, 8*(5), 481–494. doi:10.1023/A:1016542229220

Chen, S., & Xue, Q. (2011). A class-F power amplifier with CMRC. *IEEE Microwave and Wireless Components Letters, 21*, 31–33. doi:10.1109/LMWC.2010.2091265

Chen, Z. N., Xianming, Q., & Chung, H. L. (2009). A universal UHF RFID reader antenna. *IEEE Transactions on Microwave Theory and Techniques, 57*(5), 1275–1282. doi:10.1109/TMTT.2009.2017290

Chen, Z., Yuan, J., & Vucetic, B. (2001). *An improved space-time trellis coded modulation scheme on slow Rayleigh fading channels.* IEEE.

Cho, N., et al. (2005). *A 5.1-uW UHF RFID tag chip integrated with sensors for wireless environmental monitoring.* European Solid-State Circuits Conf. (ESSCIRC), Grenoble, France, Sept. 2005.

Choi, W., Son, H., Shin, C., Bae, J. H., & Choi, G. (2006, July). *RFID tag antenna with a meandered dipole and inductively coupled feed.* Paper presented at IEEE Antennas and Propagation Society International Symposium.

Choi, W., Kwon, S., & Lee, B. (2001). Ceramic chip antenna using meander conductor lines. *Electronics Letters, 37*(15), 933–934. doi:10.1049/el:20010645

Chu, L., J. (1948). Physical limitations of omni-directional antennas. *Journal of Applied Physics, 19*, 1163–1175. doi:10.1063/1.1715038

Cipriani, E., Colantonio, P., Giannini, F., & Giofre, R. (2010). Theoretical and experimental comparison of Class F vs. Class F^{-1} PAs. *Microwave Integrated Circuits Conference* (pp. 428-431).

Colantonio, P., Giannini, F., & Limiti, E. (2009). *High efficiency RF and microwave solid state power amplifiers.* John Wiley & Sons, Publication.

Collin, R. E., & Rothchild, S. (1964). Evaluation of antenna Q. *IEEE Transactions on Antennas and Propagation, 12*, 23–27. doi:10.1109/TAP.1964.1138151

Colpitts, B. G., & Boiteau, G. (2004). Harmonic radar transceiver design: Miniature tags for insect tracking. *IEEE Transactions on Antennas and Propagation, 52*(11), 2825–2832. doi:10.1109/TAP.2004.835166

Cool Magnet Man. (n.d.). Retrieved 26 August, 2011, from http://www.coolmagnetman.com/magcondb.htm

Cripps, S. C. (1999). *RF power amplifiers for wireless communication.* Artech House.

Culler, D., Estrin, D., & Srivastava, M. (2004). Overview of sensor networks. *IEEE Computer, 37*(8), 41–49. doi:10.1109/MC.2004.93

Cumming, D. R. S., & Drysdale, T. D. (2003, March). *Security tag.* (Patent, GB 0305606.6).

Dakeya, Y., Suesada, T., Asakura, K., Nakajima, N., & Mandai, H. (2000, June). *Chip multilayer antenna for 2.45 GHz-band application using LTCC technology.* Paper presented at the IEEE MTT-S International Microwave Symposium, Boston, MA.

Dam, T. V., & Langendoen, K. (2003). An adaptive energy-efficient MAC protocol for wireless sensor networks. *Proceedings of SenSys, 2003*, 171–180.

Deal, W. R., Radisic, V., Qian, Y. X., & Itoh, T. (1999). Integrated-antenna push-pull power amplifiers. *IEEE Transactions on Microwave Theory and Techniques, 47*(8), 1418–1425. doi:10.1109/22.780389

Deepu, V., Vena, A., Perret, E., & Tedjini, S. (2010, June). *New RF identification technology for secure applications.* IEEE International Conference on RFID-Technology and Applications 2010.

Deligeorgis, G., Dragoman, M., Neculoiu, D., Dragoman, D., Konstantinidis, G., & Cismaru, A. (2009). Microwave propagation in graphene. *Applied Physics Letters, 95*(7). doi:10.1063/1.3202413

Demirbas, M., Xuming, L., & Singla, P. (2009). An in-network querying framework for wireless sensor networks. *IEEE Transactions on Parallel and Distributed Systems, 20*(8), 1202–1215. doi:10.1109/TPDS.2008.217

Deshmukh, A., & Kumar, G. (2005). Compact broadband U-slot loaded rectangular microstrip antennas. *Microwave and Optical Technology Letters, 46*(6), 556–559. doi:10.1002/mop.21049

Diels, W., Vaesen, K., Wambacq, P., Donnay, S., De Raedt, W., & Engels, M. (2001). Single-package integration of RF blocks for a 5 GHz WLAN application. *IEEE Transactions on Advanced Packaging, 24*(3), 384–391. doi:10.1109/6040.938307

Dobkin, D. M. (Ed.). (2008). *The RF in RFID-passive UHF RFID in practice.* Oxford, UK: Elsevier press.

Donnay, S., Pieters, P., Vaesen, K., Diels, W., Wambacq, P., & De Raedt, W. (2000). Chip-package codesign of a low-power 5-GHz RF front end. *Proceedings of the IEEE, 88*(10), 1583–1597. doi:10.1109/5.888997

Do, V. H., Subramanian, V., Keusgen, W., & Boeck, G. (2008, March). A 60 GHz SiGe-HBT power amplifier with 20% PAE at 15 dBm output power. *IEEE Microwave and Wireless Components Letters, 18*(3), 209–211. doi:10.1109/LMWC.2008.916816

Dragoman, M., Neculoiu, D., Dragoman, D., Deligeorgis, G., Konstantinidis, G., & Cismaru, A. (2010). Graphene for microwaves. *IEEE Microwave Magazine, 11*(7), 81–86. doi:10.1109/MMM.2010.938568

Dullaert, W., & Reichardt, L., & H., R. (2011). Improved detection scheme for chipless RFIDs using prolate spheroidal wave function-based noise filtering. *IEEE Antennas and Wireless Propagation Letters, 10*, 472–475. doi:10.1109/LAWP.2011.2155023

Efthymiou, C., Nikoletseas, S., & Rolim, J. (2002). The cougar approach to innetwork query processing in sensor networks. *SIGMOD, 31*(3), 9–18. doi:10.1145/601858.601861

Elisabeth, I.-Z., Kemeny, Z., Egri, P., & Monostori, L. (2006). The RFID technology and its current applications. *Proceedings of the Modern Information Technology in Innovation Process of the Industrial Enterprises-MITIP*, (pp. 29-36).

Elo, M. (2007). *SISO to MIMO: Moving communications from single-input single-output to multiple-input multiple-output.* White Paper, Keithley Instruments.

Endo, T., Sunahara, Y., Satoh, S., & Katagi, T. (2000). Resonant frequency and radiation efficiency of meander line antennas. *Electronics and Communications in Japan (Part II Electronics), 83*, 52–58. doi:10.1002/(SICI)1520-6432(200001)83:1<52::AID-ECJB7>3.0.CO;2-7

Engel, J. (2002). DSP for RFID, circuits and systems. *45th Midwest Symposium on MWSCAS-2002,* Vol. 2, (pp. II-227- II-230).

Faizulkhakov, Y. R. (2007). Time synchronization methods for wireless sensor networks: A survey. *Programming and Computer Software, 33*(4), 214–226. doi:10.1134/S0361768807040044

Fante R., L. (1969). Quality factor of general ideal antennas. *IEEE Transactions on Antenna Propagation, 17,* 151-157

Faschinger, M., Sastry, C. R., Patel, A. H., & Tas, N. C. (2007). An RFID and wireless sensor network-based implementation of workflow optimization. *IEEE International Symposium on a World of Wireless, Mobile and Multimedia Networks (WoWMoM07)*, (pp. 1-8).

Felemban, E., Lee, C.-G., & Ekici, E. (2006). MMSPEED: Multipath multi-SPEED protocol for QoS guarantee of reliability and timeliness in wireless sensor networks. *IEEE Transactions on Mobile Computing, 5*(6), 738–754. doi:10.1109/TMC.2006.79

Finkenzeller, K. (2004). *RFID handbook: Fundamentals and applications.* Wiley.

Fodor, I. K. (2002). *A survey of dimension reduction techniques*. Technical Report UCRL-ID-148494, Lawrence Livermore National Laboratory, Center for Applied Scientific Computing, USA.

Foschini, G. J. (1996). Layered space-time architecture for wireless communication in a fading environment when using multi-element antennas. *Bell Labs Technical Journal, 1*(2), 41–59. doi:10.1002/bltj.2015

Foschini, G. J., & Gans, M. J. (1998). On limits of wireless communications in a fading environment when using multiple antennas. *Wireless Personal Communications, 6*(3), 311–335. doi:10.1023/A:1008889222784

Foster, P. R., & Burberry, R. A. (1999). Antenna problems in RFID systems. *IEE Colloquium on RFID Technology: Microwave and Antenna Systems,* (pp. 3/1-3/5).

Fudem, H., Stenger, P., Niehenke, E. C., Sarantos, M., & Schwerdt, C. (1997, June). *A low cost miniature MMIC W-band transceiver with planar antenna.* Paper presented at the IEEE MTT-S International Microwave Symposium, Denver, CO.

Gao, S., Butterworth, P., Sambell, A., Sanabria, C., Xu, H., Heikman, S., et al. (2006). Microwave Class-F and inverse Class-F power amplifiers designs using GaN technology and GaAs pHEMT. *36th European Microwave Conference* (pp. 1719-1722).

Gao, T., Chi, B., Zhang, C., & Wang, Z. (2008). Design and analysis of a highly integrated CMOS power amplifier for RFID reader. *11th IEEE Singapore International Conference on Communication Systems* (pp. 1480-1483).

Gao, B., & Yuen, M. F. (2011). Passive UHF RFID packaging with electromagnetic band gap (EBG) material for metallic objects tracking. *IEEE Transactions on Components. Packaging and Manufacturing Technology, 1,* 1140–1146. doi:10.1109/TCPMT.2011.2157150

Gao, Q. (2006). Radio range adjustment for energy efficient wireless sensor networks. *Ad Hoc Networks, 4,* 75–82. doi:10.1016/j.adhoc.2004.04.007

Gao, S. (2006). High-efficiency class F RF/microwave power amplifiers. *IEEE Microwave Magazine, 7,* 40–48. doi:10.1109/MMW.2006.1614233

Garg, R. (Eds.). (2001). *Microstrip antenna design handbook*. Norwood, MA: Artech House.

GBPPR. (n.d.). Retrieved 26 August, 2011, from http://www.qsl.net/n9zia/why_silver_plate.html

Gesbert, D., Shafi, M., Shiu, D. S., Smith, P., & Naquib, A. (2003). From theory to practice: An overview of MIMO space-time codes. *IEEE Journal on Selected Areas in Communications, 21*(3), 281–302. doi:10.1109/JSAC.2003.809458

Gimpilevich, Y. B., & Shirokov, I. B. (2006). The homodyne frequency transformation at inexact installation of phase shift change range of probing signal. *News of Higher Educational Institutions. Radio Electronics Kiev, 49*(10), 54–63.

Gimpilevich, Y. B., & Shirokov, I. B. (2007). Generalized mathematical model of homodyne frequency conversion method under a periodic variation in sounding signal phase shift. *Telecommunications and Radio Engineering, 66*(12), 1057–1065. doi:10.1615/TelecomRadEng.v66.i12.20

Giusto, D., Iera, A., Morabito, G., & Atzori, L. (2010). *The internet of things*. 20th Tyrrhenian Workshop on Digital Communications. Springer.

Glinsky, A. (2005, February). Theremin: Ether music and espionage. Urbana, IL: University of Illinois Press.

Glover, B., & Bhatt, H. (2006). *RFID essentials* (1st ed.). O'Reilly Media, Inc.

Goel, S., Abawajy, J. H., & Kim, T. (2010). Performance analysis of receive diversity in wireless sensor networks over GBSBE models. *Sensors (Basel, Switzerland), 10,* 11021–11037. doi:10.3390/s101211021

Goldsmith, A. (2005). *Wireless communications*. Cambridge University Press.

Goodrum, P., McLaren, M., & Durfee, A. (2006). The application of active radio frequency identification technology for tool tracking on construction job sites. *Automation in Construction, 15*(3), 292–302. doi:10.1016/j.autcon.2005.06.004

Goubau, G., & Schwering, F. (1961). On the guided propagation of electromagnetic wave beams. *IRE Transactions on Antennas and Propagation, 9,* 248–256. doi:10.1109/TAP.1961.1144999

Griffin, J. D., & Durgin, G. D. (2008). Gains for RF tags using multiple antennas. *IEEE Transactions on Antennas and Propagation*, *56*(2), 563–570. doi:10.1109/TAP.2007.915423

Grouchko, M., Kamyshny, A., & Magdassi, S. (2009). Formation of air-stable copper-silver core-shell nanoparticles for inkjet printing. *Journal of Materials Chemistry*, *19*(19), 3057–3062. doi:10.1039/b821327e

Guey, J. C., Fitz, M. P., Bell, M. R., & Kuo, W. Y. (1996). Signal design for transmitter diversity wireless communication systems over Rayleigh fading channels. *IEEE Vehicular Technology Conference*, (pp. 136–140).

Gupta, S., Nikfal, B., & Caloz, C. (2010, 7-10 Dec. 2010). *RFID system based on pulse-position modulation using group delay engineered microwave C-sections*. In The Asia-Pacific Microwave Conference Proceedings (APMC), 2010.

Hakim, H., Renouf, R., & Enderle, J. (2006). Passive RFID asset monitoring system in hospital environments. *IEEE 32nd Annual Northeast Bioengineering Conference*, (pp. 217-218).

Han, D., Bo, L., & Sminchisescu, C. (2009). Selection and context for action recognition. *IEEE 12th International Conference on Computer Vision (ICCV09)*, (pp. 1933-1940).

Han, J., Kim, Y., Park, C., Lee, D., & Hong, S. (2006). *A fully-integrated 900-MHz CMOS power amplifier for mobile RFID reader applications*. IEEE Radio Frequency Integrated Circuit Symposium.

Han, S., Kim, M., Kim, H., & Yang, Y. (2011). A CMOS power amplifier for UHF RFID reader systems. *13th International Conference on Advanced Communication Technology* (pp. 57-60).

Hanna, V. F. (1980). Finite boundary corrections to coplanar stripline analysis. *Electronics Letters*, *16*(15), 604–606. doi:10.1049/el:19800419

Härmä, S., Plessky, V. P., Li, X., & Hartogh, P. (2009, April). Feasibility of ultra-wideband SAW RFID tags meeting FCC rules. *IEEE Transactions on Ultrasonics, Ferroelectrics, and Frequency Control*, *56*(4). doi:10.1109/TUFFC.2009.1104

Harrington, R. F. (1960). Effect of antenna size on gain, bandwidth and efficiency. *Journal of Research of the National Bureau of Standards*, *64*, 1–12.

Hartmann, C. S. (2002, October). A global SAW ID tag with large data capacity. *IEEE Ultrasonics Symposium*, (Vol. 1, pp. 65–69), Munich, Germany.

Hartmann, C., Hartmann, P., Brown, P., Bellamy, J., Claiborne, L., & Bonne, W. (2004). *Anti-collision methods for global SAW RFID tag systems*. IEEE Ultrasonics Symposium.

Haykin, S. (2009). *Communication systems* (5th ed.). Wiley Publishing.

He, W., Huang, Y., Wang, Z., & Zhao, Y. (2008). A circular polarization MIMO antenna system applied for RFID management. *8th International Symposium on Antennas, Propagation and EM Theory, ISAPE 2008*, (pp. 225-228).

Heidemann, W., Ye, J., & Estrin, D. (2004). Medium access control with coordinated adaptive sleeping for wireless sensor networks. *IEEE/ACM Transactions on Networking*, *12*(3), 493–506. doi:10.1109/TNET.2004.828953

Heinzelman, W., Kulik, J., & Balakrishnan, H. (1999). Adaptive protocols for information dissemination in wireless sensor networks. *Proceedings 5th ACM/IEEE Mobicom*, 1999, (pp. 174–85).

Heinzelman, W. R., Chandrakasan, A., & Balakrishnan, H. (2002). An application-specific protocol architecture for wireless microsensor networks. *IEEE Transactions on Wireless Communications*, *1*(4), 660–670. doi:10.1109/TWC.2002.804190

Higgs™-2. (2008). *Product overview*. Retrieved from http://www.alientechnology.com/docs/products/DSH2.pdf

Hohlt, B., Doherty, L., & Brewer, E. (2004). Flexible power scheduling for sensor networks. *Proceedings of the ISPN*, *2004*, 205–214.

Hosrt (2007, August 19). *GPS accuracy – Myth and truth – Part 1 – DGPS / Shadowing effect*. Sensors - GPS, http://autocom.wordpress.com/2007/08/19/genaues-gps-mythos-und-wahrheit-teil-1-dgps-abschattung/

Hourani, H. (2004/2005). *An overview of diversity techniques in wireless communication systems. Postgraduate Course in Radio Communications*. Helsinki University of Technology.

Hou, Y. T., Shi, Y., & Sherali, H. D. (2006). Optimal base station selection for anycast routing in wireless sensor networks. *IEEE Transactions on Vehicular Technology, 55*(3), 813–821. doi:10.1109/TVT.2006.873822

Hsiao, K., Vosoughi, S., Tellex, S., Kubat, R., & Roy, D. (2008). Object schemas for responsive robotic language use. In the *Proceedings of the 3rd ACM/IEEE International Conference on Human Robot Interaction*, (pp. 233-240).

Huang, D., Liao, F., et al. (2003). *Plastic-compatible low resistance printable gold nanoparticle conductors for flexible electronics* (Vol. 150). Pennington, NJ: ETATS-UNIS: Electrochemical Society.

Huebler, A., Hahn, U., Beier, W., Lasch, N., & Fischer, T. (2002, 2002). *High volume printing technologies for the production of polymer electronic structures*. In The 2nd International IEEE Conference on Polymers and Adhesives in Microelectronics and Photonics, 2002. POLYTRONIC 2002.

Hu, S., Law, C. L., & Dou, W. (2008). A balloon-shaped monopole antenna for passive UWB-RFID tag applications. *IEEE Antennas and Wireless Propagation Letters, 7*, 366–368. doi:10.1109/LAWP.2008.928462

Hwang, S. H., Moon, J. I., Kwak, W. I., & Park, S. O. (2004). Printed compact dual band antenna for 2.4 and 5GHz ISM band applications. *Electronics Letters, 40*(25), 1568–1569. doi:10.1049/el:20046579

IDTechEx. (2006). *Chipless RFID - The end game*. Retrieved 3/10/2011, from http://www.idtechex.com/research/articles/chipless_rfid_the_end_game_00000435.asp

IDTechEx. (2009). *RFID forecasts, players and opportunities 2009-2019*. Executive Summary and Collusions.

Ingram, M. A., Demirkol, M. F., & Kim, D. (2001). *Transmit diversity and spatial multiplexing for RF links using modulated backscatter*. International Symposium on Signals, Systems, and Electronics (ISSSE'01).

Ingruber, B., Pritzl, W., Smely, D., Wachutka, M., & Magerl, G. (1998). High efficiency harmonic-controlled amplifier. *IEEE Transactions on Microwave Theory and Techniques, 46*, 857–863. doi:10.1109/22.681213

Intanagonwiwat, C., Govindan, R., & Estrin, D. (2000). Directed diffusion: A scalable and robust communication paradigm for sensor networks. *Proceedings of the ACM MobiCom, 2000*, (pp. 56–67).

Ishikawa, R., & Honjo, K. (2011). Distributed class-F/inverse class-F circuit considering up to arbitrary harmonics with parasitics compensation. *IEEE MTT-S International Microwave Workshop Series on Innovative Wireless Power Transmission: Technologies, Systems, and Applications* (pp. 29-32).

ISO/IEC 14496-2. (1999). *Information technology—Coding of audio-visual objects. Part 2: Visual*.

ITU. (2005). The Internet of things. *ITU Internet Reports, 2005*. Retrieved from http://www.itu.int/osg/spu/publications/internetofthings/

Jaffe, J. S., & Mackey, R. C. (1965). Microwave frequency translator. *IEEE Transactions on Microwave Theory and Techniques, 13*, 371–378. doi:10.1109/TMTT.1965.1126002

Jalaly, I., & Robertson, D. (2005, June). RF barcodes using multiple frequency bands. *IEEE MTT-S Digest*.

Jalaly, I., & Robertson, I. D. (2005). Capacitively tuned microstrip resonators for RFID barcodes. *Proceedings of the 35th EUMC* (pp 1161-1164)

Jalaly, I., & Robertson, I. D. (2005, 12-17 June). *RF barcodes using multiple frequency bands*. In The IEEE MTT-S International Microwave Symposium Digest, 2005.

Jamali, B. (2010). A development platform for SDR-based RFID reader. In Karmakar, N. C. (Ed.), *Handbook of smart antennas for RFID systems* (pp. 123–137). New Jersey: Wiley Microwave and Optical Engineering Series. doi:10.1002/9780470872178.ch5

Jandieri, G. V., Shirokov, I. B., & Gimpilevich, Yu. B. (2007). The analysis of metrological features of homodyne method of frequency transformation. *Georgian Engineering News, 2*, 38–45.

Jang, H.-S., Lim, W.-G. Oh, K.-S., Moon, S-.M., & Yu, J.-W. (November, 2010). Design of low-cost chipless system using printable chipless tag with electromagnetic code. *IEEE Microwave and Wireless Components Letters, 20.*

Jang, H.-S., Lim, W.-G., Oh, K.-S., Moon, S.-M., & Yu, J.-W. (2010). Design of low-cost chipless system using printable chipless tag with electromagnetic code. *IEEE Microwave and Wireless Component Letters, 20*(11), 640–642. doi:10.1109/LMWC.2010.2073692

Jenshan, L. (1998). Chip-package codesign for high-frequency circuits and systems. *IEEE Micro, 18*(4), 24–32. doi:10.1109/40.710868

Jiao, Y.-Y., Zhen, Y.-W., & Jiao, R. J. (2008). Hospital linens inventory control re-engineering based on RFID. *IEEE Conference on Cybernetics and Intelligent Systems* (pp. 612-617).

Joo, T., Lee, H., Shim, S., & Hong, S. (2010A). CMOS RF power amplifier for UHF stationary RFID reader. *IEEE Microwave and Wireless Components Letters, 20,* 106–108. doi:10.1109/LMWC.2009.2038552

Kamol, P., Nikolaidis, S., Ueda, R., & Arai, T. (2007). RFID based object localization system using ceiling cameras with particle filter. In the *Proceedings of the Future Generation Communication and Networking (FGCN07), 2,* (pp. 37 – 42).

Karmakar, G. C., Rahman, S. M., & Bignall, R. J. (1999). Composite features extraction and object ranking. In the *Proceedings of the International Conference on Computational Intelligence for Modeling, Control and Automation (CIMCA'99),* Vienna, Austria, (pp. 134 – 139).

Karmakar, N. C., et al. (2006). Development of low-cost active RFID tag at 2.4 GHz. *36th European Microwave Conference, 2006,* (pp. 1602-1605).

Karmakar, N. C., Zakavi, P., & Kumbukage, M. (2010). Development of a phased array antenna for universal UHF RFID reader. *Proceedings of IEEE 2010 AP-S International Symposium on Antennas and Propagation and 2010 USNC/CNC/URSI Meeting,* (pp. 1-4).

Karmakar, G. C. (2003). *An integrated fuzzy rule based image segmentation framework.* Australia: Gippsland School of Information Technology, Monash University.

Karmakar, G. C., Dooley, L., & Rahman, S. M. (2001). Review on fuzzy image segmentation technique. In Rahman, S. M. (Ed.), *Design and management of multimedia information systems: Opportunities and Challenges* (pp. 282–313). Hershey, PA: Idea Group Publishing. doi:10.4018/978-1-930708-00-6.ch014

Karmakar, N. C. (2008). Smart antennas for automatic radio frequency identification readers. In Sun, C., Cheng, J., & Ohira, T. (Eds.), *Handbook on advancements in smart antenna technologies for wireless networks* (pp. 449–472). Hershey, PA: IGI Global. doi:10.4018/978-1-59904-988-5.ch021

Karmakar, N. C. (2010). *Handbook of smart antennas for RFID systems* (pp. 13–52). doi:10.1002/9780470872178

Karmakar, N. C., & Bialkowski, M. E. (1999). An L-band 90° hybrid-coupled phase shifter using UHF band p-i-n diodes. *Microwave and Optical Technology Letters, 21*(1), 144–148. doi:10.1002/(SICI)1098-2760(19990405)21:1<51::AID-MOP15>3.0.CO;2-H

Karmakar, N. C., Zakavi, P., & Kambukage, M. (2010). FPGA-controlled phased array antenna development for UHF RFID reader. In Karmakar, N. C. (Ed.), *Handbook of smart antennas for RFID systems* (pp. 57–82). New Jersey: Wiley Microwave and Optical Engineering Series. doi:10.1002/9780470872178.ch3

Karp, B., & Kung, H. T. (2000). GPSR: Greedy perimeter stateless routing for wireless sensor networks. *Proceedings of the MobiCom, 2000,* 243–254. doi:10.1145/345910.345953

Keinhorst, T., Tsigkourakos, P., Yaqoob, M. A., van Driel, W. D., & Zhang, G. Q. (2007). Test and health monitoring of microelectronics using RFID. *The 8th International Conference on Electronic Packaging Technology* (pp. 1-6).

Keith, R. C., Cooper, W. K., & Stutzman, W. L. (1973, March). Beam-pointing errors of planner-phased arrays. *IEEE Transactions on Antennas and Propagation, 21*(2), 199–202. doi:10.1109/TAP.1973.1140434

Khannur, P. B., Chen, X., Yan, D. L., Shen, D., Zhao, B., & Raja, M. K. (2008). A universal UHF RFID reader IC in 0.18-µm CMOS technology. *IEEE Journal of Solid-state Circuits, 43,* 1146–1155. doi:10.1109/JSSC.2008.920355

Kim, B., Derickson, D., & Sun, C. (2007). A high power, high efficiency amplifier using GaN HEMT. *Asia-Pacific Microwave Conference* (pp. 1-4).

Kim, J.-Y., Oh, D.-S., & Kim, J.-H. (2007). Design of a harmonically tuned class-F power amplifier. *Asia-Pacific Microwave Conference* (pp. 1 – 4).

Kim, S. W., Ha, D. S., & Kim, J. H. (2001). *Performance gain of smart antennas with diversity combining at hand-sets for the 3GPP WCDMA system*. 13th International Conference on Wireless Communications (Wireless 2001), Calgary, Alberta, Canada.

Kim, J. H., Jo, G. D., Oh, J. H., Kim, Y. H., Lee, K. C., & Jung, J. H. (2011). Modeling and design methodology of high-efficiency Class-F and Class-F^{-1} power amplifiers. *IEEE Transactions on Microwave Theory and Techniques, 59*, 153–165. doi:10.1109/TMTT.2010.2090167

Kipphan, H. (2001). *Handbook of print media: Technologies and production methods*. Springer.

Kirousis, L. (2000). Power consumption in packet radio networks. *Theoretical Computer Science, 243*(1-2), 289–305. doi:10.1016/S0304-3975(98)00223-0

Kirsch, N. J., Vacirca, N. A., Plowman, E. E., Kurzweg, T. P., Fontecchio, A. K., & Dandekar, K. R. (2009). *Optically transparent conductive polymer RFID meandering dipole antenna*. In The IEEE International Conference on RFID.

Kitayoshi, H., & Sawaya, K. (2005). *Long range passive RFID-tag for sensor networks*. IEEE Vehicular Technology Conference.

Klair, D. K., Chin, K. W., & Raad, R. (2010). A survey and tutorial of RFID anti-collision protocols. *IEEE Comm. Surveys & Tutorials, 12*(3), 400–421. doi:10.1109/SURV.2010.031810.00037

Klaus, F. (2003). *RFID handbook: Fundamentals and applications in contactless smart cards and identification*. New York, NY: John Wiley & Sons.

Ko, Y., Roy, S., Smith, J. R., Lee, H., & Cho, C. (2007). An enhanced RFID multiple access protocol for fast inventory. *IEEE Global Telecommunications Conference (GLOBECOM07)*, (pp. 4575 – 4579).

Kobayashi, Y., & Sakuraba, T. (2008). Silica-coating of metallic copper nanoparticles in aqueous solution. *Colloids and Surfaces A: Physicochemical and Engineering Aspects, 317*(1-3), 756–759. doi:10.1016/j.colsurfa.2007.11.009

Koshurinov, E. I. (2003). The homodyne radar with optimal signal processing. *IEEE Proceedings of Crimean Microwave Conference*, (pp. 737-739). Sevastopol, Ukraine, 8-12 Sep. 2003.

Koswatta, R., & Karmakar, N. C. (2010). Moving average filtering technique for signal processing in digital section of UWB chipless RFID reader. *Microwave Conference Proceedings (APMC), Asia-Pacific*, (pp. 1304-1307).

Koswatta, R., & Karmakar, N. C. (2010, Dec. 2010). *Development of digital control section of RFID reader for multi-bit chipless RFID tag reading*. Paper presented at the International Conference on Electrical and Computer Engineering (ICECE), Dhaka, Bangladesh.

Kulik, J., Heinzelman, W. R., & Balakrishnan, H. (2002). Negotiation-based protocols for disseminating information in wireless sensor networks. *Wireless Networks, 8*, 169–185. doi:10.1023/A:1013715909417

Kuroda, K., Ishikawa, R., & Honjo, K. (2010). Parasitic compensation design technique for a C-band GaN HEMT Class-F amplifier. *IEEE Transactions on Microwave Theory and Techniques, 58*, 2741–2750. doi:10.1109/TMTT.2010.2077951

Kyutae, L., Obatoyinbo, A., Sutono, A., Chakraborty, S., Chang-Ho, L., Gebara, E., et al. (2001, May). *A highly integrated transceiver module for 5.8 GHz OFDM communication system using multi-layer packaging technology*. Paper presented at the IEEE MTT-S International Microwave Symposium, Phoenix, AZ.

Landt, J. (2001, October). *Shrouds of time: The history of RFID*. Retrieved from www.aimglobal.org/technologies/rfid/resources/shrouds_of_time.pdf

Langwieser, R., Angerer, C., & Scholtz, A. L. (2010). A UHF frontend for MIMO applications in RFID. *Radio and Wireless Symposium (RWS), IEEE*, (pp. 124-127).

Lau, P.-Y., Yung, K. K.-O., & Yung, E. K.-N. (2010). A low-cost printed CP patch antenna for RFID smart bookshelf in library. *IEEE Transactions on Industrial Electronics, 57*, 1583–1589. doi:10.1109/TIE.2009.2035992

Lazaro, A., Ramos, A., & Girbau, D., & R., V. (2011). Chipless UWB RFID tag detection using continuous wavelet transform. *IEEE Antennas and Wireless Propagation Letters, 10*, 520–523. doi:10.1109/LAWP.2011.2157299

Lee, C., & Park, Y. (2009). Design of compact-sized class-F PA for wireless handset applications. *IEEE MTT-S International Microwave Symposium Digest* (pp. 405 – 408).

Lee, J., Park, K., Hong, S., & Cho, W. (2007). Object tracking based on RFID coverage visual compensation in wireless sensor network. *IEEE International Symposium on Circuits and Systems (ISCAS),* (pp. 1597 – 1600).

Lee, J. (2009). Global positioning/GPS. In Kitchen, R., & Thrift, N. (Eds.), *International encyclopedia of human geography* (pp. 548–555). doi:10.1016/B978-008044910-4.00035-3

Lee, K. F. L., Chebolu, S. R., Chen, W., & Lee, R. Q. (1994). On the role of substrate loss tangent in the cavity model theory of microstrip patch antennas. *IEEE Transactions on Antennas and Propagation, 42*(1), 110–112. doi:10.1109/8.272308

Lee, T. H. (1999). *The design of CMOS radio-frequency integrated circuits*. Cambridge.

Lee, Y. W., Lee, K. Y., & Kim, M. H. (2009). Energy-efficient multiple query optimization for wireless sensor networks. *Proceedings of the SENSORCOMM, 2009*, 531–538.

Lee, Y.-S., & Jeong, Y.-H. (2007). A high-efficiency Class-E GaN HEMT power amplifier for WCDMA applications. *IEEE Microwave and Wireless Components Letters, 17*, 622–624. doi:10.1109/LMWC.2007.901803

Lee, Y.-S., Lee, M.-W., & Jeong, Y.-H. (2008). High-efficiency Class-F GaN HEMT amplifier with simple parasitic-compensation circuit. *IEEE Microwave and Wireless Components Letters, 18*, 55–57. doi:10.1109/LMWC.2007.912023

Li, R. L., DeJean, G., Tentzeris, M. M., Laskar, J., & Papapolymerou, J. (2003, June). *LTCC multilayer based CP patch antenna surrounded by a soft-and-hard surface for GPS applications.* Paper presented at the IEEE International Antennas and Propagation Society Symposium and URSI National Radio Science Meeting, Columbus, OH.

Lindskog, E., & Paulraj, A. (2000). A transmit diversity scheme for channels with intersymbol interference. In *Proceedings of International Conference on Communications,* (p. 1).

Linlin, Z., et al. (2008). Design and implementation of a fully reconfigurable chipless RFID tag using Inkjet printing technology. *IEEE International Symposium on Circuits and Systems, USA, 2008*, (pp. 1524-1527).

Lin, P., Qiao, C., & Wang, X. (2004). Medium access control with a dynamic duty cycle for sensor networks. *Proceedings of the WCNC, 2004*, 1534–1539.

Liu, J., Bowyer, K., Goldgof, D., & Sarkar, S. (1997). *A comparative study of texture measures for human skin treatment*. Presented at International Conference on Information, Communications and Signal Processing (ICICS '97), Singapore.

Liu, S. (2006). A lifetime-extending deployment strategy for multi-hop wireless sensor networks. *IEEE Proceedings of the Communication Networks and Services Research Conference*, May 2006, (pp. 1-8).

Liu, H. (2008). Taxonomy and challenges of the integration of RFID and wireless sensor networks. *IEEE Network, 22*(6), 26–35. doi:10.1109/MNET.2008.4694171

Liu, H., Wan, P., & Jia, X. (2007). Maximal lifetime scheduling for sensor surveillance systems with K sensors to one target. *IEEE Transactions on Parallel and Distributed Systems, 15*(2), 334–345.

Li, X., Yang, L., Gong, S.-X., Yang, Y.-J., & Liu, J.-F. (2009). A compact folded printed dipole antenna for UHF RFID reader. *Progress in Electromagnetics Research Letters, 6*, 47–54. doi:10.2528/PIERL08121303

Loo, C. H., Elmahgoub, K., Yang, F., Elsherbeni, A., Kajfez, D., & Kishk, A. (2008). Chip impedance matching for UHF RFID tag antenna design. *Progress in Electromagnetics Research, 81*, 359–370. doi:10.2528/PIER08011804

Lou, J., Liu, Q., Tan, T., & Hu, W. (2002). Semantic interpretation of object activities in a surveillance system. *International Conference on Pattern Recognition, 3*, 777 – 780.

Lowe, D. G. (2004). Distinctive image features from scale-invariant keypoints. *International Journal of Computer Vision, 60*(2), 91–110. doi:10.1023/B:VISI.0000029664.99615.94

Luechinger, N. A., Athanassiou, E. K., & Stark, W. J. (2008). Graphene-stabilized copper nanoparticles as an air-stable substitute for silver and gold in low-cost ink-jet printable electronics. *Nanotechnology, 19*(44). doi:10.1088/0957-4484/19/44/445201

Lunglmayr, M., & Huemer, M. (2010). Least squares equalization for RFID. *2nd International Workshop on Near Field Communication,* (pp. 90-94).

Magdassi, S. (Ed.). (2009). *The chemistry of inkjet inks.* Hackensack, NJ: World Scientific. doi:10.1142/9789812818225

Magdassi, S., Grouchko, M., Berezin, O., & Kamyshny, A. (2010). Triggering the sintering of silver nanoparticles at room temperature. *ACS Nano, 4*(4), 1943–1948. doi:10.1021/nn901868t

Magdassi, S., Grouchko, M., & Kamyshny, A. (2010). Copper nanoparticles for printed electronics: routes towards achieving oxidation stability. *Materials, 3*(9), 4626–4638. doi:10.3390/ma3094626

Mahfoudh, S., & Minet, P. (2008).Survey of energy efficient strategies in wireless ad hoc and sensor networks. *Proceedings of the International Conference on Networking,* 2008, (pp. 1-7).

Mahtre, V. (2005). A minimum cost heterogeneous sensor network with a lifetime constraint. *IEEE Transactions on Mobile Computing, 4*(1), 4–15. doi:10.1109/TMC.2005.2

Mailloux, R. J. (1994). *Phased array antenna handbook.* Artech House.

Maleki, M., & Pedram, M. (2005). QoM and lifetime-constrained random deployment of sensor networks for minimum energy consumption. *IEEE Proceedings of Information Processing in Sensor Networks, 2005,* 293–300.

Maloratsky, L. G. (2004). *Passive RF & microwave integrated circuits.* USA: Elsevier.

Mandel, C., Schüßler, M., Maasch, M., & Jakoby, R. (2009). *A novel passive phase modulator based on LH delay lines for chipless microwave RFID applications.* International Microwave Workshop Series, Croatia.

Manteghi, M. (2010, 11-17 July 2010). *A novel approach to improve noise reduction in the Matrix Pencil Algorithm for chipless RFID tag detection.* Paper presented at the Antennas and Propagation Society International Symposium (APSURSI), 2010 IEEE Toronto, Ontario.

Mäntysalo, M., & Mansikkamäki, P. (2009). An inkjet-deposited antenna for 2.4 GHz applications. *AEÜ. International Journal of Electronics and Communications, 63*(1), 31–35. doi:10.1016/j.aeue.2007.10.004

Marrocco, G. (2003). Gain-optimized self-resonant meander line antennas for RFID applications. *IEEE Antennas and Wireless Propagation Letters, 2,* 302–305. doi:10.1109/LAWP.2003.822198

Marrocco, G. (2008). The art of UHF RFID antenna design: Impedance-matching and size-reduction techniques. *IEEE Antennas and Propagation Magazine, 50*(1), 66–79. doi:10.1109/MAP.2008.4494504

Mazurek, G. (2008). Experimental RFID system with active tags. *International Conference on Signals and Electronic Systems, ICSES '08,* (pp. 507-510).

McFate, M. (2005). Iraq: The social context of IEDs. *Military Review,* (May-June): 2005.

McVay, J., Hoorfar, A., & Engheta, N. (2006). Theory and experiments on Peano and Hilbert curve RFID tags. *Proceedings of the Society for Photo-Instrumentation Engineers, 6248*(1).

Merilampi, S., Laine-Ma, T., & Ruuskanen, P. (2009). The characterization of electrically conductive silver ink patterns on flexible substrates. *Microelectronics and Reliability, 49*(7), 782–790. doi:10.1016/j.microrel.2009.04.004

Mhatre, V., & Rosenberg, C. (2004). Homogeneous vs heterogeneous clustered sensor networks: A comparative study. *IEEE ICC,* 2004, (pp. 646-651).

Mhatre, V., & Rosenberg, C. (2004). Design guidelines for wireless sensor networks: communication, clustering and aggregation. *Ad Hoc Networks*, *2*, 45–63. doi:10.1016/S1570-8705(03)00047-7

Michishita, N., Yamada, Y., & Nakakura, N. (2004). *Miniaturization of a small meander line antenna by loading a high ε_r material*. Paper presented at the 5th International Symposium on Multi-Dimensional Mobile Communications and The 2004 Joint Conference of the 10th Asia-Pacific Conference.

Midrio, M., Boscolo, S., Sacchetto, F., Someda, C. G., Capobianco, A. D., & Pigozzo, F. M. (2009). Planar, compact dual-band antenna for wireless LAN applications. *IEEE Antennas and Wireless Propagation Letters*, *8*, 1234–1237. doi:10.1109/LAWP.2009.2035647

Milanés, V., Llorca, D. F., Vinagre, B. M., González, C., & Sotelo, M. A. (2010). Clavileño: Evolution of an autonomous car. *International IEEE Annual Conference on Intelligent Transportation Systems,* Madeira Island, Portugal, (pp. 1129-1134).

Mukherjee, S. (2007). Chipless radio frequency identification by remote measurement of complex impedance. *Proceedings of Wireless Technologies, European Conference* (pp. 249-252).

Mukherjee, S., & Chakraborty, G. (2009, December). *Chipless RFID using stacked multiple patches*. Applied Electromagnetics Conference, Kolkata.

Murphy, K. P. (2002). *Dynamic Bayesian networks: Representation, inference and learning*. Berkeley, USA: University of California.

Nakano, H., Tagami, H., Yoshizawa, A., & Yamauchi, J. (1984). Shortening ratios of modified dipole antennas. *IEEE Transactions on Antennas and Propagation*, *32*(4), 385–386. doi:10.1109/TAP.1984.1143321

NanoPrintTech. (n.d.). Retrieved 24 August, 2011, from http://www.nanoprinttech.com/old/printing-technology.html

Narkcharoen, P., & Pranonsatit, S. (2011, 17-19 May). *The applications of fill until full (FuF) for multiresonator-based chipless RFID system*. In The 8th International Conference on Electrical Engineering/Electronics, Computer, Telecommunications and Information Technology (ECTI-CON), 2011.

Natarajan, A., Komijani, A., Guan, X., Babakhaniand, A., & Hajimiri, A. (2006, December). A 77-GHz phased-array transceiver with on-chip antennas in silicon: Transmitter and local LO-path phase shifting. *IEEE Journal of Solid-state Circuits*, *41*(12), 2807–2819. doi:10.1109/JSSC.2006.884817

Neculoiu, D., Deligeorgis, G., Dragoman, M., Dragoman, D., Konstantinidis, G., Cismaru, A., et al. (2010). *Electromagnetic propagation in graphene in the mm-wave frequency range*. In The 40th European Microwave Conference, EuMC 2010, Paris.

Negra, R., Sadeve, A., Bensmida, S., & Ghannouchi, F. M. (2008). Concurrent dual-band Class-F load coupling network for applications at 1.7 and 2.14 GHz. *IEEE Transactions on Circuits and Systems*, *2*(55), 259–263.

Nemati, H. M., Fager, C., Thorsell, M., & Herbert, Z. (2009). High-efficiency LDMOS power-amplifier design at 1 GHz using an optimized transistor model. *IEEE Transactions on Microwave Theory and Techniques*, *5*, 1647–1654. doi:10.1109/TMTT.2009.2022590

Nemati, H. M., Saad, P., Fager, C., & Andersson, K. (2011). High-efficiency power amplifier. *IEEE Microwave Magazine*, *12*(1), 81–84. doi:10.1109/MMM.2010.939314

Nikitin, P. V., & Rao, K. V. S. (2007). Performance of RFID tags with multiple RF ports. In *Proceedings of the IEEE Antennas and Propagation Society International Symposium*, (pp. 5459–5462).

Nikitin, P. V., & Rao, K. V. S. (2008). Antennas and propagation in UHF RFID systems. *IEEE International Conference on RFID*, Las Vegas, NV, 2008, (pp. 277-288).

Nikitin, P. V., & Rao, K. V. S. (2010a). Helical antenna for handheld UHF RFID reader. *Proceedings of IEEE RFID Conference*, (pp. 166-172).

Nikitin, P. V., & Rao, K. V. S. (2010b). Compact Yagi antenna for handheld UHF RFID reader. *Proceedings of IEEE APSURSI*, (pp. 1-4).

Nikitin, P. V., Rao, K. V. S., & Lazar, S. (2007). An overview of near field UHF RFID. *IEEE International Conference on RFID,* 2007, (pp. 167-174).

Nikitin, P. V., & Seshagiri, K. V. (2009). LabVIEW-based UHF RFID tag test and measurement system. *IEEE Transactions on Industrial Electronics, 56*(7), 2374–2381. doi:10.1109/TIE.2009.2018434

Occhiuzzi, C., & Marrocco, G. (2010). The RFID technology for neurosciences: feasibility of limbs' monitoring in sleep diseases. *IEEE Transactions on Information Technology in Biomedicine, 14,* 37–43. doi:10.1109/TITB.2009.2028081

Oldenzijl, R., Gaitens, G., & Dixon, D. (2010). *Conduct radio frequencies with Inkpp.* Retrieved from http://www.intechopen.com/articles/show/title/conduct-radio-frequencies-with-inks

Pal, N. R., & Pal, S. K. (1993). A review on image segmentation techniques. *Pattern Recognition, 26,* 1277–1294. doi:10.1016/0031-3203(93)90135-J

Pan, C. Y., & Horng, T. S. (2001, December). *Miniaturized dielectric chip antenna in a C-shaped configuration with an array of shorting pins.* Paper presented at the Asia-Pacific Microwave Conference, Taipei, Taiwan.

Pan, J. (2005). Optimal base-station locations in two-tiered wireless sensor networks. *IEEE Transactions on Mobile Computing, 4*(5), 458–473. doi:10.1109/TMC.2005.68

Park, S., Kim, K., Park, S., & Park, M. (2006). Object entity-based global localization in indoor environment with stereo camera. *Proceedings of SICE,* (pp. 2681-2686).

Paruchuri, V., et al. (2004). Random asynchronous wakeup protocol for sensor networks. *Proceedings of the IEEE International Conference on Broadband Networks,* 2004, (pp. 710-717).

Pawar, S. A., Raj Kumar, K., Elia, P., Vijay Kumar, P., & Sethuraman, B. A. (2009). Space–time codes achieving the DMD tradeoff of the MIMO-ARQ channel. *IEEE Transactions on Information Theory, 55*(7). doi:10.1109/TIT.2009.2021332

Pellerano, S., Alvarado, J., & Palaskas, Y. (2010). A mm-wave power-harvesting RFID tag in 90 nm CMOS. *IEEE Journal of Solid-state Circuits, 45*(8), 1627–1637. doi:10.1109/JSSC.2010.2049916

Pentland, A. (1999). Modelling and prediction of human behaviour. *Neural Computation, 11,* 229–242. doi:10.1162/089976699300016890

Perret, E., Tedjini, S., Deepu, V., Vena, A., Garet, & F., Duvillaret, L. (2010, February). *Etiquette RFID passive sans puce.* Patent, FR, 10/50971, N. REF: B100087.

Perret, E., Hamdi, M., Vena, A., Garet, F., Bernier, M., Duvillaret, L., & Tedjini, S. (2011). *RF and THz identification using a new generation of chipless RFID tags. The Radioengineering Journal - Towards EuCAP 2012: Emerging Materials.* Methods, and Technologies in Antenna & Propagation.

Pfeiffer, U. R., & Goren, D. (2007, July). A 20 dBm fully-integrated 60 GHz SiGe power amplifier with automatic level control. *IEEE Journal of Solid-state Circuits, 42*(7), 1455–1463. doi:10.1109/JSSC.2007.899116

Pham, D. L., & Prince, J. L. (1999). An adaptive fuzzy c-means algorithm for image segmentation in the presence of intensity inhomogeneities. *Pattern Recognition Letters, 20,* 57–68. doi:10.1016/S0167-8655(98)00121-4

Philipose, M. (2005). Battery-free wireless identification and sensing. *IEEE Pervasive Computing / IEEE Computer Society and IEEE Communications Society, 4*(1), 37–45. doi:10.1109/MPRV.2005.7

Polastre, J., Hill, J., & Culler, D. (2004). Versatile low power media access for wireless sensor networks. *Proceedings of the Sensys, 2004,* 95–107. doi:10.1145/1031495.1031508

Preradovic, S., & Karmakar, N. C. (2007B). Modern RFID readers – A review. *The 4th Conference on Electrical and Computer Engineering,* (pp. 100-103).

Preradovic, S., & Karmakar, N. C. (2009, December). *Design of short range chipless RFID reader prototype.* Paper presented at the Intelligent Sensors, Sensor Networks and Information Processing (ISSNIP), Melbourne, Australia.

Preradovic, S., & Karmakar, N. C. (2009, Sept.). *Design of fully printable planar chipless RFID transponder with 35-bit data capacity.* Paper presented at the European Microwave Conference, 2009. EuMC 2009., Rome.

Preradovic, S., & Karmakar, N. C. (2010). Multiresonator based chipless RFID tag and dedicated RFID reader. *Digest of 2010 International Microwave Symposium,* Anaheim, California. (CD-ROM)

Preradovic, S., Balbin, I., Karmakar, N. C., & Swiegers, G. (2008, April 16 - 17). *A novel chipless RFID system based on planar multiresonators for barcode replacement.* Paper presented at the IEEE International Conference on RFID, 2008, Las Vegas, Nevada.

Preradovic, S., Balbin, I., Karmakar, N. C., & Swiegers, G. (2008, October). Chipless frequency signature based RFID transponders. *38th European Microwave Conference,* (pp. 1723-1726). Amsterdam, Netherlands.

Preradovic, S., Balbin, I., Karmakar, N. C., & Swiegers, G. (Apr. 2008). *A. P. P. Application.*

Preradovic, S., Balbin, I., Roy, S. M., Karmakar, N. C., & Swiegers, G. (2008). *Radio frequency transponder.* (Australian Provisional Patent Application P30228AUPI, Apr. 2008).

Preradovic, S., Roy, S., & Karmakar, N. (2009, 7-10 Dec. 2009). *Fully printable multi-bit chipless RFID transponder on flexible laminate.* Paper presented at the Microwave Conference, 2009. APMC 2009. Asia Pacific.

Preradovic, S., Roy, S., & Karmakar, N. C. (2009). *Fully printable multi-bit chipless RFID transponder on flexible laminate.* Asia Pacific Microwave Conference.

Preradovic, S. (Ed.). (2011). *Advanced radio frequency identification design and applications in RFID Tags.* Rijeka, Croatia: INTECH.

Preradovic, S., & Karmakar, N. C. (2010). Chipless RFID: Bar code of the future. *Microwave Magazine, 11*(7), 87–97. doi:10.1109/MMM.2010.938571

Preradovic, S., & Karmakar, N. C. (2010). RFID readers-review and design. In Karmakar, N. C. (Ed.), *Handbook of smart antennas for RFID systems* (pp. 85–121). New Jersey: Wiley Microwave and Optical Engineering Series. doi:10.1002/9780470872178.ch4

Preradovic, S., Karmakar, N., & Balbin, I. (2008, October). RFID transponders. *IEEE Microwave Magazine, 9*(5), 90–103. doi:10.1109/MMM.2008.927637

Preradovic, S., Karmakar, N., & Balbin, I. (2010, December). Chipless RFID, Bar code of the future. *IEEE Microwave Magazine, 11*(7), 87–97. doi:10.1109/MMM.2010.938571

Pudas, M., Halonen, N., Granat, P., & Vähäkangas, J. (2005). Gravure printing of conductive particulate polymer inks on flexible substrates. *Progress in Organic Coatings, 54*(4), 310–316. doi:10.1016/j.porgcoat.2005.07.008

Pursula, P., Karttaavi, T., Kantanen, M., Lamminen, A., Holmberg, J., & Lahdes, M. (2011). 60-GHz millimeter-wave identification reader on 90-nm CMOS and LTCC. *IEEE Transactions on Microwave Theory and Techniques, 59*(4), 1166–1173. doi:10.1109/TMTT.2011.2114200

Pursula, P., Vaha-Heikkila, T., Muller, A., Neculoiu, D., Konstantinidis, G., Oja, A., & Tuovinen, J. (2008). Millimeter-wave identification—A new short-range radio system for low-power high data-rate applications. *IEEE Transactions on Microwave Theory and Techniques, 56*(10), 2221–2228. doi:10.1109/TMTT.2008.2004252

Raab, F. H. (1996). An introduction to class-F power amplifiers. *R.F. Design, 19,* 79–84.

Raab, F. H. (1997). Class-F amplifiers with maximally flat waveforms. *IEEE Transactions on Microwave Theory and Techniques, 45,* 2007–2012. doi:10.1109/22.644215

Raab, F. H. (2001). Maximum efficiency and output of class-F power amplifiers. *IEEE Transactions on Microwave Theory and Techniques, 49,* 1162–1166. doi:10.1109/22.925511

Rahman, M. Z., Karmakar, G. C., & Dooley, L. S. (2005). Passive source localization using power spectral analysis and decision fusion in wireless distributed sensor networks. *International Conference on Information Technology: Coding and Computing* (ITCC-05), Las Vegas, USA, (pp. 260–264).

Rajendran, V., Garcia-Luna-Aceves, J., & Obraczka, K. (2005). Efficient application-aware medium access for sensor networks. *Proceedings of the MASS, 2005,* 630–637.

Rajendran, V., Obraczka, K., & Garcia-Luna-Aceves, J. J. (2006). Energy efficient, collision-free medium access control for wireless sensor networks. *Wireless Networks, 12*(1), 63–78. doi:10.1007/s11276-006-6151-z

Ramadan, A., Reveyrand, T., Martin, A., Nebus, J. M., Bouysse, P., & Lapierre, L. (2010). Experimental study on effect of second-harmonic injection at input of classes F and F-1 GaN power amplifiers. *Electronics Letters, 46,* 570–572. doi:10.1049/el.2010.0392

Ramadan, A., Reveyrand, T., Martin, A., Nebus, J.-M., Bouysse, P., & Lapierre, L. (2011). Two-stage GaN HEMT amplifier with gate–source voltage shaping for efficiency versus bandwidth enhancements. *IEEE Transactions on Microwave Theory and Techniques, 59,* 699–706. doi:10.1109/TMTT.2010.2095033

Ramamurthy, H. (2007). Wireless industrial monitoring and control using a smart sensor platform. *IEEE Sensors Journal, 7*(5), 611–618. doi:10.1109/JSEN.2007.894135

Rashed, J., & Tai, C. T. (1982, May). *A new class of wire antennas.* Paper presented at Antennas and Propagation Society International Symposium.

Ray, K. P., & Krishna, D. D. (2006). Compact dual band suspended semi-circular microstrip antenna with half U-slot. *Microwave and Optical Technology Letters, 48*(10), 2021–2024. doi:10.1002/mop.21844

Rhee, I. (2008). Z-MAC: A hybrid MAC for wireless sensor networks. *IEEE/ACM Transactions on Networking, 16*(3), 511–524. doi:10.1109/TNET.2007.900704

Rida, A., Li, Y., Reynolds, T., Tan, E., Nikolaou, S., & Tentzeris, M. M. (2009, 1-5 June 2009). *Inkjet-printing UHF antenna for RFID and sensing applications on liquid crystal polymer.* In The Antennas and Propagation Society International Symposium, APSURSI '09.

Roy, S. M., & Karmakar, N. C. (2010). RFID planar antenna-smart design approach at UHF band. In Karmakar, N. C. (Ed.), *Handbook of smart antennas for RFID systems* (pp. 141–171). New Jersey: Wiley Microwave and Optical Engineering Series. doi:10.1002/9780470872178.ch6

Saad, P., Nemati, H. M., Thorsell, M., Andersson, K., & Fager, C. (2009). An inverse class-F GaN HEMT power amplifier with 78% PAE at 3.5 GHz. *EuMC 2009 European.* (pp. 496-499).

Sandhu, S., Heath, R., & Paulraj, A. (2001). *Space-time block codes versus space-time trellis codes.* IEEE.

Sangoi, R., Smith, C. G., Seymour, M. D., Venkataraman, J. N., Clark, D. M., & Kleper, M. L. (2005). Printing radio frequency identification (RFID) tag antennas using inks containing silver dispersions. *Journal of Dispersion Science and Technology, 25*(4), 513–521. doi:10.1081/DIS-200025721

Sanjay Monie, P. (n.d.). *Developments in conductive inks.* Retrieved 29 August, 2011, from http://www.vorbeck.com/news.html

Sankir, N. D. (2008). Selective deposition of PEDOT/PSS on to flexible substrates and tailoring the electrical resistivity by post treatment. *Circuit World, 34*(4), 32–37. doi:10.1108/03056120810918105

Sanming, H. (2010). Study of a uniplanar monopole antenna for passive chipless UWB-RFID localization system. *IEEE Transactions on Antennas and Propagation, 58,* 271–278. doi:10.1109/TAP.2009.2037760

Schmelzer, D., & Long, S. I. (2007). A GaN HEMT class F amplifier at 2 GHz with > 80% PAE. *IEEE Journal of Solid-state Circuits, 42,* 2130–2136. doi:10.1109/JSSC.2007.904317

Scholkopf, B., Smola, A., & M̈uller, K. (1999). Kernel principal component analysis. In Scholkopf, B., Burges, C. J. C., & Smola, A. J. (Eds.), *Advances in kernel methods–Support vector learning* (pp. 327–352). Cambridge, MA: MIT Press.

Schurgers, C., Tsiatsis, V., & Srivastava, M. B. (2002). STEM: topology management for energy efficient sensor networks. *IEEE Aerospace Conference,* 2002, (pp. 1099-1108).

Schurgers, C., & Srivastava, M. B. (2001). Energy efficient routing in wireless sensor networks. *Proceedings of the IEEE MILCOM,* 2001, 357–361.

Shaker, G., Safavi-Naeini, S., Sangary, N., & Tentzeris, M. M. (2011). Inkjet printing of ultra-wideband (UWB) antennas on paper-based substrates. *Antennas and Wireless Propagation Letters, 99,* 1–1.

Shao, B., Chen, Q., Amin, Y., David, S. M., Liu, R., & Zheng, L. R. (2010, Nov). *An ultra-low-cost rfid tag with 1.67 gbps data rate by ink-jet printing on paper substrate.* Paper presented at the IEEE Asian Solid-State Circuits Conference (A-SSCC) Beijing

Sharma, R., Chakravarty, T., & Bhattacharyya, A. B. (2009). Analytical model for optimum signal integrity in PCB interconnects using ground tracks. *IEEE Transactions on Electromagnetic Compatibility, 51*(1), 66–77. doi:10.1109/TEMC.2008.2010054

Sharma, R., Chakravarty, T., & Bhattacharyya, A. B. (2010). Reduction of signal overshoots in high-speed interconnects using adjacent ground tracks. *Journal of Electromagnetic Waves and Applications, 24,* 941–950. doi:10.1163/156939310791285218

Sharpe, C. A. (1995). *Wireless automatic vehicle identification* (pp. 39–58). Applied Microwave & Wireless FALL.

Shechtman, E., & Irani, M. (2007). Space-time behaviour based correlation or How to tell if two underlying motion fields are similar without computing them? *IEEE Transactions on Pattern Analysis and Machine Intelligence, 29*(11), 2045–2056. doi:10.1109/TPAMI.2007.1119

Sheikh, A., Roff, C., Benedikt, J., Tasker, P. J., Noori, B., Wood, J., & Aaen, P. H. (2009). Peak class F and inverse Class F drain efficiencies using Si LDMOS in a limited bandwidth design. *IEEE Microwave and Wireless Components Letters, 19,* 473–475. doi:10.1109/LMWC.2009.2022138

Sheng, X., & Yu-Hen Hu, Y. (2005). Maximum likelihood multiple-source localization using acoustic energy measurements with wireless sensor networks. *IEEE Transactions on Image Processing, 53*(1), 44–53.

Shim, S., Han, J., & Hong, S. (2008). CMOS RF polar transmitter of a UHF mobile RFID reader for high power efficiency. *IEEE Microwave and Wireless Components Letters, 18,* 635–637. doi:10.1109/LMWC.2008.2002490

Shin, D.-Y., Lee, Y., & Kim, C. H. (2009). Performance characterization of screen printed radio frequency identification antennas with silver nanopaste. *Thin Solid Films, 517*(21), 6112–6118. doi:10.1016/j.tsf.2009.05.019

Shirasaka, Y., Yairi, T., Kanazaki, H., Junichi Shibata, J., & Machida, K. (2006). Supervised learning for object classification from image and FDID data. *SICE-ICASE International Joint Conference,* Bexco, Busan, Korea, (pp. 5940-5944).

Shirokov, I. B. (2010, September 10). *The method of radio frequency identification.* (Patent of Ukraine, #91937, MPC G01R 29/08, 6 p).

Shirokov, I. B. (2011, January 5). *The method of operation range increasing of RFID system.* (Patent Application of Ukraine, #a201100198, MPC G01R 29/08, 7 p.).

Shirokov, I. B., & Durmanov, M. A. (2007). Multi-Abonent RFID system. *IEEE Proceedings of Crimean Microwave Conference* (pp. 756-757). Sevastopol, Ukraine, 10-14 Sep. 2007.

Shirokov, I. B., & Isofatov, V. V. (2003). The increasing of noise immunity of automatic system of RFID of vehicles and cargos. *IEEE Proc. of Crimean Microwave Conference* (pp. 718-719). Sevastopol, Ukraine, 8-12 Sep. 2003.

Shirokov, I. B., & Polivkin, S. N. (2004). The selection of the number of phase shifter steps in tasks of investigations of channel characteristics, carried out with homodyne methods (in Russian). *Radio Engineering: All-Ukrainian Interdependent Science-Technology Magazine, 137,* 36-43. Kharkov 2004.

Shirokov, I. B., Gimpilevich, Y. B., & Polivkin, S. N. (2010). The generalized mathematical model of homodyne frequency converter at discrete change of probing signal phase. *Radio Engineering: All-Ukrainian Interdependent Science-Technology Magazine, 161,* 119-125. Kharkov 2010.

Shirokov, I. B. (2009). The multitag microwave RFID system. *IEEE Transactions on Microwave Theory and Techniques, 57*(5), 1362–1369. doi:10.1109/TMTT.2009.2017315

Shirokov, I. B., & Gimpilevich, Yu. B. (2006). Influence of phase shift change nonlinearity of probing signal on an error of homodyne frequency converter. *News of Higher Educational Institutions. Radio Electronics Kiev, 49*(12), 20–28.

Shirokov, I. B., Jandieri, G. V., & Sinitzyn, D. V. (2006). Multipath angle-of-arrival, amplitude and phase progression measurements on microwave line-of-sight links. *European Space Agency, (Special Publication). ESA SP, 626.*

Shrestha, S., Balachandran, M., Agarwal, M., Phoha, V. V., & Varahramyan, K. (2009). A chipless RFID sensor system for cyber centric monitoring applications. *IEEE Transactions on Microwave Theory and Techniques, 57*(5), 1303–1309. doi:10.1109/TMTT.2009.2017298

Siden, J., Fein, M. K., Koptyug, A., & Nilsson, H. E. (2007). Printed antennas with variable conductive ink layer thickness. *Microwaves. Antennas & Propagation, 1*(2), 401–407.

Simon, M. K., & Alouini, M. S. (2000). *Digital communication over fading channels: A unified approach to performance analysis.* New York, NY: Wiley-Interscience Series in Telecommunications and Signal Processing. doi:10.1002/0471200697

Sim, S. H., Kang, C. Y., Yoon, S. J., Yoon, Y. J., & Kim, H. J. (2002). Broadband multilayer ceramic chip antenna for handsets. *Electronics Letters, 38*(5), 205–207. doi:10.1049/el:20020176

Skinner, B. F. (1953). *Science and human behavior.* New York, NY: The Free Press.

Smith, J. R., Fishkin, K. P., Jiang, W. B., Philipose, M., Rea, A. D., Roy, S., & Rajan, K. S. (2005). RFID-based techniques for human-activity detection. *Communications of the ACM, 48*(9), 39–44. doi:10.1145/1081992.1082018

Snider, D. M. (1967). A theoretical analysis and experimental confirmation of the optimally loaded and overdriven RF power amplifier. *IEEE Transactions on Electron Devices, 14*, 851–857. doi:10.1109/T-ED.1967.16120

Sohrabi, K., & Pottie, J. (2000). Protocols for self-organization of a wireless sensor network. *IEEE Personal Communication, 7*(5), 16–27.

Song, X., & Fan, G. (2006). Joint key-frame extraction and object segmentation for content-based video analysis. *IEEE Transactions on CSVT, 16*(7), 904–914.

Song, Y., Blostein, S. D., & Cheng, J. (2003). Exact outage probability for equal gain combining with cochannel interference in Rayleigh fading. *IEEE Transactions on Wireless Communications, 2*, 865–870. doi:10.1109/TWC.2003.816796

Stockman, H. (1948). *Communication by means of reflected power* (pp. 1196–1204). IRE Proceedings.

Strang, G. (2009). *Linear algebra and its applications* (4th ed.). Wellesley-Cambridge Press.

Stutzman, W. L., & Thiele, G. A. (1997). *Antenna theory and design* (2nd ed.). New Jersey: John Wiley & Sons, Inc.

Subramanian, V., Liao, F., & Huai-Yuan, T. (2010, 27-28 September). *Printed RF tags and sensors: The confluence of printing and semiconductors.* In The European Microwave Integrated Circuits Conference (EuMIC), 2010.

Sung, D., de la Fuente Vornbrock, A., & Subramanian, V. (2010). Scaling and optimization of gravure-printed silver nanoparticle lines for printed electronics. *IEEE Transactions on Components and Packaging Technologies, 33*(1), 105–114. doi:10.1109/TCAPT.2009.2021464

Tanidokoro, H., Konishi, N., Hirose, E., Shinohara, Y., Arai, H., & Goto, N. (1998, June). *1-wavelength loop type dielectric chip antennas.* Paper presented at the IEEE Antennas and Propagation Society International Symposium and URSI National Radio Science Meeting, Atlanta, GA.

Tarokh, V., Jafarkhani, H., & Calderbank, A. R. (1999). Space-time block codes from orthogonal designs. *IEEE Transactions on Information Theory, 45*, 1456–1467. doi:10.1109/18.771146

Tarokh, V., Seshadri, N., & Calderbank, A. R. (1998). Space-time codes for high data rate wireless communication: Performance criteria and code construction. *IEEE Transactions on Information Theory, 44*(2), 744–764. doi:10.1109/18.661517

Taylor, D. (2009, April). Introducing SAR code – A unique chipless RFID technology. *RFID Journal, 7th Annual Conference,* Orlando.

Tedjini, S., Perret, E., Deepu, V., & Bernier, M. (2009, September). *Chipless tags, the RFID next frontier.* Invited paper, 20th Tyrrhenian Workshop on Digital Communications, Sardina, Italy.

Telatar, I. E. (1999). Capacity of multi-antenna Gaussian channels. *European Transactions in Telecommunications, 10*(6), 585–595. doi:10.1002/ett.4460100604

Tentzeris, E., Li, R. L., Lim, K., Maeng, M., Tsai, E., DeJean, G., et al. (2002, June). *Design of compact stacked-patch antennas on LTCC technology for wireless communication applications.* Paper presented at the IEEE Antennas and Propagation Society International Symposium and URSI National Radio Science Meeting, San Antonio, TX.

Thomas, K. G., & Sreenivasan, M. (2010). Compact CPW-fed dual-band antenna. *Electronics Letters, 46*(1), 13–14. doi:10.1049/el.2010.1729

Thouroude, D., Himdi, M., & Daniel, J. P. (1990). CAD-oriented cavity model for rectangular patches. *Electronics Letters, 26*(13), 842–844. doi:10.1049/el:19900552

Tseng, Y. C., Pan, M. S., & Tsai, Y. Y. (2006). Wireless sensor networks for emergency navigation. *IEEE Computer, 39*(7), 55–62. doi:10.1109/MC.2006.248

Turkmani, A. M. D., Arowogolu, A. A., Jefford, P. A., & Kellent, C. J. (1995). An experimental evaluation of performance of two-branch space and polarization diversity schemes at 1800 MHz. *IEEE Transactions on Vehicular Technology, 44*(2), 318–326. doi:10.1109/25.385925

Tyler, V. J. (1958). A new high-efficiency high power amplifier. *Marconi Review, 2*, 96–109.

Uddin, J., Reaz, M. B. I., Hasan, M. H., Nordin, A. N., Ibrahimy, M. I., & Ali, M. A. M. (2010). UHF RFID antenna architectures and applications. *Scientific Research and Essays, 5*(10), 1033–1051.

Vanjari, S. V., Krogmeier, J. V., & Bell, M. R. (2007). Remote data sensing using SAR and harmonic reradiators. *IEEE Transactions on Aerospace and Electronic Systems, 43*(4), 1426–1440. doi:10.1109/TAES.2007.4441749

Veltkamp, R. C., & Tanase, M. (2000). *Content-based image retrieval systems: A survey.* Technical Report UU-CS-2000-34, Department of Computing Science, Utrecht University.

Vena, A., Perret, E., & Tedjini, S. (2011, June). *RFID chipless tag based on multiple phase shifters.* IEEE MTT-S IMS, Baltimore, USA.

Vena, A., Perret, E., & Tedjini, S. (2011, March). Novel compact RFID chipless tag. *Progress in Electromagnetics Research Symposium 2011*, Marrakesh, Morocco.

Venguer, A. P., Medina, J. L., Chávez, R. A., & Velázquez, A. (2002). Low noise one-port microwave transistor amplifier. *Microwave and Optical Technology Letters, 33*(2), 100–104. doi:10.1002/mop.10236

Venguer, A. P., Medina, J. L., Chávez, R. A., Velázquez, A., Zamudio, A., & Il'in, G. N. (2004). The theoretical and experimental analysis of resonant microwave reflection amplifiers. *Microwave Journal, 47*(1), 80–93.

Verma, A., Bo, W., Shepherd, R., Fumeaux, C., Van-Tan, T., Wallace, G. G., et al. (2010, 20-24 September). *6 GHz microstrip patch antennas with PEDOT and polypyrrole conducting polymers.* In The International Conference on Electromagnetics in Advanced Applications (ICEAA), 2010.

Volkman, S. K., Pei, Y., Redinger, D., Yin, S., & Subramanian, V. (2004). Ink-jetted silver/copper conductors for printed RFID applications. In *The Materials Research Society Symposium Proceedings.*

Vornbrock, A. F. (2009). *Roll printed electronics: Development and scaling of gravure printing techniques.* Berkeley, CA: University of California.

Vucetic, B., & Yuan, J. (2005). Space-time trellis codes. In *Space-time coding.* Chichester, UK: John Wiley & Sons, Ltd.

Wakejima, A., Asano, T., Hirano, T., Funabashi, M., & Matsunaga, K. (2005). C-band GaAs FET power amplifiers with 70-W output power and 50% PAE for satellite communication use. *IEEE Journal of State Circuits, 40*, 2054–2060. doi:10.1109/JSSC.2005.854596

Wambacq, P., Donnay, S., Pieters, P., Diels, W., Vaesen, K., De Raedt, W., et al. (2000, February). *Chip-package co-design of a 5 GHz RF front-end for WLAN.* Paper presented at the IEEE International Solid-State Circuits Conference, San Francisco, CA.

Wang, J. J. M., Winters, J., & Warner, R. (2007). *RFID system with an adaptive array antenna.* US Patent, No. 7212116.

Wang, J., Luo, Z., Wong, E. C., & Tan, C. J. (2007). RFID assisted object tracking for automating manufacturing assembly lines. *IEEE International Conference on e-Business Engineering*, (pp. 48-53).

Wang, D. (2008). An energy-efficient clusterhead assignment scheme for hierarchical wireless sensor networks. *International Journal of Wireless Information Networks*, *15*(2), 61–71. doi:10.1007/s10776-008-0079-4

Wang, J. J., Xue, Y., & Zhang, Y. P. (2005). Frequency-band selection for an integrated-circuit package antenna using LTCC technology. *Microwave and Optical Technology Letters*, *44*(5), 439–441. doi:10.1002/mop.20660

Wang, Q. (2008). Lightning: a hard real-time, fast, and lightweight low-end wireless sensor election protocol for acoustic event localization. *IEEE Transactions on Mobile Computing*, *7*(5), 570–584. doi:10.1109/TMC.2007.70752

Wang, W., & Zhang, Y. P. (2004). 0.18-um CMOS push-pull power amplifier with antenna in IC package. *IEEE Microwave and Wireless Components Letters*, *14*(1), 13–15. doi:10.1109/LMWC.2003.821489

Want, R. (2004). The magic of RFID. *ACM Queue; Tomorrow's Computing Today*, *2*(7), 40–48. doi:10.1145/1035594.1035619

Want, R. (2006). An introduction to RFID technology. *IEEE Pervasive Computing / IEEE Computer Society and IEEE Communications Society*, *5*(1), 25–33. doi:10.1109/MPRV.2006.2

Warr, P. A., Morris, K. A., Watkins, G. T., Horseman, T. R., Takasuka, K., & Ueda, Y. (2009). A 60% PAE WCDMA handset transmitter amplifier. *IEEE Transactions on Microwave Theory and Techniques*, *57*, 2368–2377. doi:10.1109/TMTT.2009.2029021

Weinstein, R. (2005). RFID: A technical overview and its application to the enterprise. *IT Professional*, *7*(3), 27–33. doi:10.1109/MITP.2005.69

Wenyi, C., et al. (2009). Analysis, design and implementation of semi-passive Gen2 tag. *IEEE International Conference on RFID*, 2009, (pp. 15-19).

Wheeler, H. A. (1947). Fundamental limitations of small antennas. *Proceedings of IRE*, *35*, 1479–1484. doi:10.1109/JRPROC.1947.226199

Wikipedia. (2011). *Computer vision*. Retrieved October 1, 2011, http://en.wikipedia.org/wiki/Computer_vision

Winters, J. H. (1987). On the capacity of radio communication systems with diversity in a Rayleigh fading environment. *IEEE Journal on Selected Areas in Communications*, *5*, 871–878. doi:10.1109/JSAC.1987.1146600

Woo, Y. Y., Yang, Y., & Kim, B. (2006). Analysis and experiment for high-efõciency class-f and inverse class-f power amplifiers. *IEEE Transactions on Microwave Theory and Techniques*, *54*, 1969–1974. doi:10.1109/TMTT.2006.872805

Wu, D. Y.-T., & Boumaiza, S. (2009). *10W GaN inverse class F PA with input/output harmonic termination for high efficiency WiMAX transmitter*. 10th Annual IEEE Wireless and Microwave Technology Conference.

Wu, J., Osuntogun, A., Choudhury, T., Philipose, M., & Rehg, J. M. (2007). A scalable approach to activity recognition based on object use. *IEEE 11th International Conference on Computer Vision*, Rio de Janeiro, Brazil, (pp. 1–8).

Xiao, Y., Yu, S., Wu, K., Ni, Q., Janecek, C., & Nordstad, J. (2007). Radio frequency identification: Technologies, applications, and research issues: Research articles. *Wireless Communication and Mobile Computing*, *7*(4), 457–472. doi:10.1002/wcm.365

XPEDX. (n.d.). Retrieved 24 August, 2011, from http://xpedx.edviser.com/default.asp?req=knowledge/article/151

Xue, Q., Shum, K. M., & Chan, C. H. (2000). Novel 1-D microstrip PBG cells. *IEEE Microwave Wireless Components Letters*, *10*, 403–405.

Yamada, Y., & Michishita, N. (2005, March). *Antenna efficiency improvement of a miniaturized meander line antenna by loading a high ε_r material*. Paper presented at IEEE International Workshop on Antenna Technology: Small Antennas and Novel Metamaterials.

Yang, X., & Vaidya, N. (2004). A wakeup scheme for sensor networks: achieving balance between energy saving and end-to-end delay. *Proceedings of the IEEE Real-Time and Embedded Technology and Applications Symposium*, 2004, (pp. 19-26).

Yao, Y., & Gehrke, J. (2002). The cougar approach to in-network query processing in sensor networks. *SIGMOD*, *31*(3), 9–18. doi:10.1145/601858.601861

Yilmaz, A., Javed, O., & Shah, M. (2006). Object tracking: A survey. *ACM Computing Surveys*, *38*(4), 1–45. doi:10.1145/1177352.1177355

Yi, S. (2007). PEACH: Power-efficient and adaptive clustering hierarchy protocol for wireless sensor networks. *Computer Communications*, *30*(14-15), 2841–2852. doi:10.1016/j.comcom.2007.05.034

Yoon, C. S., Jeon, K. Y., & Cho, S. H. (2008). The performance enhancement of UHF RFID reader in multi-path fading environment using antenna diversity. In *Proceedings of the 23rd International Technical Conference on Circuits/Systems, Computers and Communications*, (pp. 1749–1752).

Younis, O., & Fahmy, S. (2004). HEED: A hybrid, energy-efficient distributed clustering approach for ad hoc sensor networks. *IEEE Transactions on Mobile Computing*, *3*(4), 366–379. doi:10.1109/TMC.2004.41

Yu, C., & Ballard, D. H. (2002). Learning to recognize human action sequences. *International Conference on Development and Learning*, (pp. 28-33).

Yu, Y., Estrin, D., & Govindan, R. (2001). *Geographical and energy-aware routing: A recursive data dissemination protocol for wireless sensor networks*. UCLA Comp. Sci. Dept. tech. rep., 2001.

Zabin, F. (2008). REEP: data-centric, energy-efficient and reliable routing protocol for wireless sensor networks. *IET Communication*, *2008*, 995–1008. doi:10.1049/iet-com:20070424

Zhang, L., Rodriguez, S., Tenhunen, H., & Zheng, L. R. (2006). An innovative fully printable RFID technology based on high speed time-domain reflections. *HDP'06 Conference*, (pp. 166-170).

Zhang, Y. P., Lo, T. K. C., & Hwang, Y. M. (1995, June). *A dielectric-loaded miniature antenna for microcellular and personal communications*. Paper presented at the IEEE Antennas and Propagation Society International Symposium, Newport Beach, Calif.

Zhang, C. C., Liu, J. J., & Zhang, Y. P. (2003). ICPA for highly integrated concurrent dual-band wireless receivers. *Electronics Letters*, *39*(12), 887–889. doi:10.1049/el:20030582

Zhang, H., & Shen, H. (2011). Energy-efficient beaconless geographic routing in wireless sensor networks. *IEEE Transactions on Parallel and Distributed Systems*, *21*(6), 881–896. doi:10.1109/TPDS.2009.98

Zhang, Y. P. (2002). Integration of microstrip antenna on cavity-down ceramic ball grid array package. *Electronics Letters*, *38*(22), 1307–1308. doi:10.1049/el:20020937

Zhang, Y. P. (2004a). Finite-difference time-domain analysis of integrated ceramic ball grid array package antenna for highly integrated wireless transceivers. *IEEE Transactions on Antennas and Propagation*, *52*(2), 435–442. doi:10.1109/TAP.2004.823889

Zhang, Y. P. (2004b). Integrated circuit ceramic ball grid array package antenna. *IEEE Transactions on Antennas and Propagation*, *52*(10), 2538–2544. doi:10.1109/TAP.2004.834427

Zhang, Y. P., & Wang, J. J. (2006). Theory and analysis of differentially-driven microstrip. *IEEE Transactions on Antennas and Propagation*, *54*(4), 1092–1099. doi:10.1109/TAP.2006.872597

Zhang, Y. P., Wang, J. J., Li, Q., & Li, X. J. (2008). Antenna-in-package and transmit-receive switch for single-chip radio transceivers of differential architecture. *IEEE Transactions on Circuits and Systems. I, Regular Papers*, *55*(11), 3564–3570. doi:10.1109/TCSI.2008.925822

Zhang, Z., Ma, M., & Yang, Y. (2004). Energy-Efficient multihop polling in clusters of two-layered heterogeneous sensor networks. *IEEE Transactions on Computers*, *57*(2), 231–245. doi:10.1109/TC.2007.70774

Zheng, L., Rodriguez, S., Zhang, L., Shao, B., & Zheng, L. R. (2008, May). Design and implementation of a fully reconfigurable chipless RFID tag using inkjet printing technology. *IEEE International Symposium on Circuits and Systems*, (pp. 1524 – 1527).

Zheng, R., Hou, J., & Sha, L. (2003). Asynchronous wakeup for ad hoc networks. *Proc. of the ACM MobiHoc*, 2003, (pp. 35–45).

Zorzi, M., & Rao, R. R. (2003). Geographic random forwarding (GeRaF) for ad hoc and sensor networks: multihop performance. *IEEE Transactions on Mobile Computing*, *2*(4), 337–348. doi:10.1109/TMC.2003.1255648

About the Contributors

Nemai Chandra Karmakar obtained his PhD in Information Technology and Electrical Engineering from the University of Queensland, St. Lucia, Australia, in 1999. He has about twenty years of teaching, design, and research experience in smart antennas, microwave active and passive circuits, and chipless RFIDs in both industry and academia in Australia, Canada, Singapore, and Bangladesh. He has published more than 180 refereed journal and conference papers and many book chapters. He holds two patents in the field. Currently, he is an Associate Professor in the Department of Electrical and Computer Systems Engineering at Monash University.

* * *

Jemal H. Abawajy is an Associate Professor, Deakin University, Australia. Dr. Abawajy is the director of the "Pervasive Computing & Networks" research groups at Deakin University and a Senior Member of IEEE. Dr. Abawajy is actively involved in funded research in robust, secure, and reliable resource management for pervasive computing and networks. He has published more than 200 research articles in refereed international conferences and journals as well as a number of technical reports. Dr. Abawajy has given keynote/invited talks at many conferences. Dr. Abawajy has guest-edited several international journals and served as an associate editor of international conference proceedings. In addition, he is on the editorial board of several international journals. Dr. Abawajy has been a member of the organizing committee for over 150 international conferences serving in various capacity including chair, general co-chair, vice-chair, best paper award chair, publication chair, session chair and program committee. He is also a frequent reviewer for international research journals (e.g., FGCS, TPDS, and JPDC), research grant agencies, and PhD examinations.

Emran Md Amin received the B.Eng. degree in Electrical and Electronics Engineering Department from Bangladesh University of Engineering and Technology (BUET) in 2009. Currently he is pursuing his PhD at Electrical and Computer Systems Engineering Department of Monash University. His research area is Chipless RFID Sensors and Radiometric Partial Discharge detection of High Voltage Equipments. He is working on a project titled, "Smart Information Management of Partial Discharge in Switchyards Using Smart Antennas," funded by ARC and SP-AusNet.

Rubayet-E-AzimAnee completed her Bachelor of Science in Electrical and Electronic Engineering (EEE) from Bangladesh University of Engineering and Technology (BUET), Dhaka, Bangladesh in 2009. She is currently a Graduate Student and pursuing research in Antenna and RFID Research Group in Monash University. Her areas of research interests include signal processing, RFID, and Antenna.

AKM Azad completed his PhD from the Faculty of Information Technology, Monash University, Australia, on May, 2010, and Master and Bachelor of Science degrees in Computer Science and Engineering from the Dept. of Computer Science and Engineering, Bangladesh University of Engineering and Technology (BUET), Dhaka, Bangladesh, in 1998 and 2005, respectively. Dr. Azad is currently working as a research engineer at the Antenna and RFID research lab of the department of Electrical and Computer Systems Engineering, Monash University, Australia. His research interest includes RFID and Sensors, ad hoc and sensor networks, signal processing, control system design, graph theory, et cetera.

A. K. M. Baki received B.Sc. in Electrical and Electronic Engineering degree from Bangladesh University of Engineering and Technology (BUET), Bangladesh in 1992 and his M. Sc. degree from the University of Bolton (former Bolton Institute of Higher Education), UK in 2003 having followed a postgraduate program in Electronic System and Engineering Management at the South-Westphalia University of Applied Science, Germany. He received Ph. D. degree in Electrical Engineering from Kyoto University, Kyoto, Japan in 2007. From 2007 to 2009, he worked as an Assistant Professor in the School of Engineering and Computer Science (SECS) of the Independent University, Bangladesh. From 2009 to 2011, he worked as a Research Fellow in Monash University, Australia.

Uditha Wijethilaka Bandara obtained his Bachelor of Science in Electrical and Electronics Engineering degree from University of Peradeniya, Sri Lanka in 2007. He has gained industrial experience in implementation and operation of various wireless communication networks. Currently he is pursuing a PhD in Engineering at Monash University. His research interests include design of smart antenna systems for radiometric detection and localization of partial discharge sources in switchyards using radiated signals.

Shivali G. Bansal was born in Punjab, India in 1980. She received her B. Tech. degree in Electrical Engineering from Punjab Technical University, Punjab, India in 2002. She received her M.Tech. degree from the Electronics and Communication Engineering Department (ECED) from Thapar Institute of Technology (TIET), Deemed University (now known as Thapar University), Punjab, India in 2004. Shivali G. Bansal also served as a Lecturer in ECED in TIET for the 2004 – 2005 term before starting with her PhD. Presently she is a PhD student with the School of Information Technology, Deakin University, Waurn Ponds, Geelong, Vic, Australia. She received the Deakin University International Research Scholarship in 2005 for the duration of her PhD. She has authored publications during her research in the areas of wireless communication. Her research interests include space-time coding and modulation, MIMO systems, relays, channel estimation, and antenna selection.

Tapas Chakravarty is a Senior Scientist (R & D) in TCS Innovation Lab at Tata Consultancy Services Ltd (TCS), Bangalore. He holds a Ph.D (Science) from Jadavpur University awarded for his work on compact & tunable microstrip antennas. His current major area of research is in the field of wireless platforms, solutions including sensor networks. Prior to joining TCS, he was associated with Society for Applied Microwave Electronics Engineering & Research (SAMEER) and multiple academic institutes like Jaypee University of Information Technology, Solan, and HP India. Tapas is a Senior Member of IEEE (USA). Tapas has over 20 years of research and development experience in RF & Microwave design, Wireless Communications, RFID based auto-identification systems, Sensor networks etc. He has specialization in design of microstrip antennas. He has over 100 publications in various international journals, conferences as well as 2 book chapters and holds a few invention disclosures. He has lead & executed many customer sponsored R&D projects in embedded systems. He continues to deliver key note addresses in national/international conferences & symposia. Tapas has also supervised one Ph.D thesis (antennas) and a number of M.Tech thesis from frontline Indian universities.

Shichang Chen was born in Zhejiang, China, in 1987. He received B.S. degree in Electronic Engineering from Nanjing University of Science and Technology, Nanjing, China, in 2009, and is currently working toward the Ph.D. degree at the City University of Hong Kong, Kowloon, Hong Kong SAR. His research interests include high efficiency power amplifiers, Doherty power amplifier and RFIC design.

Peter Cole has the degrees of B.Sc., B.E., and Ph.D. from the University of Sydney. He has interests in radio frequency identification, electromagnetic engineering, microcircuit design, and signal processing. He has taught at the Universities of Adelaide and Sydney, and is now Professor of Radio Frequency Identification Systems in the School of Electrical and Electronic Engineering at The University of Adelaide. He was a founder and was from 1984 to 1999 the Chairman of Directors of Integrated Silicon Design Pty. Ltd., a South Australian Company specialising in Radio Frequency Identification Systems. In 2002 he was invited to join the Auto-ID Center established by MIT, and is now Director of the Auto-ID Laboratory at Adelaide, one of seven Auto-ID research laboratories throughout the word supported by EPCglobal.

Laurence S. Dooly was awarded his B.Sc. (Hons), M.Sc. and Ph.D. degrees in Electrical and Electronic Engineering from the University of Wales/Cymru (Swansea) in 1981, 1983, and 1987 respectively. He is *Chair of Information and Communication Technologies* in the Department of Communication and Systems at The Open University, UK, where his research interests include: cognitive radio systems, distributed source coding, multimodal medical imaging, MANET and LTE/4G security, educational technologies, and SME technology/knowledge transfer. He has co-edited one book and published 215 peer-reviewed scientific journals, book chapters, monographs and conference papers, with 3 papers being awarded international research prizes/nominations. He received the 2010 *IEEE Certificate of Award* for promoting international exchange in recognition of his services to the IEEE international conference series on signal processing. He has supervised 18 PhD/MPhil students to completion together with being a recipient of significant public and private sector funding to support his multifaceted research. He is a Chartered Engineer, a Fellow of the British Computer Society, and a Senior Member of the IEEE, as well as being a Vice President of the Crawshays Rugby Football Club.

Christophe Fumeaux received the Diploma and Ph.D. degrees in Physics from the ETH Zurich, Switzerland, in 1992 and 1997, respectively. From 1998 to 2000, he was a Postdoctoral Researcher with the School of Optics, University of Central Florida, Orlando. In 2000, he joined the Swiss Federal Office of Metrology, Bern, Switzerland. From 2001 to 2008, he was a Research Group Leader with the Laboratory for Electromagnetic Fields and Microwave Electronics (IFH), ETH Zurich. During Fall 2005, he was a Visiting Scientist with the Laboratory of Sciences and Materials for Electronics and Automation (LASMEA), University Blaise Pascal, Clermont-Ferrand, France. In 2008, he joined the School of Electrical and Electronic Engineering, The University of Adelaide, Australia, as an Associate Professor. Since 2011, he is a Future Fellow of the Australian Research Council. His current main research interest concerns computational electromagnetics, antenna engineering, terahertz technology, and the modeling of optical microstructures/nanostructures. Dr. Fumeaux is an Associate Editor for the *IEEE Transactions on Microwave Theory and Techniques* since 2010. He was the recipient of the ETH Silver Medal of Excellence for his doctoral dissertation. He was the corecipient of the 2004 Outstanding Paper Award of the Applied Computational Electromagnetics Society (ACES).

Liming Gu was born in Jiangsu, China, in 1987. He received B.S. degree in Electronic Engineering from Nanjing University of Science and Technology, Nanjing, China, in 2009, and is currently working toward the Ph.D. degree at the Nanjing University of Science and Technology, Nanjing, China. His research interests include microstrip passive circuit designs, power amplifiers designs, phase shifter designs, and MMIC designs.

Jyun-Yan He was a graduate student in the Department of Computer Science and Information Engineering at National Dong Hwa University, Taiwan. His research includes wireless network and cryptography.

Zhonghao Hu received his Bachelor degree in Electronic Information Engineering from Northwestern Polytechnical University, China in 2006 and Ph.D. degree from The University of Adelaide, Australia in 2011. He was also a Research Assistant in the Auto-ID Lab, Adelaide during his Doctorate study. He is now the Senior Product Development Engineer in Confidex, Finland. His current research interest includes the passive RFID tag antenna design for detecting metallic items, active tags, and analog circuits for RFID reader.

Darine Kaddour received in 2003 the B.S. degree in Physics from the Lebanese University, Faculty of Sciences, Tripoli, Lebanon and the M.S. degree in "Optics & Radiofrequency" from the Grenoble Institute of Technology (Grenoble-INP) France, in 2004. Then, she prepared her Ph.D. degree at the Institut de Microélectronique, Electromagnétisme et Photonique (IMEP) at Grenoble-INP. Since September 2009, she joined the 'Laboratoire de Conception de Intégration des Systèmes (LCIS) located in Valence, France, as an Assistant Professor in Electrical Engineering. Her research interests include the design, realization and test of passive microwave devices such as filters, couplers, and antennas.

Prasanna Kalansuriya received the B.Sc. degree (with first-class honours) in Electronic and Telecommunication Engineering from the University of Moratuwa, Sri Lanka, in 2005, the M.Sc. degree in Wireless Communications from the University of Alberta, Edmonton, Canada in 2009 and is currently pursuing a PhD in Electrical and Computer Systems Engineering at Monash University, Australia. He worked as an electronic engineer in Sri Lanka from 2005 – 2007. He also served as a Lecturer in the Department of Electronic and Telecommunication Engineering, University of Moratuwa in 2007, and as a Research Assistant in the iCORE Wireless Communication Laboratory, University of Alberta from 2008 - 2010. His research interests include wireless communication, signal processing, and detection mechanisms for chipless RFID.

Joarder Kamruzzaman received a B.Sc. and M.Sc. in Electrical Engineering from Bangladesh University of Engineering & Technology, Dhaka, Bangladesh in 1986 and 1989, respectively, and a PhD in Information System Engineering from Muroran Institute of Technology, Japan, in 1993. Currently, he is a faculty member in the Faculty of Information Technology, Monash University, Australia. His research interests include computer networks, computational intelligence, and bioinformatics. He has published over 150 peer-reviewed publications which include 40 journal papers and 6 book chapters, and edited two reference books on computational intelligence theory and applications. He is the recipient of Best Paper award in two IEEE sponsored international conferences. He is currently serving as a program committee member of a number of international conferences and an editor of international journal.

Gour Chandra Karmakar received the B.Sc. Eng. degree in Computer Science and Engineering from Bangladesh University of Engineering and Technology in 1993 and Master's and Ph.D. degrees in Information Technology from the Faculty of Information Technology, Monash University, in 1999 and 2003, respectively. He is currently a Senior Lecturer at the Gippsland School of Information Technology, Monash University. He has published over 90 peer-reviewed research publications including 13 international peer reviewed reputed journal papers and was awarded 3 best papers in reputed international conferences. His research interest includes image and video processing, mobile ad hoc, and wireless sensor networks.

Subhas Chandra Mukhopadhyay graduated from the Department of Electrical Engineering, Jadavpur University, Calcutta, India with a Gold medal and received the Master of Electrical Engineering degree from Indian Institute of Science, Bangalore, India. He has PhD (Eng.) degree from Jadavpur University, India and Doctor of Engineering degree from Kanazawa University, Japan. Currently he is working as a Professor of Sensing Technology with the School of Engineering and Advanced Technology, Massey University, Palmerston North, New Zealand. He has over 21 years of teaching and research experiences. His fields of interest include sensors and sensing technology, electromagnetics, control, electrical machines and numerical field calculation, et cetera. He has authored/co-authored over 240 papers in different international journals, conferences and book chapter. He has edited nine conference proceedings. He has also edited nine special issues of international journals as lead guest editor and eleven books out of which nine are with Springer-Verlag. He was awarded numerous awards throughout his career and attracted over NZ $3.5 M on different research projects. He is a Fellow of IEEE (USA), a Fellow of IET (UK), an Associate Editor of *IEEE Sensors Journal* and *IEEE Transactions on Instrumentation and Measurements*. He is in the editorial board of *e-Journal on Non-Destructive Testing, Sensors and Transducers, Transactions on Systems, Signals and Devices (TSSD), Journal on the Patents on Electrical Engineering*, et cetera. He is the co-Editor-in-chief of the *International Journal on Smart Sensing and Intelligent Systems* (www.s2is.org). He is in the technical programme committee of IEEE Sensors Conference, IEEE IMTC Conference, and numerous other conferences. He was the Technical Programme Chair of ICARA 2004, ICARA 2006 and ICARA 2009. He was the General chair/co-chair of ICST 2005, ICST 2007, IEEE ROSE 2007, IEEE EPSA 2008, ICST 2008, IEEE Sensors 2008, ICST 2010 and IEEE Sensors 2010. He has organized the IEEE Sensors conference 2009 at Christchurch, New Zealand during October 25 to 28, 2009 as General Chair. He is the Chair of the IEEE Instrumentation and Measurement Society New Zealand Chapter. He is a Distinguished Lecturer of the IEEE Sensors Council.

Balamuralidhar P. is heading TCS Innovation Lab at Tata Consultancy Services Ltd (TCS), Bangalore. Major area of research is in applications of Networked Embedded Systems. Before TCS his research careers were with Society for Applied Microwave Electronics Engineering & Research (SAMEER) and Sasken Communications Ltd. Balamuralidhar has over 22 years of research and development experience in signal processing, embedded systems and wireless communications. He has over 50 publications in various international journals and conferences and several invention disclosures. He had lead many R&D programs in wireless communications involving national and international collaborations. He is also a member of IET and CII. His areas of research interest are in networked embedded systems, cognitive networking, and internet of things.

Etienne Perret was born in Albertville, France, on October 30, 1979. He received the Eng. Dipl. in Electrical Engineering from the ENSEEIHT, Toulouse, France, in 2002, and the M.Sc. and Ph.D. degrees from the Toulouse Institute of Technology, France, in 2002 and 2005, respectively, all in Electrical Engineering. From 2005 to 2006, he held a post-doc position at the Institute of Fundamental Electronics, Orsay, France. Since September 2006, Dr. Perret is Assistant Professor in electronic with the Grenoble Institute of Technology. He is authored and co-authored of more than 50 technical conferences, letters, and journal papers. He is member IEEE and Technical Program Committee member of IEEE-RFID. He was keynote speaker and the chairman of the 11th Mediterranean Microwave Symposium MMS'2011. He also served as the co-chair of the 29th PIERS 2011 in Marrakesh, Morocco. His research activities cover electromagnetic modeling of passive devices for millimeter and submillimeter-wave applications.

Sushim Mukul Roy is a Senior Research Engineer working at Monash University on Chipless RFID for Australian Polymer Banknote ARC Linkage Project supported by Securency and SatNet. Dr. Roy obtained PhD in June 2009 in the same filed from Monash University. He did his Bachelor of Engineering in Electronics and Telecommunication Engineering from Bengal Engineering College in 2003 from India. His research interests include microwave circuits and systems.

Igor B. Shirokov obtained his PhD degree in Radio Engineering from the Sevastopol Instrument Making Institute, Sevastopol, USSR, in 1984. Since 2007 he is a member of IEEE. He has thirty years of teaching, design, and research experience in antennas and microwave propagation, searching for people under avalanches, remote sensing, technological equipment sensors, and power electronics in both industry and academia in Ukraine and Russia. He has published more than 300 refereed journal and conference papers, as well as two referred book chapters. He holds 40 patents in applied problems. Currently, he is an Associate Professor in the Chair of Radio Engineering at Sevastopol National Technical University, Sevastopol, Ukraine.

Bala Srinivasan is a Professor of Information Technology and Head of Clayton School of Information Technology in the Faculty of Information Technology, Monash University, Australia. He was formerly an academic staff member of the Department of Computer Science and Information Systems at the National University of Singapore, Singapore and the Indian Institute of Technology, Kanpur, India. He has more than 30 years of experience in academia, industries and research organizations. He has authored and jointly edited 7 technical books and more than 300 refereed publications in international journals and conferences in the areas of multimedia databases, data communications, data mining and distributed systems, and has attracted a number of research grants. His substantial contribution towards research and training has been recognized by Monash University by awarding him the Vice-Chancellor's Medal for excellence in supervision. He is a founding chairman of the Australasian Database Conference.

Gerry Swiegers is a Professor in the Intelligent Polymer Research Institute (IPRI). He is also an Adjunct Fellow of the ARC Centre of Excellence for Electromaterials Science (ACES). He actively collaborates with various members of the centre, including Prof Gordon Wallace, Prof Leone Spiccia, and Dr. Jun Chen. Gerry is concurrently Vice-President R&D at Datatrace-DNA Pty Ltd, a company formed from his research in 2005. He was previously leader of the Security Devices research group at CSIRO Molecular and Health Technologies in Melbourne.

Smail Tedjini, Doctor in Physics from Grenoble University in 1985. Since 1996, he is Professor at Grenoble Institute of Technology. His teaching topics concern electromagnetism, RF, wireless, and optoelectronics. He serves as Director of the ESISAR Eng. Dept. Past research concerns the modeling of devices and circuits at both RF and optoelectronic domains. He is the founder and past Director of the LCIS Lab. Now, he is ORSYS group leader. Current research concerns wireless systems with specific attention to RFID. He supervised 27 PhD and he has more than 250 publications. He is Member of several TPC and serves as expert/reviewer for national and international scientific committees such as Piers, IEEE, URSI, ISO, ANR, OSEO, FNQRT… He organized several conferences/workshops. Senior Member IEEE, Past-President and founder of the IEEECPMT French Chapter, Vice-President of IEEE France Section since 2008, Chair of URSI Commission D "Electronics & Photonics" since 2011.

Arnaud Vena was born in Monaco in 1982 . He received the Engineer degree from the Grenoble Institute of Technology (Grenoble-INP) in 2005. From July 2005 to September 2009 he served as a R&D Engineer at ACS Solution France SAS, solution provider for the management of urban transport network. He was in charge of RFID Contactless card reader development and contributed to the evolution of ISO/IEC 14443 regulation. Since October 2009 he is a PhD student at the Laboratoire de Conception et d'intégration des Systèmes (LCIS) in Valence, France, under supervision of Smail Tedjini and Etienne Perret. He has many recent publications in the field of RFID. His current researches are mainly focused on RFID chipless tag development.

Emanuele Viterbo received his degree (Laurea) in Electrical Engineering in 1989 and his Ph.D. in 1995 in Electrical Engineering, both from the Politecnico di Torino, Torino, Italy. From 1990 to 1992 he was with the European Patent Office, The Hague, The Netherlands, as a patent examiner in the field of dynamic recording and error-control coding. Between 1995 and 1997 he held a post-doctoral position in the Dipartimento di Elettronica of the Politecnico di Torino. In 1997-98 he was a post-doctoral research fellow in the Information Sciences Research Center of AT&T Research, Florham Park, NJ, USA. He became first Assistant Professor (1998) then Associate Professor (2005) in Dipartimento di Elettronica at Politecnico di Torino. In 2006 he became Full Professor in DEIS at University of Calabria, Italy. From September 2010 he is Full Professor in the ECSE Department at Monash University, Melbourne, Australia. Prof. Emanuele Viterbo is a 2011 Fellow of the IEEE, a ISI Highly Cited Researcher and Member of the Board of Governors of the IEEE Information Theory Society (2011-2013). He is Associate Editor of *IEEE Transactions on Information Theory, European Transactions on Telecommunications,* and *Journal of Communications and Networks,* and Guest Editor for *IEEE Journal of Selected Topics in Signal Processing: Special Issue Managing Complexity in Multiuser MIMO Systems.* In 1993 he was visiting researcher in the Communications Department of DLR, Oberpfaffenhofen, Germany. In 1994 and 1995 he was visiting the Ecole Nationale Suprieure des Telcommunications (E.N.S.T.), Paris. In 2003 he was visiting researcher at the Maths Department of EPFL, Lausanne, Switzerland. In 2004 he was visiting researcher at the Telecommunications Department of UNICAMP, Campinas, Brazil. In 2005, 2006, and 2009 he was visiting researcher at the ITR of UniSA, Adelaide, Australia. In 2007 he was visiting fellow at the Nokia Research Center, Helsinki, Finland.

Christina Junjun Wang received the B.E. degree from Shandong University of Technology, Shandong, China, and M.E. degree from Shanghai University, Shanghai, China, in 1999 and 2002 respectively, and the Ph.D. degree at School of Electrical Engineering, Nanyang Technological University, Singapore, in 2006, all in Electronic Engineering. From 2005 to 2008, she worked at Sony Electronics (Singapore) as a Senior R&D Engineer. Since 2009, she has been an Associate Professor with the School of Electronic and Information Engineering, Beihang University, Beijing, China. Her research interests include design of integrated in-package and on-chip antennas, analysis of electromagnetic compatibility, and design of radio frequency circuits.

Yuexian Wang received the B.E. and M.Eng.Sc. degrees in Electronic and Information Engineering and Pattern Recognition and Intelligent Systems from Northwestern Polytechnical University, Xi'an, China in 2006 and 2009, respectively. He is currently pursuing his M.Eng.Sc. degree in the School of Electrical and Electronic Engineering at University of Adelaide. His interests include RF circuit design, signal processing, and wireless sensor networks.

Yang Yang was born in Inner Mongolia, China, in 1982. He received B.E. degree from Dalian Nationalities University, Dalian, China, in 2005, M.E. degree in Telecommunication Eng. from Monash University in 2007, M.Sc. Degree in Digital Communications and Protocols from Monash University in 2008 and is currently working toward the Ph.D. degree in Electrical and Computer Systems Engineering at Monash University, Melbourne, Victoria, Australia. His research interests include microstrip passive filter and antenna designs for portable wireless communications system, transceivers designs, MMIC designs, and bio-sensors development. His publications mainly focus on microstrip filter designs and portable wireless monitoring system designs.

Ramprakash Yerramilli is presently holding a position of R&D Scientist at Securency International Pty Ltd, at its location in Craigieburn, Victoria, Australia. In the past, he has been actively associated with research projects in semiconductor based photo electrochemical devices including dye-sensitized solar cells. He contributed extensively to the research leading to the technology development of solid oxide fuel cells in 1993-2002 at Ceramic Fuel Cells Ltd. He is now engaged in a collaborative research project with Monash University to develop a robust high speed printing technology for RFID tag printing on polymer banknote substrate. Dr Yerramilli has published a number of research papers in the past in his field of research and is deeply interested in the science of RFID printing to manufacture commercial products.

Index